GREATER CHINA'S QUEST FOR INNOVATION

Edited by
Henry S. Rowen, Marguerite Gong Hancock,
and William F. Miller

SHORENSTEIN APARC STANFORD | THE WALTER H. SHORENSTEIN ASIA-PACIFIC RESEARCH CENTER

THE WALTER H. SHORENSTEIN ASIA-PACIFIC RESEARCH CENTER (Shorenstein APARC) is a unique Stanford University institution focused on the interdisciplinary study of contemporary Asia. Shorenstein APARC's mission is to produce and publish outstanding interdisciplinary, Asia-Pacific-focused research; educate students, scholars, and corporate and governmental affiliates; promote constructive interaction to influence U.S. policy toward the Asia-Pacific; and guide Asian nations on key issues of societal transition, development, U.S.-Asia relations, and regional cooperation.

The Walter H. Shorenstein Asia-Pacific Research Center
Freeman Spogli Institute for International Studies
Stanford University
Encina Hall
Stanford, CA 94305-6055
tel. 650-723-9741
fax 650-723-6530
http://APARC.stanford.edu

Greater China's Quest for Innovation may be ordered from:
The Brookings Institution
c/o DFS, P.O. Box 50370, Baltimore, MD, USA
tel. 1-800-537-5487 or 410-516-6956
fax 410-516-6998
http://www.brookings.edu/press

Walter H. Shorenstein Asia-Pacific Research Center Books, 2008.
Copyright © 2008 by the Board of Trustees of the Leland Stanford Junior University.

All rights reserved. No part of this publication may be reproduced, stored in a retrieval system, or transmitted in any form or by any means, electronic, mechanical, photocopying, recording, or otherwise, without written permission of the publisher.

First printing, 2008.
13-digit ISBN 978-1-931368-12-4

GREATER CHINA'S QUEST FOR INNOVATION

PRODUCED AS PART OF THE
STANFORD PROGRAM ON
REGIONS OF INNOVATION
AND ENTREPRENEURSHIP

THE WALTER H. SHORENSTEIN
ASIA-PACIFIC RESEARCH CENTER

CONTENTS

Preface	VII
Introduction	9
Henry S. Rowen	

RESEARCH INSTITUTES   33
1. Success in "Pasteur's Quadrant"? The Chinese Academy of Sciences and Its Role in the National Innovation System   35
   Richard P. Suttmeier and Bing Shi
2. ITRI's Role in Developing the Access Network Industry in Taiwan   57
   Kristy H. C. Sha, Tzung-Pao Lin, Chih-Young Hung, and Bao-shuh Paul Lin
3. Restructuring China's Research Institutes   67
   Gary Jefferson, Bangwen Cheng, Jian Su, Paul Duo Deng, Haoyuan Qin, and Zhaohui Xuan

R&D BY MULTINATIONAL COMPANIES   83
4. The Political Economy of Technological Development: China's Information Technology Sector   85
   Douglas B. Fuller
5. Multinational R&D in China: Myths and Realities   109
   Lan Xue and Zheng Liang
6. Linking Local and National Innovation Networks: Multinational Companies' R&D Centers in Taiwan   129
   Yuan-chieh Chang, Chintay Shih, and Yi-Ling Wei
7. R&D Globalization and Silicon Triangle Dynamics   147
   Kung Wang and Yi-Ling Wei
8. Cooperation in Chinese Innovation Systems   157
   Ingo Liefner and Stefan Hennemann
9. Cross-border R&D Networks and International R&D: A Study of Taiwanese Firms   169
   Meng-chun Liu and Shin-horng Chen

TALENT AND INNOVATIVE CAPACITY   179
10. China's Emerging Science and Technology Talent Pool: A Quantitative and Qualitative Assessment   181
   Denis Fred Simon and Cong Cao
11. Can Chinese IT Firms Develop Innovative Capabilities Within Global Knowledge Networks?   197
   Dieter Ernst
12. Toward a Better Understanding of China's Scientific Elite   217
   Cong Cao
13. China's Return Migrants and Its Innovative Capacity   231
   Claudia Müller and Rolf Sternberg

STATISTICAL INDICATORS: PATENTS AND JOURNALS	253
14. What Do They Patent in China, and Why? *Albert Guangzhou Hu*	255
15. A Comparative Analysis of Chinese and International Aggregated Citation Relations among Journals *Ping Zhou and Loet Leydesdorff*	269
16. The Role of Global Multinational Corporations versus Indigenous Firms in the Rapid Growth of East Asian Innovation: Evidence from U.S. Patent Data *Poh Kam Wong*	281
HIGH-TECH REGIONS	309
17. Knowledge Spillovers and Growth in the Hsinchu Science-based Industry Park *Yih-Luan Chyi and Yee-Man Lai*	311
18. The Development of Innovative Capacity in Hong Kong *Erik Baark*	317
19. China and the Emerging Regional System of Technological Entrepreneurship *Adam Segal*	337
CHANGING INDUSTRY DYNAMICS: INTELLECTUAL PROPERTY AND THE NEW MEDIA	357
20. Managing Intellectual Property in the Chinese Semiconductor Industry *Xiaohong Quan, Henry Chesbrough, and Jihong Wu Sanderson*	359
21. China's New Media: Shaping New Industries with New Policy Regimes *F. Ted Tschang and Seng-Su Tsang*	371
About the Contributors	389

Preface

The origins of this book were in a workshop held in Beijing in May 2006, sponsored by the China Institute for Science and Technology Policy (CISTP) of Tsinghua University, Beijing; the Industrial Technology Research Institute (ITRI), Hsinchu; and the Administrative Committee of Zhongguancun Science Park, Beijing. Later workshops that developed some of these themes were held at Stanford University in November 2006 and in Taipei in September 2007.

We are grateful to the contributors this volume, to the organizers of the meeting and especially to Professor Lan Xue of Tsinghua University, Madame Mulan Zhao of the Administrative Committee of Zhongguancun Science Park, and Tze-Chen Tu, general director of ITRI's Industrial Economics and Knowledge Center (IEK). We are also grateful that Naushad Forbes came from India to chair one of our sessions in Beijing.

The Stanford Program on Regions of Innovation and Entrepreneurship (SPRIE) was fortunate during this period to have three postdoctoral scholars—Dan Breznitz, Douglas Fuller, and Xiaohong Quan—two of whom contributed chapters to this book. We also were fortunate to have Gu Ming, a talented Stanford predoctoral student, who did much valuable research toward this topic.

Among those who helped on the administrative side to make the Beijing meeting a success were Chen Ling and others of the CISTP staff. Our editors, Victoria Tomkinson, Barbara Milligan, and Fayre Makeig, have struggled with sometimes-recalcitrant texts and tables. And for all of our activities during the past two years, the Walter H. Shorenstein Asia-Pacific Research Center (Shorenstein APARC) in Stanford's Freeman Spogli Institute for International Studies has provided us with a home, and its faculty and staff have provided both support and intellectual engagement.

Throughout, Chintay Shih, dean of the College of Technology Management at National Tsing Hua University, has been an inspiring force. Thanks also to ITRI's Yi-Ling Wei and Oliver Hung, and to Jen-Chang Chou, a visiting scholar here at SPRIE. John Seely Brown, Richard Walker, David Wang, and Kyung Yoon have each provided advice and support.

None of this would have been possible without the financial support of the Ministry of Economic Affairs of Taiwan and ITRI, the Semiconductor Research Corporation, and the Hewlett-Packard Corporation.

Finally, our program coordinator, George Krompacky, has been essential to this enterprise.

Henry S. Rowen
Marguerite Gong Hancock
William F. Miller

Introduction

Henry S. Rowen

Mainland China is the world's second-largest producer of electronic and information technology (IT) products, after the United States. In 2006 its high-tech exports were a remarkable $342 billion, up $80 billion from 2005. That one-year *increase* is almost the equal of Chile's *total* national output. No wonder it is said that China is taking over world manufacturing.

However, much reporting misleads on China's high-tech trade. Did China really "make" $342 billion of these products in 2006? In reality, its IT industry is part of a global supply chain in which components (such as microprocessors, screens, disk drives, and memory chips) are shipped from Japan, Taiwan, the United States, Korea, and elsewhere, assembled on the Mainland, and then (mostly) exported. The value of its imports in this global system is about 75 percent of the value of exports, with the difference being the value added in China.[1] So China added approximately $85 billion of value in 2006. That is still an impressive number, but less than that added by Japan, the fifteen original members of the European Union (EU-15), and the United States (also in 2006).

The story is similar on integrated circuits (ICs). China's IC "consumption" reportedly went up fivefold from 1999 to 2005—reaching $56 billion, or one-fourth of global output. But much of that wasn't really consumption. Two-thirds of these chips were imported and then promptly exported in assembled goods. Although its domestic IC production is growing rapidly, China is likely to remain by far the world's largest importer of semiconductors for many years to come—and to continue to be a large exporter of them in finished products.

An interesting analysis has recently been published on where value is added on the iPod.[2] Most striking is the finding that of its $150 factory cost, which is recorded as contributing $150 to the U.S. trade deficit with China, only about $4 of value is actually added there. Further, of the iPod's $299 retail cost, Apple (especially) and other American firms captured most of the value. It is not only Apple's brand that is of value here but its conception of the system that comprises the iPod, iTunes, and computers.

China's pattern of production is similar to that of late-developing countries, at least in Asia. At first, their main advantage is in low-cost labor and their main challenges are acquiring technologies from abroad, developing management skills, and learning about world markets. The focus is on reducing costs by adopting better manufacturing processes; domestic companies often produce for foreign firms on an original equipment manufacturing (OEM) basis. With experience, more attention is paid to improving products for global and domestic markets, if the latter are large enough. This leads to original design manufacturing (ODM), in which design activities that had been performed by downstream, overseas, buyers are gradually assumed by the (formerly) OEM suppliers. Some of them "orchestrate" products in the sense that they play a larger role in conceiving their designs and managing their supply. An increasing number of Taiwanese firms have reached this stage. Eventually, some firms make the transition to original brand manufacturing (OBM) and become known to consumers globally. Taiwan's Acer (the world's second-largest desktop PC seller in the fourth quarter of 2006) has achieved this status. China's Lenovo jump-started the process by buying IBM's PC business, an event that Dieter Ernst discusses in chapter 11 of this book. Likewise, Haier products are being marketed around the world.

The histories of Taiwan and Hong Kong are instructive. Taiwan went down the path of becoming a leading manufacturer of IT products (among others) based on links with multinational corporations (MNCs) in the developed countries, especially the United States. It set up an industrial structure with many fast-reacting, competitive manufacturers—one that differs markedly from those of Korea, Hong Kong, Singapore, and the Mainland, with its many state-owned companies.

Rising wages then led Taiwan's companies to move their manufacturing operations to the Mainland. This has produced what has been called, with some oversimplification, the "Silicon Triangle," with nodes in Silicon Valley, Hsinchu, and Shanghai. Several chapters of this book deal with these linkages, and Kung Wang and Yi-Ling Wei explicitly deal with the "Triangle" in chapter 7. In 2003 Taiwanese firms accounted for 90 percent of China's $41 billion in exports of computers, components, and peripherals, and 70 percent of its exports of electronics and telecommunications equipment. In chapter 4, Douglas Fuller discusses the important role of "foreign-invested enterprises" (FIEs) in China, many of which are Taiwanese.

Hong Kong evolved very differently. Twenty years ago it was home to IT-manufacturing activities, but high wages and land costs made it uncompetitive as a manufacturing hub. It evolved instead into a major financial center, based on the crucial advantage of operating under the rule of law, and acquired skills in managing supply chains, acting as a middleman between low-cost manufacturing sites in nearby Guangdong and distributors worldwide. In chapter 18, Erik Baark discusses the case of Hong Kong, arguing that the special administrative region (SAR) should view innovation broadly to encompass its learned skills.

Investments by foreign firms and their research and development (R&D) activities are increasingly motivated by prospects in the Mainland China market. It has the most mobile-phone users in the world, the largest TV market, the second-largest PC market, and a thriving Internet services industry. The high- and middle-income classes have created new demands for products and services in head-count-based industries such as telecommunications, e-commerce, and entertainment.

Recent e-commerce growth exemplifies the distinctiveness of this market. Between 2000 and 2006, Internet users in China increased at a cumulative annual growth rate of 40 percent, to a total of 123 million. Before the end of the decade, China is likely to overtake the United States on this score. A set of local dot-coms has emerged. Often cofounded by returnees to China and locals, these companies tend to adopt—although with some variations—proven business models from the United States.[3]

What is most evident is what can be called a paradigm of execution: a focus on exploiting existing knowledge. Often this is labor-intensive manufacturing with knowledge embedded in machines, tools, standard operating procedures, and product designs in industries where competition has not yet required firms to innovate or die. This produces hierarchical organizations with little variance in operations. It relies on workers who are good at carrying out orders. The rewards are usually tangible, immediate, and certain, with factor inputs going in one end of the process and products coming out the other.

In a paradigm of innovation the problem is to come up with something better (process or product). Individuals are encouraged to think outside the box and to challenge existing rules and practices. Organizations are usually less rigidly hierarchical, with enough flexibility to help vertical (including bottom-up) as well as horizontal information flows. The undertaking seems riskier and the rewards uncertain. These two concepts are shown in Table 1.

Table 1 The Paradigm of Execution versus the Paradigm of Innovation

Characteristics	Paradigm of execution	Paradigm of innovation
Attitude toward knowledge	Exploit existing knowledge as a user	Create new knowledge and own it
Attitude toward rules	Accept and execute them	Challenge, modify, and create them
Organizational structure	Structured	Structured chaos
Communication within firms	Mostly vertical (give orders, report results)	Vertical and horizontal (exchange ideas and information)
Time horizon	Focus on the present	Look further ahead
Reward	Soon, highly likely	Delayed, risky
Competition	Compete by efficiency: "Do it better"	Compete by differentiation: "Find the better thing"

Source: This table and the distinction between the two paradigms were contributed by Ming Gu, a researcher on this book.

As the table indicates, China's current stage of development is weighted toward execution, which nonetheless can lead to significant process innovations. The best example of this is the Toyota system of manufacturing, which revolutionized world manufacturing. The importance of execution has also been demonstrated by Taiwanese companies' relentless driving down of costs. In the case of Toyota, process advances eventually led to a dominant position in products, whereas Taiwanese firms' process advances are leading toward product innovations. Process innovations, too, may be occurring in some Mainland companies, but, if so, they are hard to detect because they typically do not involve patents or other external manifestations.

The flurry of start-ups in various IT sectors in Mainland China, such as those in e-commerce and IC design, began with proven ideas, business models, or technologies from the West. However, from portal to search, and auction to online community, local entrepreneurs quickly adapted these features to local markets. With several celebrated exceptions—such as FocusMedia (which produces LCD screens in elevators), Shanda (online games), and Ctrip (travel)—few have demonstrated something new.

The transition to a more innovative society presents a systemic challenge at the institutional, organizational, and individual level. Most institutions adapt slowly. The educational system may be the slowest to change, since it emphasizes learning from history, where wisdom and memorization serve as the main mechanisms for knowledge transfer. The process is widely called "stuffing the duck": facts, concepts, and formulae are known, but little attention is paid to how they can be applied and implemented.

At the organizational and individual levels, learning to innovate also involves serious and interlocking challenges. Entrepreneurial spirit abounds in Greater China, but there is a deficiency of knowledge about how to build a team capable of innovating. This reflects a widespread shortage of experienced engineers and technical managers in various IT industries. Building competent teams—let alone innovative firms—is Greater China's first and most immediate challenge.

The Challenge of Measuring Innovations

With respect to innovation in Greater China—which is defined in these pages as Mainland China, Hong Kong, and Taiwan—this book addresses four main themes. First, the quest to be more innovative is a powerful force in all parts of Greater China. It has not yet become a significant source of new technologies, nor of more than a few global companies. But several indicators show progress, particularly increasing numbers of trained people, R&D activities (including by MNCs), and patents and papers published in international journals (along with citations in both categories). The most progress so far has occurred less in products than in processes (manufacturing in Taiwan), managing logistics chains (Hong Kong), and business models (Mainland).

Second, almost all of the technologies used in Greater China originate from outside the region. Technologies and management skills are being transferred from the countries of the Organisation for Economic Co-operation and Development (OECD) and within Greater China. Several mechanisms are at work: trade in goods and services, direct investments, academic collaborations, and students going abroad—some returning not only well educated but also with valuable job skills. Many MNCs focus on design work in the IT sector. They are motivated to serve the ever-growing China market, and to develop products for global markets. In the IC industry, the most effective transfers of technology are made by MNCs that combine foreign finance with a commitment to local operations.

Third, research institutes, too, have played important roles, especially the Industrial Technology Research Institute (ITRI) in Taiwan and the Chinese Academy of Sciences (CAS). ITRI's mission is to help commerce. The CAS has many missions, but until recently, helping commerce was not prominent among them. Relative at least to the United States, such institutes produced more important research than universities.

Fourth and finally, the talent pool is enormous in Greater China, but on the Mainland it is not very experienced. Key institutions—including state-owned companies, the CAS, and universities—have not yet demonstrated world-class research abilities. Nevertheless, Chinese scholars are increasingly contributing to science and technology (S&T), as measured by numbers of papers in international journals. The quality of their research is also measurably improving, judging by citations to these papers.

In addition to these four themes, this book focuses not on making things but on introducing new things and services. The economist most famous on this topic, Joseph Schumpeter, took a broad view. He identified five types of innovations:

1. A new product or a qualitative change in an existing product
2. A process new to an industry
3. The opening of a new market
4. New sources of supply
5. A change in industrial organization

Most of these kinds of innovations are illustrated in this book. Together, along with several types of institutions, these five types define what has become known as systems of innovation. These usually exist at the national level (national innovation systems, NISs), but regional innovation systems, RISs) have also been identified, as Claudia Müller and Rolf Sternberg observe in chapter 13.

The systems concept is useful in that it provides a framework for understanding differences in the processes by which innovations are made. For instance, France has a long-established centralized system, while the U.S. system is much more decentralized. It is also interesting to consider the top-down versus the bottom-up method of allocating national resources to S&T. The Mainland has had a top-down, state-run system (which is now being altered), while Hong Kong has had a bottom-up, private system (which is also now undergoing change). One important aspect of a NIS is the connection—or the absence of a connection—among institutions. In chapter 10, Denis Fred Simon and Cong Cao, and in chapter 3, Gary Jefferson, Bangwen Cheng, Jian Su, Paul Duo Deng, Haoyuan Qin, and Zhaohui Xuan consider the case of the CAS. Before restructuring, it was largely disconnected from commerce—its great successes having occurred in the military sphere—and thus contributed little to it. ITRI, by contrast, is a very different kind of research institute, whose mission is to help industries.

Much interest in innovation focuses on R&D, and this book is no exception. R&D is an indicator of an important input to the advancement of commercial innovations. But, as this book's chapters show, formal R&D is not always necessary—many innovations have been made on the shop floor in Taiwanese companies. Another limitation is that reported R&D can encompass a range of activities, from trying to discover something new about the universe to tweaking the features of some existing product. Accepting a few (near) truths—commercial firms do not undertake basic research, for example, and universities and science institutes do not make incremental changes in products—still leaves room for a wide range of activities called "R&D."

Making progress requires defining criteria and making measurements, and here there are difficulties. It is easy to agree on big scientific/technical breakthroughs, but we would like to measure less cosmic ones by identifying and comparing novel products or services across companies within an industry and then relating them to the inputs used to make them, such as money and people. This is challenging even for narrowly defined products, and it is increasingly difficult the more heterogeneous the products under consideration. The authors of three chapters in this book undertake this task: Fuller, by estimating outputs from China's IC design sector; Jefferson and his coauthors, by estimating technology-generated revenues from China's research institutes; and Ted Tschang and Seng-Su Tsang, by estimating the market for China's new media, such as games and animation.

Measuring inputs is less difficult, although still not easy. Favorite inputs are money spent on R&D and the numbers of scientists and engineers engaged in R&D. Nor is it hard to identify possible limitations: Do tax or political considerations play a role in reported R&D? Is innovative work done in small firms adequately reported? How should new business models be considered? As Baark remarks in chapter 18: "Quantitative R&D statistics, patent statistics, and citations reflect only formal aspects of the processes of innovation. If there are few or no formal R&D expenditures, R&D statistics do not exist or, where they do exist, are of little use. Despite low levels portrayed in formal R&D statistics, there may still be extensive innovative activity taking place."

Nevertheless, reported R&D spending is a useful indicator. China's spending in this arena has been rising rapidly and promises to keep doing so. It quintupled in real terms between 1995 and 2004, doubling as a percentage of gross domestic product, GDP (from 0.6 to 1.3 percent), while the reported number of researchers increased by 77 percent. A 2005 survey of the largest R&D spenders worldwide, conducted by the United Nations Conference on Trade and Development (UNCTAD), rated China third in R&D spending globally.[4]

The Importance of Research Institutes

Intermediate outputs from the innovation process can be measured through several useful indicators. One is scientific publications. In chapter 15, Ping Zhou and Loet Leydesdorff measure the quality of such outputs by assessing both the reputation of the journals in which papers appear and the citations to them in the global literature.

From an American perspective at least, the importance of research done in institutes in Taiwan and Mainland China is striking, relative to its scale in both companies and universities. This book discusses two important research institutes, ITRI in Taiwan and the CAS in China. Baark also comments on the recent creation of institutes and science parks in Hong Kong.

ITRI is justifiably famous for its role in adapting and developing technologies for use by companies. ITRI was responsible for creating the IC foundry industry, and with its help Taiwan has become the world's most important supplier of OEM and ODM, with such products as PCs, semiconductors, mobile phones, LCDs, and more.

In chapter 2, Kristy Sha, Tzung-Pao Lin, Chih-Young Hung, and Bao-shuh Paul Lin offer a case study of how ITRI's Information and Communications Research Laboratories (ICLs) helped Taiwan to build a broadband technology equipment sector. Given that its companies were relatively weak in R&D, and that foreign firms dominated the long-distance and metropolitan city markets, the ICLs focused on broadband local loops, where it perceived a niche opportunity. As a result, Taiwan's digital subscriber line (DSL) equipment was first in the world in 2005, with 78 percent of the global market. Taiwanese vendors have advanced to the stage where they can partner with major manufacturers such as Nokia, Alcatel, and Samsung.

The Mainland's research institute structure is enormous, with about five thousand institutes nationwide, including those that function at the local level. This structure, copied from the Soviet Union, is a legacy of the planned economy. As Richard P. Suttmeier and Bing Shi put it in chapter 1, the CAS is the "backbone" of the NIS. It has enjoyed some major achievements, largely in national security, but the CAS has also had serious flaws: a decoupling from markets, a confused set of missions, and a lack of competition for research funds.

Its commercial inadequacy is captured in the observation that before reforms began in 1979, it had made forty thousand inventions but commercialized none.[5] CAS institutes retained too many nonresearch workers and scientists who had passed their peak productivity and who lagged behind international research frontiers.

Suttmeier and Shi pay special attention to the CAS reform program, known as the Knowledge Innovation Program (KIP), one of whose goals is creating thirty internationally recognized research institutes by 2010, with five of them being world leaders. Under this program, the numbers of administrative staff members have been streamlined, and more professors have been hired. The average age of scientists and managers has been reduced. Competition for funds has intensified. The number of peer-reviewed papers in *Science Citation Index* (*SCI*) journals has skyrocketed (up 148 percent between 1998 and 2004), and there has been an eighteenfold jump in patents granted. The authors report that CAS-industry relations have, correspondingly, been transformed.

The CAS is unique in having many goals. It aims, first, to be a preeminent center of basic research, performing cutting-edge, high-tech R&D; conducting research in agriculture, health, energy, the environment, and national defense; training graduates; and promoting high-tech entrepreneurship, industrial extension, and economic development in cooperation with local governments. Second, it seeks to be an honorific organization whose elite academicians play an important science advisory function.

Having so many missions raises questions. One relates to the perceived neglect of commercial innovation. Suttmeier and Shi report that many experts in China believe that commercial innovation needs to come principally from company R&D, but companies have been weak in this arena precisely because so much of this activity had been assigned to government research institutes. Managers of state-owned companies often find it easier to seek government bailouts—that is, they have soft budget constraints—than to do the hard work of carrying out R&D and introducing new technologies. In any case, more than 60 percent of China's reported national R&D is now done by industry, a large increase from the past. As the authors put it, this is not entirely compatible with a CAS-centered view of innovation, nor with the increased role now intended for university research. In short, the CAS is experiencing increased competition from the industrial sector on the one hand, and from universities on the other.

In chapter 3, Jefferson and his coauthors use three criteria to discuss outcomes from the restructuring of the research institutes: revenue generation (especially, given that most of these institutes were supposed to become self-financing), patent production, and financial return on patents. The authors find that moving institutes to S&T enterprises—the main kind of conversion—has substantially improved their research productivity. Those converted to nonprofit, nonresearch status showed some improvement, while nonprofit institutes that remained under government supervision actually declined somewhat, as measured by revenues earned from their technology.

Research and Development by Multinational Companies

In their chapter on the spread of technological competencies in *The Dynamic Firm: The Role of Technology, Strategy, Organization, and Regions*, Pari Patel and Keith Pavitt offer two main reasons why MNCs tend to do their research in-house. The first is the fact that much knowledge about technology is tacit, person-embodied, and nontransferable.[6] The second is that firms are shaped by the specialties, accumulated research, and labor-force skills of their home countries. In Asia, however, where very large numbers of people are involved, the growth of skills, lower costs of telecommunications, and growing markets with distinctive demands are causing this pattern to change. Estimated R&D expenditures by U.S.-owned subsidiaries in China went from $7 million in 1994 to $650 million in 2002 and have doubtless grown since then.[7] A recent study highlights some of the perceived advantages of doing R&D in China—notably tapping the vast pool of talent and staying abreast of competitors in China and elsewhere in Asia—and predicts that these R&D labs will use China's emerging talent pools and technologies to shift their focus from support and adaptation to full-scale R&D work.[8]

David Michael, senior partner and managing director of the Beijing office of the Boston Consulting Group (BCG), offers three main reasons for foreign firms doing R&D in China.[9] First, growth in the domestic market drives companies to customize products to suit its needs, which means having more R&D capability on the ground. That China is the world's largest mobile-phone market offers a case in point. Second, Michael notes that "a critical mass of manufacturing and sourcing activity is emerging in China, and R&D is complementary to these activities . . . you need R&D to help those companies comply with your standards, to understand how they fit into your development and production processes." The third driver is talent. "There's a global war for talent," he observes, "and you can't find the talent you need in sufficient numbers just by getting it from traditional sources in the West."

Dependence on technologies from elsewhere fits the pattern for developing regions, which Taiwan and Hong Kong were until recently, and the Mainland is still. This should not be surprising, given that over 90 percent of all R&D in the world is carried out in the developed countries. Therefore, it is eminently rational for Chinese companies to acquire their technologies from abroad. They do this by following world market trends and licensing, hiring returnees with expertise, "me-too" copying, and exploiting various work-arounds, both legitimate and illegitimate. According to one Chinese venture capitalist, "Why go to the trouble to innovate when there is so much low-hanging fruit out there?"

This is not to assert that China lacks the capacity to carry out major technology projects. This capacity was evident in its nuclear weapons and space accomplishments from the 1960s on, often expressed in the phrase "two bombs and a missile." At its current state of development, however, it is not feasible for China to create technologies across a wide spectrum—and it would be wasteful to try. Nevertheless, the government finds this dependence intolerable and is determined to end it.

Many foreign firms have established R&D centers in Mainland China and Taiwan. According to official Chinese statistics, 750 foreign centers had been established by the end of 2004; most of them were set up after 2001. Eight of the world's top ten R&D–spending international companies had centers in China or India (Microsoft, Pfizer, DaimlerChrysler, General Motors, Siemens, Matsushita Electric, IBM, and Johnson & Johnson). By 2004 China had become the third most important offshore R&D location after the United States and the United Kingdom, followed by India (sixth) and Singapore (ninth). Much of foreign firms' R&D offshoring to Asia is concentrated in the electronics industry, which China dominates in the area of hardware.

What is actually going on inside the MNC centers is not easy to answer. This is because R&D activities are sensitive for competitive reasons and because foreign companies may exaggerate the technical complexity of what they are doing to gain favor with the authorities. Broadly, there is little of what is properly called "research" being done, and not a great deal of "development," but much design activity (giving a different meaning to the D in R&D from the usual one). Nonetheless, significant activities are under way.

Several chapters of this book address R&D activities by foreign companies in Greater China. Fuller's main message in chapter 4 is that foreign firms differ greatly in both the transfer and creation of technology. Based primarily on 342 interviews with government and business participants in China's IT industry from 1998 to 2006, Fuller found that a particular kind of foreign firm stands out in this respect: what he calls the "hybrid FIE." The hybrid FIE combines finances from the advanced countries with a commitment to China-based operations. Firms with foreign financing face hard budget constraints, whereas state-owned ones do not. A commitment to China-based operations implies a willingness to do serious R&D operations there, as distinct from viewing China as one place among many for such activities.

In semiconductors, Fuller reports that hybrid FIEs far outperformed others in China's technological upgrading. As the number of chip designers went from fewer than two thousand in 1998 to over seven thousand in 2003, these firms led the way in training. In interviews with firms employing about half of China's chip-design engineers, he found that hybrid FIEs trained 1,200 engineers in real design skills, whereas ordinary foreign firms trained only 100, and domestic firms at most trained 488. Furthermore, only two of the 26 hybrids were primarily doing reverse engineering, whereas 11 of the 19 domestic firms were.

In chapter 5, Lan Xue and Zheng Liang focus on foreign firms' research in Beijing. Using survey data through early 2005, the authors found low wages to be the first motivating factor, especially in software. These R&D centers have close connections within company networks, but not locally in Beijing. Although many of them have ties with universities and research institutes, Xue and Liang found these links to be weak; the centers are a kind of enclave. The authors identified both "knowledge-exploiting activities" (that is, bringing technology from outside and adapting it locally) and "knowledge-exploring" ones (creating new knowledge for global use) with the domestic market becoming increasingly important over time. However, few of these centers had applied for patents, and they were not deeply embedded in the Chinese innovation system.

In chapter 6, Yuan-chieh Chang, Chin-tay Shih, and Yi-Ling Wei distinguish between demand-oriented and supply-oriented forces in stimulating the globalization of R&D. The demand-oriented forces include integrating with production plants overseas, local ambitions among subsidiaries, host government policies, and the need for closeness to local customers. Supply-oriented forces include access to low-cost talent and access to foreign S&T. Based on interviews and a survey, the authors conclude that the R&D centers of foreign firms in Taiwan have been mainly exploiting technologies brought from overseas, but that they are gradually evolving into coordinators of regional R&D activities in Greater China and global markets. According to the authors, the main reasons that MNCs set up these centers are to support government policy and to tap local research networks. They argue that Taiwan should focus on specialized areas of S&T and emphasize the domestic market, such as in information products and services. MNCs might then want to learn from Taiwan's experience by using their R&D centers as testers of lead markets. The best examples are in mobile phones and computers. This implies a need for Taiwan to (1) have a large supply of R&D experts, strong niche research bases, and strong intellectual property (IP) protection; (2) encourage R&D–related foreign direct investment via government promotion, R&D incentives, and the creation of science parks; and (3) support stronger MNC-local R&D networks.

Wang and Wei address the links that connect Silicon Valley, Hsinchu, and Shanghai—the so-called "Silicon Triangle." Silicon Valley carries out product and technology innovation, Taiwan takes care of product development and logistics management, and China handles manufacturing. (Obviously it isn't quite this simple: companies from Europe and Japan—such as Ericsson, Alcatel, Sony, and NEC—also participate.) The authors consider the motivations and strategies of the MNCs with R&D activities in Taiwan as part of this system. Because Taiwanese manufacturers have moved their production to China it is no longer a mass-production base, and making innovations has therefore become an urgent matter. They find these centers to be highly focused on current market demand with a one- to two-year time horizon. Most are sales or regional headquarters, not R&D headquarters, and contribute little to Taiwan's technology.

Wang and Wei, in chapter 7, anticipate a tripartite future in which, first, Silicon Valley continues to focus on innovation, including creating new industries; integrating technologies; and establishing new industrial standards, marketing, and services. Second, Taiwan will become a product-design R&D center and a global supply center while sustaining its innovative low-cost model. Third, China, with low labor and land costs and market advantages, looks set to gradually become a capital- and labor-intensive manufacturing center, which will also develop its own market-oriented R&D and product designs and brands. These authors conclude that the pattern of technology division in the "Silicon Triangle" is unlikely to change much in the next five years. Taiwan's enterprises need to cooperate closely to keep their leading position globally. They need to maintain close links with U.S. companies and to bolster those connections by forging additional links to companies in Europe and Japan. Furthermore, the growing importance of services requires that Taiwan develop this sector.

Ingo Liefner and Stefan Hennemann, in chapter 8, investigate factors that determine cooperation between foreign and local firms in China, connections among high-tech firms there, and the impacts of such cooperation. Together with academic and research institute partners in China, they surveyed high-tech companies in Beijing and Shanghai, and found foreign ownership, company size, and a commitment to developing products for the China market to be the key factors. In an inquiry into public research organizations and universities, they found that the most innovative firms had close connections both with local research organizations and with foreign companies.

In chapter 9, Meng-chun Liu and Shih-horng Chen report on R&D investments by Taiwanese companies overseas. These firms began to make such investments in the late 1970s, first in Southeast Asia. In the late 1980s, investments shifted to Hong Kong's Pearl River Delta, in such industries as apparel, umbrellas, and footwear, followed by PC assembly and notebook computers. By 2005 all Taiwanese company notebook PCs were assembled in China and investment had shifted to the Yangtze River Delta region. The Taiwanese IC foundry service business began in Shanghai in 2003. Seventy percent of all Taiwanese foreign direct investment is in China, mainly in manufacturing, with the top five in 2002–2004 being electronics, basic metals, chemicals, precision machinery, and nonmetals. In contrast, Taiwanese firms' direct investments in other parts of the world are concentrated more in services. Liu and Chen's survey shows that a large majority (84 percent) of Taiwanese firms mainly use technology originated in Taiwan; only 25 percent report doing R&D locally. Those firms that do so are motivated by developing products, accessing the market, and lowering costs. Catching up with rivals' technology gets a low rating. Outside China, these firms prefer to have clients as R&D partners, but not inside. Liu and Chen highlight the general importance of MNCs having ties with local innovation networks. Promoting scientific/technological cooperation can help to foster these connections, but—they speculate—Taiwanese firms find building ties locally to be difficult.

Talent and Innovative Capacity

Anyone who doubts the depth and scale of China's talents should consider the following statement, made in chapter 10 by Simon and Cao:

Between 1999 and 2003, Beijing University and the University of Science and Technology of China in Hefei were the two largest baccalaureate-origin institutions of U.S. doctorates in physical science (558 and 461 doctorate recipients, respectively), surpassing both the Massachusetts Institute of Technology (MIT) and the University of California–Berkeley by well over 100. In engineering, for the same period, Qinghua [Tsinghua] University was the largest baccalaureate-origin institution, with more than twice as many graduates earning U.S. doctorates than the largest U.S.-origin institution, MIT (863 versus 344).

One possible reaction to this news is that all numbers in China are big. Even so, these numbers are striking. Another response, salient to current Chinese government concerns, is that only a small proportion of these talented and well-educated people have returned to China so far.

A huge expansion is taking place in Chinese higher education, with the number of students admitted to tertiary education going from 1.5 million in 1999 to 7.5 million in 2005. The number of doctoral degrees awarded in 2000 reached 7,300, more than Japan and second only to the United States (which awarded 26,200 doctoral degrees that year).[10] Much of this expansion is in scientists and engineers; 33 percent of university students in China studied engineering, compared with 20 percent in Germany and just 4 percent in India. Official statistics for 2003 show that higher-education graduates in S&T reached 800,000—more than double the figure of a decade earlier.[11] The world has never seen such a combination of scale and speed—albeit at a cost in quality.

Vivek Wadhwa, founder and CEO of Relativity Technologies and an active member of the influential nonprofit network The Indus Entrepreneurs, raises questions about these numbers. He finds that Chinese (and Indian) official data include graduates of two- and three-year programs.[12] Particularly in China the label "engineer" is used more loosely than in the United States. Looking at all computer science and IT degrees from four-year schools in 2004, Wadhwa originally came up with a figure of 137,000 engineering graduates for the United States, compared with 112,000 for India and 351,000 for China. After further inquiry, the only clear conclusion he reached was that engineering numbers are increasing in both India and China.

For a long time, academic positions in China were relatively unattractive. The Cultural Revolution did much damage to universities, which have been slowly recovering. Although academics have traditionally been held in high esteem, pay became so low that many faculty members were forced to engage in outside business to make ends meet. This situation is improving with higher pay and more research support, but there is still a long way to go.

Cong Cao, in chapter 12, notes that the level of much of China's scientific and technical elite is not very high from an international perspective. The members of the CAS perform many roles: doing—and managing—research, training students, providing expert opinion on a large variety of national issues, and participating in international exchanges. Their professional freedoms and their influence have fluctuated widely. From the low point of the Cultural Revolution these have risen greatly; still, few exhibit signs of political nonconformity—the Party is wary of people for whom scientific autonomy is a crucial value. Cao asserts, "It is certain that current CAS members are not at the same level as members in some of the most advanced countries." He also writes that the quality of members appointed since 1980 is inferior to those appointed earlier, a decline he attributes to the fact that more recent members were trained at home during a period in which higher education was in a sorry state.

There is also a serious shortage of experienced workers, especially in management and research, which is to be expected given the high expansion rate of graduates.[13] According to a 2005 McKinsey report, of the 1.6 million young engineers in the country, only about 160,000 have the practical and language skills necessary to work for a multinational—an amount no larger than the United Kingdom's talent pool.[14] McKinsey also reports that China will need 75,000 managers with some form of global experience in a decade; currently it has only about 5,000 such people. On-the-job experience will correct many of these deficiencies.

There is now a trickle of returnees. They have the advantage of being among the most skilled of their generation, with good education and research experience. Although their management skills are valuable, some have lost touch with changes in China, especially those who have been away a long time, and this limits their usefulness.

Simon and Cao address the demand and supply of scientific and engineering talent. Their model suggests that upward pressure in this market will be moderated over time through experience and the return of experts. They also assert that China lacks a pool of skilled people capable of breakthroughs in scientific research and technology, a perception that government officials evidently share, given their efforts to improve higher education and to recruit overseas Chinese.

Ernst asks if China will be able to move beyond being a low-cost, export-oriented global factory to becoming innovative. He sees China as having a unique combination of advantages: the world's largest pool of low-cost and trainable knowledge workers, a booming market for electronics products and services, sophisticated lead users and test-bed markets, and strong policy support for its innovation system. As a late-latecomer, China has the additional advantage of learning from the achievements and mistakes of others. Most important, China can take advantage of global knowledge networks that now extend beyond markets for goods and finance to those for technology. Ernst argues that China is more integrated into these networks than were Japan and Korea at a similar stage of their development. Corporate networks link Chinese firms to customers, investors, technology suppliers, and strategic partners through foreign direct investment, venture capital, private equity investment, and contract-based alliances. Informal global social networks connect it to overseas innovation systems through the circulation of students and researchers. He shows how integration into global knowledge networks enabled the computer company Lenovo to jump into global markets by buying IBM's PC business.

Müller and Sternberg focus on a key aspect of talent in an innovation system, and one in which the Chinese have a strong and justified reputation: entrepreneurship. They focus on Shanghai's RIS, and the role of returnees in the semiconductor and software sectors there. Through interviews, they find that returnee entrepreneurs serve an important educational function, both with workers and customers. And the activities of returnees produce knowledge spillovers that help high-tech industries in the region more broadly.

Statistical Indicators: Patents and Publications

More foreign firms are securing patents in China, and more Chinese firms are securing them abroad. China's inventors moved from fifty-seventh in the U.S. patent system in 1985 to eighteenth in 2003. Among the outsiders within China's patent system, Japanese, Taiwanese, and Korean firms are in the lead.

Fuller's research has found that the international hybrid firms, also called hybrid FIEs, stand out as a source of patents. He notes, "Examining U.S. utility patent data within the broader IT sector and in the IC sector, from 1997 through 2004, global hybrid firms (those with foreign financing and China operations) created 503 of China's 616 corporate U.S. IT utility patents. Domestic firms created only eleven and the remainder were created by foreign multinationals. Their outsized role is all the more impressive because they were competing against large-scale MNCs and growing domestic giants, such as Huawei."[15]

To discern the direction and growth of innovative activities over time in (non-Japan) Asia, in chapter 16 Poh Kam Wong traces the five hundred largest patent-owning companies in the world. He finds that the four Asian newly industrialized economies (NIEs), plus China and India, have increased their share of influential patents and also that the share of foreign patents has been growing rapidly, especially those by inventors in China, India, Taiwan, Korea, and Singapore. Taiwan has become the third-largest patenting economy in the world, behind the United States and Japan. The (non-Japanese) Asian ownership of U.S. patents granted between 2000 and 2004 varied greatly. Inventors from Taiwan received 53 percent, from Korea 32 percent, from Hong Kong 5 percent, and from both Singapore and China 3 percent. The growth rates of their patenting also ranged widely, with Singapore and China growing much faster than the others, though from lower bases. The four NIEs increased their share of influential patents worldwide and, within the top 5 percent of the most cited patents, Taiwan had overtaken the United Kingdom and closed in on Germany. Korea, meanwhile, had overtaken France and approached Switzerland.

Wong makes some observations about what he calls the "Global IP 500" companies, that is, those that own the most patents. He finds that most of these companies' patents are home-based, but that they are gradually becoming more widely distributed. Thus, the less advanced countries naturally depended on foreign knowledge when their local innovation systems were weak, but this becomes less necessary as local knowledge strengthens.

In chapter 14, Albert Guangzhou Hu focuses on China's "explosion" in patents. In 2004 outsiders accounted for two-thirds of invention patents, a higher share than in the United States and much higher than in Japan. Invention patents granted to Chinese applicants have been growing annually at almost 26 percent, but those to inventors in the United States, Japan, Germany, Taiwan, and Korea have grown even faster—with Korean ones growing at 58 percent a year.

Why is there such aggressive acquisition of IP rights when IP enforcement is so weak? Is this surge driven by foreigners' need to protect proprietary technologies against Chinese imitators, or is it spurred instead by competition among foreign investors in the Chinese market? Observing that foreign patenting in China has been growing three to five times the rate of foreign patenting in the United States, Hu infers that competition among foreign firms and between foreign and domestic Chinese firms is the driving force.

According to Zhou and Leydesdorff, the scientific production of China, as measured by publications, has grown exponentially for over a decade. Specifically, it advanced from seventeenth in the world in 1993 to fifth in 2004. China strives to be an innovative country, and the role of its scientific publications is a crucial indicator. However, sciento-metricians regard the number of citations that publications receive as more significant than publications alone, because citations show the visibility or impact of scientific output. Although higher than earlier, in 2004 China's publications ranked only fourteenth in the share of citations. Zhou and Leydesdorff ask: Why is this, and can Chinese S&T institutions make a larger contribution to knowledge?

The authors judge that too much emphasis is placed on counts of publications, which leads Chinese authors to focus on the number of publications, rather than the quality. The high share of publications and relatively low share of citations illustrate this tendency. It is not that citations are seen as unimportant in China; on the contrary, inclusion in the *SCI* is a major aim of Chinese journals' editorial boards. However, Zhou and Leydesdorff find that inclusion in the *SCI* does not necessarily increase visibility and conclude that research institutes, authors, and editorial boards need to try harder. Chinese authors, they suggest, must focus more on original and innovative research; producing papers in English particularly enhances visibility.

High-Tech Regions

Industry concentrations, or clusters, often emerge naturally. For example, national capitals attract telecommunications companies because that is where government regulators are located. Other clusters tend to form around major universities, or develop in places with nice weather or in cultural centers—or in locations that combine such factors. Silicon Valley boasts two of these three features, as does Shanghai—but not the same two. Much of Greater China's high-tech industry is geographically concentrated in such cities as Hsinchu, Shanghai, Beijing, Shenzhen, and Suzhou.

Of particular interest here is government action—sometimes successful, sometimes not—to create clusters.[16] Government created the Hsinchu Science-based Industrial Park (HSIP); Beijing and Shanghai, too, are natural sites for high-tech clusters, but governments have boosted the process. Beijing's Torch program supports fifty-three high-tech regions that are widely distributed throughout the country.

Yih-Luan Chyi and Yee-Man Lai, in chapter 17, focus on the workings of HSIP, which is probably the most successful government-created high-tech park in the world. Set up in 1980 to attract high-tech firms, including start-ups, the HSIP's goal was to become a Silicon Valley of the East. Nearby were ITRI and two major universities, National Tsing Hua University and National Chiao Tung University. By the end of 2004, HSIP had 384 tenants, and it had grown at an annual rate of 12 percent over the previous two decades. As of 2004, returnees owned almost one-third of these firms. Total sales were NT$11 trillion (US$350 billion),with an annual growth rate of 38 percent. The number of employees had increased more than ten times, from 8,275 in 1986 to 113,000 in 2004.

The success of HSIP firms is due in part to the fact that technology spills over into the knowledge networks in which they operate. This means that without paying any cost, a firm can benefit from other firms' research—that is, research performed in one firm can stimulate the creation of new knowledge or combine ideas from other firms. This type of knowledge spillover does not require direct input-output connections among firms or industries.

With patent data from HSIP and Silicon Valley firms, Chyi and Lai construct measures of knowledge spillovers, both within HSIP and between HSIP and Silicon Valley. (Their measures of international knowledge spillovers pertain to the semiconductor industry.) Hsinchu's high-tech clusters display knowledge spillovers measured by R&D performed. In particular, the authors find spillover elasticities from R&D done *outside* the companies to be higher than from the companies' *own* R&D. In the semiconductor industry they find that the foreign knowledge stock has a stronger impact on net sales than the domestic knowledge stock.

In chapter 18, Baark describes the evolution of Hong Kong's high-tech policies. Since returning to China in 1997, Hong Kong has moved away from its policy of "positive noninterventionism" and toward the fostering of technology and innovation. It has set up the Innovation and Technology Fund (ITF), supplied venture capital, and created the Cyberport, the Hong Kong Science and Technology Park, and the Applied Science and Technology Research Institute (ASTRI). Baark also notes that total R&D spending in Hong Kong, both public and private, went from less than 0.30 percent of GDP in the 1990s to (a still comparatively low) 0.69 percent in 2003. Significantly, the business sector share of R&D spending went from 18 percent in 2000 to 41 percent in

2003. Moreover, economic integration with the Mainland, the strengthening of international research linkages via its universities, the growth of overseas R&D activities by Hong Kong companies, and its role as a financial center have all created a web of innovation networks. Hong Kong endeavors to be an innovation hub for both China and global markets.

Services are especially important. Most Hong Kong firms' large-scale production facilities are located in the Pearl River Delta, and their activities at home are focused on management, marketing, and development. Services in finance, insurance, communication, and logistics now contribute more than 85 percent of the SAR's value-added. These types of firms are making innovations that are not adequately reflected in available R&D statistics. Baark mentions the famous trading company, Li and Fung, as an innovative orchestrator of loosely coupled supply chains encompassing many consumer products. It has a network of more than 7,500 suppliers. He argues that the government should enhance existing sectors rather than aim to create new innovation systems, and that an economy can innovate without focusing today on creating new knowledge. Creating new knowledge is likely to become more relevant at later stages, but policymakers should try to identify organic strengths and realize that it is possible to build on traditional strengths to attain innovative excellence.

In chapter 19, Adam Segal compares efforts to foster technological entrepreneurship in China, India, and Korea, focusing on three policy arenas: university-industry collaboration and university-related start-ups, policy support for small- and medium-sized enterprises, and venture capital. He summarizes China's current strategy as complementing its traditional state directed top-down approach with a more bottom-up entrepreneurial method. The government now supports all domestic enterprises designated "high technology" and helps inventive entrepreneurs by doing more to define and protect property rights (including IP), using venture funds, and developing technology markets.

Segal identifies a similar desire to foster innovation throughout Asia. India is supporting small- and medium-sized enterprises, promoting university-industry linkages, and encouraging cooperation between state-run labs and multinationals. Korea's IT839 strategy—a government effort to introduce eight new IT services, encourage investment in three key network infrastructures, and develop nine promising sectors—is complemented by the promotion of venture companies and "inno-biz." These activities are taking place within NISs that do not change quickly. In China, the legacy of the Soviet system of S&T is still felt. In India, "mission-oriented" research institutes, especially those in defense and nuclear energy, cast a long shadow. In Korea, the promotion of small, tech-focused technology enterprises, university-industry collaborations, and regional ecosystems of innovation are intertwined with efforts to reduce the gap between the *chaebol* (South Korea's large, family-controlled conglomerate firms) and small firms, and between Seoul and the rest of the country.

There are important commonalities among these countries. All see opportunities in the globalization of R&D and the return of skilled expatriates.

Policymakers are learning from neighbors, entrepreneurs are cooperating across borders, and efforts to develop new standards in open software and home media are bringing firms together. The rise of China has created a wide concern that producers in all sectors will be squeezed, which adds to the impetus to promote local innovation. Segal quotes a venture capitalist in Seoul who noted that "all Koreans think about China all the time."

University-industry collaborations, too, are an active topic in China, India, and South Korea. In China such collaborations include patent licensing, technology service contracts, joint research projects, university-based science parks, consulting agreements between individual faculty members and commercial firms, and university-affiliated enterprises. In 2002, Zhongguancun Science Park, a technology hub situated primarily in Haidian District, Beijing, was home to more than 9,500 high-tech firms, more than 200 of them university-affiliated. In 2007 that total reached 20,000 firms. There are, however, certain negatives in blurring the lines between industry and academia, most notably the impact of commercial activities on the academic environment.

Indian universities have received proportionately less research funding than their Chinese counterparts and have been less closely linked to companies. Neither faculty nor companies have valued the connection, with many academics preferring to connect to MNCs than to local firms.

Korean universities have faced similar barriers. Often weak in research, they have faced legal and social barriers to entrepreneurship. Although these barriers are being reduced, the overarching question of the proper balance of education, university research, and commerce remains.

Segal sees some convergence in policies toward small firms across these countries, as well as similar barriers: ineffective policies, a dearth of early-stage capital, a lack of scale, weak technological capabilities and management skills in small firms, and cultural barriers to entrepreneurship. He suggests that the shift toward an innovation strategy has not gone far enough because it still posits a central role for government. Instead, he sees success depending more on the roles of civic and business associations. In India it is the growing number of successful entrepreneurs involved in business-plan competitions, entrepreneurial clubs at the Indian Institutes of Technology (IITs), and angel capitalists who will overcome cultural barriers to networking and risk-taking. Only through more active and independent involvement of such groups will entrepreneurship be fostered.

Intellectual Property: The Cases of Chips and the New Media

In addressing IP protection in China, Xiaohong Quan, Henry Chesbrough, and Jihong Wu Sanderson observe in chapter 20 that there is no lack of statutes and that case law is forming in trademarks, copyrights, and patents. Enforcement, they note, is the problem. However, the government is coming to understand that creating IP requires protecting it.

In the IC sector, the authors assert that government efforts (in money, equipment, talents, and so forth) to make a competitive manufacturing industry have failed, while design houses, with no government effort, have flourished. In 2005 there were around six hundred design houses, of which a few that are run by returnees, such as Vimicro and Spreadtrum, had global competence.

Given that IP protection is weak, Quan and her coauthors ask why there are so few infringements in this industry. Or, if there are infringements, why is this not a big concern for designers? They argue that the growing complexity of semiconductor designs makes imitation very difficult and that even successful infringers would have trouble selling their products to foundries or system companies. In any case, patents are effective in only a few industries, such as the chemical one. In other industries, lead time, learning curve advantage, secrecy, and sales and service efforts offer better means of appropriating IP.

Quan and her coauthors mention several ways in which IP is being protected. Different actors specialize in different activities along the supply chain; such specialization can be done within a company across levels and locations. For example, one MNC the authors studied carefully separated its systems-design work from specific components—the systems work is done at home, while parts of the implementation are done in the firm's China laboratory. Another firm refused to move people across processes, so that any employee leaving the company would at most know a single process.

Tschang and Tsang address the new Chinese media in chapter 21. This market has seen intense, imitative, cost-competitive competition, which makes it hard for firms to differentiate themselves on IP and other output characteristics. Foreign producers have occupied strong niches or held dominant market shares. Tschang and Tsang focus in particular on animation, video games, and mobile-phone content. In contrast to manufacturing and much software, this is a domestic market with distinctive preferences—as, for example, in animation—in which government regulators play large roles. With advanced mobile phones, the rapidly growing sales of mobile-phone games was predicted to be $3.8 billion in 2007, making it the fastest-growing market in the world. Tschang and Tsang see big challenges for suppliers, especially in anticipating consumer demands and cultivating hits in a market where tastes vary across provinces. There is also a huge shortage of marketing and other talent that may last for years.

Online games naturally have much lower piracy rates than packaged traditional games (which have a piracy rate as high as 95 percent), but because users want multiplayer games, broadband penetration is critical. Broadband connectivity in homes might stimulate multiplayer online games in the same way advanced handsets did for mobile-phone games. In 2003, 70 percent of the online games in China were made in Korea, a troubling fact that led the government to restrict foreign games. Domination (to the tune of 90 percent) of the animation market by producers from Japan, the United States, and South Korea also led the government to adopt protectionist measures in this sector.

One consequence of protecting these industries on cultural grounds is that the government has gotten serious about IP protection for domestic content. More broadly, government policies include funding (for programs and infrastructure), protection (such as bans on foreign content and reservations of space in the various channels for domestic content), training, and various forms of promotion. The decision to spend $1.8 billion to develop one hundred online games based on Chinese history and heroes means trying to pick commercial winners, an activity that governments do poorly, and one—as Tschang and Tsang observe—that is especially dubious given how hard it is to anticipate consumer demand.

Intense price competition suggests the importance of ideas for new gameplay and technology. Countries with deep traditions in creative work foster new kinds of games and gameplay. Tschang and Tsang note that this bears on the possible social consequences of government prohibitions on foreign content. While limiting influences from global entertainment might be deemed "good," such bans might deny players exposure to a variety of influences, including potentially new sources of innovation. China risks raising a generation of players not exposed to new gameplay styles.

Given the intense domestic and foreign competition in online games, it may be difficult for all but the largest Chinese firms to compete, at least in the near term. This is similar to the software industry's experience, where domestic firms were unable to match foreign multinationals at the high end of the market and were competing destructively with one another at the low end.

Looking Ahead

China's leaders find dependence on foreign technology deeply unsatisfactory. They consider it to be unseemly for a great nation. There also exists a perceived national security vulnerability, and there is resentment at having to pay royalties to foreigners. The government aims to change this pattern by turning China into a major creator of S&T. The announced goal is "self-reliance," meaning reduced use of imported technology or, more broadly, IP. The 15-Year Science and Technology Plan specifies sixteen major engineering projects, including design of large aircraft, moon exploration, and drug development. The plan further highlights four major basic research programs: protein science, quantum physics, nanotechnology, and developmental and reproductive science. Each of these four programs is to receive about $1 billion. The plan places the National Center for Nanoscience and Technology and the Beijing Protein Research Center in charge of the megaprojects in their fields.

One issue is a top-down versus bottom-up decision process for these programs. That a mix of the two methods is appropriate should not be in doubt, but, for China, a balance requires a more decentralized process than it has historically embraced. Many scientists perceive this need; for example, those in developmental and reproductive biology say they intend to establish a merit-based system to distribute funds.

R&D spending by all sources, industry included, is supposed to go from $30 billion in 2005 to $113 billion in 2020. Basic research is to climb from 6 percent of R&D expenditures in 2004 to perhaps 15 percent in fifteen years. The goal is to make China a world powerhouse of S&T.

There are challenges to achieving this goal, several of which this book addresses. However, there should be little doubt that with the talents and resources available, China will reach its objective. The growth in the numbers of better-educated young people is extraordinary, as is the government's commitment to creating technology. China will become a major source of technology, but there are questions about its path over time.

Notes

[1] See National Science Foundation (NSF), *Science and Engineering Indicators 2006*, vol. 2 (Arlington, VA: NSF, 2006), Tables 6-1 to 6-4.

[2] See Greg Linden, Kenneth L. Kraemer, and Jason Dedrick, "Who Captures Value in a Global Innovation System? The Case of Apple's iPod," Personal Computing Industry Center (PCIC), University of California, Irvine, June 2007.

[3] See John Hagel III, and J. S. Brown, *The Only Sustainable Edge: Why Business Strategy Depends on Productive Friction and Dynamic Specialization* (Cambridge, MA: Harvard Business School Press, 2005).

[4] See United Nations Conference on Trade and Development (UNCTAD), *World Investment Report 2005* (New York: United Nations, 2005), 20. But Bergsten et al. criticize this report's purchasing power parity (PPP) adjustments to R&D spending, calling them "dubious" and almost certainly leading "to substantial overstatement"—see C. Fred Bergsten, Bates Gill, Nicholas R. Lardy, and Derek Mitchel, *China: The Balance Sheet: What the World Needs to Know about the Emerging Superpower* (New York: Public Affairs, 2006), 174.

[5] After the reforms began, however, the CAS spun off Legend Computer, now Lenovo.

[6] Pari Patel and Keith Pavitt, "The Wide (and Increasing) Spread of Technological Competencies in the World's Largest Firms: A Challenge to Conventional Wisdom," in *The Dynamic Firm: The Role of Technology, Strategy, Organization, and Regions*, ed. Alfred D. Chandler, Peter Hagstrom, and Orjan Solvell (New York: Oxford University Press, 1999), 192–213.

[7] NSF, *Science and Engineering Indicators 2006*. vol. 2.

[8] UNCTAD, *World Investment Report 2005*, 26.

[9] Boston Consulting Group (BCG), "The New Global Challengers: How 100 Top Companies from Rapidly Developing Economies Are Changing the World," report, May 2006.

[10] Andrew Wyckoff and Martin Schaaper, "The Changing Dynamics of the Global Market for the Highly Skilled," Organisation for Economic Co-operation and Development (OECD), "Advancing Knowledge and the

Knowledge-Economy" conference, National Academy of Science, Washington, D.C., January 10–11, 2005.

[11] National Statistics Bureau, China, 2004.

[12] Vivek Wadhwa, Testimony to the U.S. House of Representatives Committee on Education and the Workforce, May 16, 2006.

[13] Heidrick & Struggles and the Stanford Program on Regions of Innovation and Entrepreneurship, "Getting Results in China: How China's Tech Executives Are Molding a New Generation of Leaders," special report, 2006.

[14] *The McKinsey Quarterly*, November 2005.

[15] Douglas Fuller, "China's Global Hybrid Model: A New Path to Development Under Globalization," manuscript, 2006.

[16] For an examination of failed efforts to create high-tech clusters, see Scott Wallsten's "High-tech Cluster Bombs: Why Successful Biotech Hubs Are the Exception, Not the Rule," AEI-Brookings Joint Center, March 2004.

Research Institutes

SUCCESS IN "PASTEUR'S QUADRANT"? THE CHINESE ACADEMY OF SCIENCES AND ITS ROLE IN THE NATIONAL INNOVATION SYSTEM

Richard P. Suttmeier and Bing Shi[1]

The innovative capacities of a nation or a region are affected by a number of factors—the entrepreneurial talents of the inhabitants, the quality of industrial management, the nature of financial intermediaries, the system of laws and the intellectual property rights regime, the quality of the educational system, and the supply of social capital. Especially with regard to innovation in new science-based industries, innovative capacity is also affected by the quality of the research institutions of the country or region, and their capacity for original research and development (R&D), human resource development; and linkages to industry, agricultural producers, and other users of knowledge. In terms of its size, facilities, and record of achievements, one of the most important research institutions in Greater China is the Chinese Academy of Sciences (CAS).[2]

Over the past eight years, the CAS has been engaged in a major plan of reform and revitalization known as the Knowledge Innovation Program (KIP).[3] This generously funded program is intended to help the Academy adapt to the radically transformed economic, security, and ecological conditions that China faces in the early twenty-first century, and to better use the CAS's many institutional resources—human, material, and infrastructural—to fulfill its central role as the "backbone" (*gugan*) of the nation's innovation system. As the employer of much of China's best scientific and engineering talent, and with an extensive system of research institutes and laboratories, the CAS is seen as a critical part of the nation's innovative capacity in information and communications technology (ICT) and the bio-nano revolutions. Not surprisingly, many CAS leaders see the backbone role in ways reminiscent of the Academy's role in the 1950s, when (using another metaphor) it was referred to as the "locomotive" (*huoche tou*) of the emerging national innovation system (NIS).

On the one hand, since the CAS possesses many of the nation's human and material research assets, it is reasonable to think of it as having a leading role in NIS. On the other hand, and especially in recent years, Chinese research and innovation policies have put industrial enterprises at the center of the nation's innovation system and have expanded university-based research. The newly announced Medium- and Long-Term Science and Technology Development

Plan (MLP) reinforces this "enterprise-centered" view of the NIS, thus making the question of the Academy's role and mission all the more important.

This chapter argues that although the achievements of the KIP are many, the Academy still faces significant challenges. Some of these are internal to the CAS and subject to its control, while others are more systemic, having to do with the transitional nature of China's reforms, problems with the national science and technology (S&T) policy system, and the problems and opportunities resulting from its place in the international innovation system. Donald Stokes's *Pasteur's Quadrant* provides a suggestive way of thinking about these challenges as the CAS seeks not only to sort through its sense of place in "Pasteur's Quadrant" (PQ) of "use-inspired basic research" but also to respond to the pulls toward "pure" basic research ("Bohr's Quadrant") from many of its scientists and market pressures for more "pure" applied research ("Edison's Quadrant").[4] The expansion of research in institutions of higher education and the rapid growth of industrial research further challenge the CAS to define a distinctive place in the evolving Chinese innovation system. And, as in Stokes's account, the CAS's attempts to accommodate itself to PQ are linked to a larger set of issues involving the relationships between the general structure of the research system and institutions for national S&T policy.[5]

Background

The CAS entered the 1990s with an uncertain future. Its basic institutional architecture had been established under Soviet influence in the 1950s, at a time when central planning and public ownership of the economy seemed future certainties. Geopolitically, Marxist-Leninist regimes were in strategic competition with the capitalist democracies, and China felt itself threatened by the U.S. presence in Asia. The CAS helped lay the foundations for China's modern scientific development during the 1950s and subsequently came to play an important role in its strategic weapons programs. However, over time, the Academy also fell victim to the vagaries of domestic politics, especially the Cultural Revolution, and, as that ended, of a changing international economic order in which ongoing scientific advance and technological change in the capitalist countries underscored how inappropriate China's institutions were for research and innovation.

By the late 1970s, a series of national economic reforms and foreign policy changes showed that the CAS's founding assumptions of the 1950s were becoming increasingly irrelevant. Accordingly, by 1978 the Academy faced the tasks of recovering from the damages of the Cultural Revolution and meeting the challenges of its new international environment. Reform policies for S&T, initiated in 1985, drastically reduced guaranteed state funding for the CAS—and for government research institutions more generally[6]—thus forcing the Academy and its constituent institutes into a series of commercial ventures. These ventures included Legend (later Lenovo, China's leading computer maker, which recently acquired IBM's PC business, and which in turn grew out of the commercial initiatives of the Institute of Computer Technology in the 1980s. Then-president of the CAS, Zhou Guangzhao, introduced the concept of "one academy, two systems" (as both research institution and commercial technology agent) to give the CAS a sense of direction under these new circumstances—it was a national research center, but one connected closely to the economy.

In place of guaranteed funding, the reform policies introduced a series of competitive, project-based national programs for research and institutional improvement. These included the National Natural Science Foundation of China (NSFC), the National High-Tech Research and Development Program (known as the 863 Program), and (with help from the World Bank) the National Key Laboratory Program and the National Engineering Research Center Program. Although these helped to revitalize the Academy during the 1990s, the financial pressures on the CAS nevertheless continued and became especially intense with regard to personnel issues. The Academy was saddled with an aging research force, faced increasing pension costs, and suffered from the absence of a generation that resulted from the interruption of higher education during the Cultural Revolution and the ensuing brain drain. Its resources for recruiting a new generation of research personnel and for initiating new research directions—independent of project funding from the government or revenues from industry—were thus seriously constrained. This was especially so in light of the challenges of attracting back to China the many foreign-trained students who remained abroad in response to better pay, better equipment and research conditions, and better career opportunities generally.

A National Innovation System (NIS)

In 1995 the government convened a major National Conference on Science and Technology and, under the slogan "Revitalize the Country through Science and Education," elevated scientific and technological development to a major national priority. Following this conference, work began on new plans and programs to rejuvenate higher education, expand basic research, advance reforms of the S&T system, stimulate the growth of R&D in industry, and steadily increase R&D expenditures. A major CAS report to the central leadership in 1997, "The Coming of the Knowledge-Based Economy and the Construction of the National Innovation System," further supported the concept of an NIS in China's evolving technology policy and led to the introduction of a broad range of new measures affecting industrial research, intellectual property rights, and venture capital.[7]

These policy initiatives posed persistent questions about the CAS's value to a society that was rapidly abandoning the assumptions that had prevailed when the Academy was established. The appointment in 1977 of Lu Yongxiang as president of the CAS provided the occasion for the new CAS leadership to convince the government that, with a major infusion of financial support, the Academy could transform itself into a center for research and innovation that would serve China's ambitious twenty-first-century goals. The most prominent of these was to make China a center of original basic research and creative indigenous innovations and thereby to reduce its dependence on foreign technology. The resulting KIP, introduced in 1998, was seen as a phased pilot project of reform that would lead to a remade Academy by 2010. A first experimental phase (1998–2000) was to be followed by a five-year implementation of reform measures (2001–05). The CAS is now in the third phase (2006–10), in which it hopes to leapfrog into positions of scientific leadership in key areas of research.

Achievements to Date

There is no doubt that the CAS today is a markedly different and improved institution compared with what it was in 1998. When the KIP began, the CAS still supported some 120 institutes, many with overlapping missions and research agendas inconsistent with the intellectual challenges of twenty-first-century science. Institutes were seriously overstaffed with nonresearch personnel, and lumbered with scientists who had passed their peak productivity and lagged behind international research frontiers. Research programs were often derivative of foreign science, physical facilities were typically run down, and the quality of equipment was very uneven. One of the KIP's principal goals is to attack these problems by creating some thirty internationally recognized research institutes by 2010, with five being recognized as world leaders.[8]

During the first two phases of the KIP, major progress was made. The number of institutes was scaled back to eighty-nine (the number has now inched up above ninety),[9] as some applied research institutes were hived off into commercial entities and others were reorganized. In addition, duplication was reduced and missions were rationalized, bringing focus to new intellectual opportunities and societal challenges. For instance, a new Shanghai Institute of Biological Sciences was established by merging the institutes of Biochemistry and Cell Biology, Plant Physiology and Ecology, Physiology, and Brain Research. In the city of Lanzhou, an Academy of Mathematics and System Science brings together four former mathematics institutes, and a Cold and Arid Regions Environmental and Engineering Research Institute was organized out of the former institutes of Glaciology and Cyropedology, Desert Research, and Atmospheric Physics. New institutes of genomics, neuroscience, and nutrition science have also been added to the Academy.[10]

At individual institutes, traditional disciplinary orientations and missions have been redefined and restructured to bring Chinese research into new international knowledge networks that are relevant to the IT-bio-nano revolutions. At the Institute of Automation in Shenyang, for instance, a narrow robotic engineering mission has given way to broader multidisciplinary research initiatives in intelligent machinery and advanced manufacturing. At the

Institute of Microsystem and Information Technology in Shanghai, formerly the Institute of Metallurgy, the focus has shifted to electronics and information and telecommunications engineering. At the Dalian Institute of Chemical Physics, traditional strengths in physical chemistry, chemical engineering, and catalysis have been reconfigured into an imaginative strategy of basic and applied research that taps into the various funding streams that now support Chinese science. Basic research is centered in two "state key laboratories" for catalysis and for molecular reaction dynamics. Applied research identified as relevant to high-priority national needs is performed in three laboratories for new energy resources, chemical lasers, and aerospace catalysis and new materials. Additional applied research occurs in laboratories of modern analytical chemistry and micro instruments, fine chemicals, environmental engineering, natural gas utilization and applied catalysis, and in a new program in biotechnology. The Institute of Metal Research in Shenyang merged with the former Institute of Corrosion and Protection of Metals to form an important materials science center that incorporates national laboratories in materials science and corrosion studies as well as engineering research centers for high-performance homogenized alloys and corrosion control of metals. A national drug-screening center was set up at the Institute of Materia Medica in Shanghai, with the joint support of the CAS, the Ministry of Science and Technology (MOST), and the Shanghai municipal government, taking advantage of the institute's strength in pharmaceutical research and innovation.

As these examples indicate, special programs begun in the early 1990s in support of national laboratories and engineering research centers have left the CAS with centers of excellence of considerable importance for the KIP. As of 2004, the CAS had 4 "national laboratories" (*guojia shiyanshi*), 56 "state key laboratories"(*guojia zhongdian shiyanshi*), 73 "CAS key laboratories," 24 "state engineering centers" (*guojia gongcheng zhongxin*), and 8 "CAS engineering centers."[11]

An important objective of the KIP has been to reduce the number of CAS personnel and to recruit younger, better-trained researchers. Between 1998 and 2004, the Academy's professional and technical staff was reduced from 42,693 to 29,474, with the number of senior personnel being reduced from 16,031 to 13,058, along with notable reductions in administrative and support staff. At the same time, the number of researchers who qualified as full professors increased from 4,677 to 4,871.[12] New programs have been designed to recruit a new generation of talented group and laboratory leaders, from brain-drain scientists working abroad to promising young researchers in China. Among the more prominent of these is the 100 Talents Program, which offers high salaries, responsible positions, and generous start-up research support to leading young Chinese researchers working both abroad and in China.[13] Between 1998 and 2004, 899 researchers were recruited using this mechanism, 778 of whom were working overseas. Of these, 392 had earned doctorates from foreign universities. The CAS also expanded its graduate training, with the total enrollment as of the end of 2004 reaching some 33,000 at its institutes, its graduate school, and its University of Science and Technology campus. A major new CAS university center in Beijing is now under construction.[14]

These restructurings, new programs, and new personnel have rejuvenated the CAS leadership and have also served to link the Academy's work much more strongly to international research. Whereas the average age of institute directors and deputy directors in 1991 was fifty-six, by 2003 it had been reduced to forty-seven. Between 1998 and 2003, the CAS made 14,409 new appointments, 67.8 percent of which were senior scientists under the age of forty-five.[15] The percentage of research personnel with doctorates has tripled in the years since the KIP was initiated, but remains relatively low (at only 14.6 percent, or 6,297 individuals) of the total research staff.[16] While these appointments are significant, the recruitment and retention of young talent are also of long-term significance. New appointments no longer carry promises of lifetime tenure but are subject to careful evaluations early into each researcher's CAS career. In addition, salary structures have changed and now recognize the extra responsibilities that accompany leadership positions, and they also include generous provisions for merit increases.

Over the past eight years, and particularly during the most recent, second, phase, the KIP has also supported the CAS's three broad domains: fundamental research, high technology with strategic significance, and S&T for managing resources and the environment. KIP funding—with 70 percent going directly to institutes that have scored well on CAS performance evaluations and 30 percent controlled by CAS headquarters—has given the institutes more discretion in management of research, and has made them somewhat less dependent on scouring for support from national programs and commercial activities. At the same time, however, the additions from KIP were intended to make CAS institutes more competitive, vis à vis universities and other government research institutes, for grants and contracts associated with the programs that MOST, NSFC, and the National Development and Reform Commission (NDRC) administer. Although success in this goal may be disappointing, as noted later in this chapter, research funding at the CAS is notably more robust than before the KIP. There has also been a notable increase in CAS research outputs. Peer-reviewed papers in journals catalogued in *Science Citation Index* increased by 148 percent between 1998 and 2004 (from 5,860 to 14,516), and marked rises have also occurred over this period in patent applications (3.2 times), patents granted (18.6 times), and registered copyrights.[17]

A robust evaluation system has accompanied the implementation of the KIP. CAS evaluation work involves administrative reviews to assess the consistency of institute activity with CAS policy and KIP objectives, and also entails peer review of professional work by leading Chinese and foreign scientists. Thus, this process benchmarks CAS performance against the best international standards. It has also led to major investments in upgrading facilities and equipment. These days, most CAS institutes are housed in nicely constructed new buildings filled with recently acquired equipment.

Finally, much has been made of the need to introduce a culture of innovation. This has taken a variety of forms—new attention to the ethics of research, a commitment to popularization and science-society relations, and more open and cooperative relations with universities, industry, and local governments. The strong commercial orientation into which the CAS was forced in the 1980s has evolved into a more mature approach to the operation of its own companies, and to relations with Chinese enterprises and local governments.[18] As the industrial economy is forced into international market competition, the demand for new technologies increases, placing CAS-industry relations on a different footing from fifteen years ago. Market forces are both driving and unlocking new market forces, and forging commercial ties with many new high technology companies that have appeared in China in recent years. One example of this evolution is the technology developed in robotic engineering at the CAS-founded Institute of Automation, which has been successfully transferred to a leading computer game developer, the ShanDa Company, based in Shanghai.

The Challenges of Phase 3

The CAS is now entering the third phase of the KIP, which will run until 2010. Its objectives call for significant new research initiatives and institutional changes that go beyond those in Phase 2. In particular, the CAS seeks to develop its ability to respond to emerging national policy priorities, including those identified in the national 11th Five-Year Plan and the new fifteen-year MLP. By proposing a sharper mission focus to its work, the CAS hopes to secure its backbone role in the Chinese NIS. To accomplish this, the Academy is establishing a new matrix management scheme, in which the activities of its research institutes will be linked to ten high-priority national mission areas. Supporting the entire effort is a new commitment to interdisciplinary basic research in frontier areas—defined as the "purpose-orientated cross-disciplinary research base at scientific frontiers"—in what is referred to as the "10+1" scheme.

To coordinate activities in each of the ten areas with the work of relevant institutes, this PQ-like scheme will require the CAS headquarters to initiate new mechanisms to support these innovation bases. These new mechanisms have the potential to change relationships between the institutes and the central CAS in ways that could compromise some of the other KIP objectives. Such conflicts could occur if mission-oriented objectives from above clash with the pursuit of scientific distinction in selected fields below.[19]

A new mission orientation is also evident in the CAS's efforts to cultivate relations with local governments. With local governments spending more on R&D and steering their economies toward higher-value-added production, the CAS has found willing partners in a number of newly established institutes. These include facilities in Shenzhen (advanced technology), Suzhou (nano-technology and nano-bionics), Qingdao (biomass energy and bioprocess technology), Yantai (coastal zone research for sustainable development), Ningbo (materials technology and engineering), Xiamen (urban environment), and Guangzhou (biomedicine and health). Through these new organizations unencumbered by inherited managerial cultures, the CAS hopes to create institutes with strategic research orientations that couple a commitment to advanced science with the sensitivity to national needs that is characteristic of PQ objectives.

The progress of transforming the CAS through the IP has been truly impressive, but the problems of implementing Phase 3 are not insignificant. These are summarized in the ensuing sections.

Table 1.1 "10+1 Science and Technology" Innovation Bases

Basic Research	Purpose-orientated, cross-disciplinary
	Comprehensive research based on Big Science facilities
Strategy High-Tech	Information technology
	Space science and technology
	Advanced energy science and technology
	Nano, advanced manufacturing, and new materials
Sustainable Development Research	Human health and medicine
	Industrial biotechnology
	Advanced sustainable agriculture
	Eco-environmental studies
	Resources and marine science

Human Resources

Technical talent continues to be a major concern and has a number of dimensions. First, while the KIP helped to rejuvenate the ranks of researchers, it still faces the problem of sustaining that rejuvenation since the senior ranks have been filled with recently appointed, relatively young scientists. The KIP needs to recruit excellent scientists and bring them into positions of leadership at the stages of their careers when they are not only most creative, but also most oriented toward application. Second, although the CAS has sought to recruit the very best scientific talent, most observers would agree that the results are mixed. Although the 100 Talents Program has had successes, it has not attracted those Chinese scientists working abroad who are most active at the frontiers of international science. Indeed, some of these scientists have become especially vocal in their criticisms of the Chinese research environment.[20] China continues to lose many of its top students (including those coming out of the CAS educational system), both to study and research opportunities abroad and alternative employment opportunities in China, not to mention work in universities and in the growing number of R&D facilities of multinational corporations (MNCs). Thus, in spite of many reports about the abundance of scientists and engineers in China, there is intense competition for the best of these, with the CAS having to compete with the high salaries of the industrial sector on the one hand and an improving university environment on the other. Within the CAS graduate school system—the world's largest in terms of student enrollment—a balance must be struck between quality and quantity. The steady expansion of graduate enrollment in the CAS, driven in part by its wish to ensure that it has a steady supply of young researchers, implies a need for quality control as the numbers increase. Does the Academy's large graduate program do more to build its own work force (as some in universities allege) than to train the next generation of researchers for science-based innovation?[21]

In Phase 3, the Academy thus faces the problem of being able to consistently recruit and retain top people. High-quality researchers expect a degree of stability in the research environment and worry that the new Phase 3 initiatives could threaten it. The evaluation system, especially for new group leaders, places enormous pressure to produce on young scientists, who require a few years to get their laboratories established and to recruit new graduate students to join their groups. In some cases, pressure for research achievements has caused promising scientists to leave the CAS for employment elsewhere.[22]

Considering the variety of institutes within the CAS, due regard must also be given to different types of evaluation standards and processes. The new mission orientation will clearly make developing an appropriate evaluation system more challenging. On the one hand, CAS aspirations to achieve world-class status in research will put a premium on an evaluation system that focuses on scientific and technical merit. On the other hand, the emphasis on a centrally directed and coordinated mission orientation calls for an evaluation system that focuses on the consistency of research performance with national policy and on the extent to which social needs are met. In short, developing an evaluation system suitable for programs in PQ is a major challenge.

In addition, the problems of fraud and corruption that have recently plagued Chinese science now beset some of the programs designed to improve the talent pool by enticing Chinese scientists working abroad back to China. In some cases, the high salaries and attractive material incentives used in these programs have been abused. Researchers have enjoyed the salaries without taking their research responsibilities seriously, while employing institutions have been satisfied to use their relationships with these star scientists to improve their evaluations and thereby qualify for increased funding.[23]

Institutional Mission and Focus

Many of the challenges that the CAS faces in Phase 3 are tied to the Academy's own institutional identity and how it fits into the NIS. Over the years, the CAS has evolved into what is surely a highly distinctive organization. While in some respects comparable to scientific institutions in other countries (resembling, variously, the Max Planck Society and U.S. national labs, and—in its honorific functions—the Royal Society and the U.S. National Academy of Sciences), it is unique in its size and the range of activities and functions it attempts to accomplish. Few institutions in other parts of the world incorporate, in one organizational framework, a desire to be (1) a preeminent center of basic research; (2) a leading institution for cutting-edge high-technology R&D; (3) a performer of research in support of public goods programs in agriculture, health, energy, the environment, and national defense; (4) a sponsor of institutions of higher education and graduate training; (5) a mechanism for high-technology entrepreneurship, industrial extension, and economic development in cooperation with local governments; *and* (6) an honorific organization whose elite academicians (*yuanshi*) play an increasingly important science advisory role.[24] The CAS leadership seeks to integrate and harmonize these diverse functions during Phase 3, and to do so in ways that convince China's political leaders of the CAS's indispensability. At the same time, the CAS continues to face daunting financial challenges in providing social safety nets for both its current employees and its retirees. The Academy believes that it can best perform its functions and meet its responsibilities through more generous state support and through the success of its commercial ventures.

In return for increased government support, however, the CAS faces new problems of accountability, with the government wanting assurances that it is serving national needs in a cost-effective manner. In the face of such expectations, the idea of a stronger mission orientation is being advanced in Phase 3. Yet such an orientation runs the risk of imposing excessive top-down requirements on the research community in ways that could discourage research creativity and bottom-up innovation. Some Chinese scientists working abroad have identified this problem as one that is holding back Chinese science more generally.

The Role of the CAS in China's Evolving NIS

A clear definition of the CAS mission is closely related to active policy debates in today's China over the shape of the NIS. As China has moved from a planned to a market economy, there is a growing realization among national policymakers that Chinese industry must become far more innovative if it is to move up the value chain of the international economy. A core question is where that innovation will come from. Following the model of the capitalist countries in the Organisation for Economic Co-operation and Development (OECD), many in China believe that the development of R&D in Chinese companies is the key strategic task for building the NIS. Chinese industrial enterprises have long been weak in R&D since, under the planned economy, research was centralized in government research institutes. However, in recent years, government policy has favored research in enterprises, and now more than 60 percent of the nation's R&D is performed by industry (up from less than 40 percent in the 1990s). This

"OECD-inspired" view of an enterprise-centered NIS is not entirely compatible with the CAS-centered view associated with the KIP's initiation, nor with China's institutional tradition more generally. The new orientation of national policy toward strengthening the role of enterprises is leading to revisions in national R&D programs, making partnerships with companies a necessary ingredient of successful competition for research funding.

The role of universities also figures prominently in current debates about the NIS. Again, in China's planned-economy era, universities had a very limited research role. This has changed dramatically in the reform period, as the value of research in conjunction with advanced training, associated with the Western model of the university, has taken root, and as new sources of funding for university research, especially the NSFC, have become available. The scope and quality of university research have grown rapidly, and with them recurring questions about the CAS's role in relation to universities, especially the training and subsequent employment of graduate students.[25]

Defenders of the Academy plausibly argue that given the past weaknesses of both industry- and university-based research, China has a distinct need for an institution like the CAS. Although visions of its role may differ, one version, as shown in Figure 1.1, sees the CAS playing a critical PQ-type role in the innovation system. Clearly furthering the growth of both enterprise and university research is a strength that the CAS, in the past, could have claimed as being unique. In spite of the KIP's achievements to date, some still question whether the CAS, in its present form, really has a place in a marketized and globalized China.

In the short to medium term, the Academy's defenders are likely to carry the day. Very few Chinese companies will be able to put together the combination of scientific and engineering talent, facilities, research management, and a strategic vision for innovation of the sort found in leading technology-based companies in the OECD countries. The CAS has a reservoir of assets that in many fields approach or exceed international standards. Similarly, although it increasingly competes with China's leading universities for professional staff, top graduate students, and research project funding, the CAS seems able to hold its own, as it offers more generous stipends and working conditions for graduate students, better equipped laboratories, and sophisticated equipment not available in universities. Despite the efforts being made to upgrade Chinese universities, only the distinguished few can rival the Academy.

Chinese policymakers are paying considerable attention to a more enterprise-centered NIS, but it is unlikely that many Chinese companies will develop R&D capabilities to support novel, science-based technologies in the near future. China's more entrepreneurial high-technology companies in the private sector often lack the resources to support their own R&D, while larger state-owned enterprises frequently find that short-term business objectives are better met by the less risky course of procuring advanced technology from abroad. In some cases, Chinese companies outsource their innovation needs to centers of knowledge creation in China, such as the CAS and the universities, or to research centers abroad. CAS-industry and university-industry relations, while still less than ideal, are thus likely to become considerably more important. If so, this suggests that as a matter of national policy, and as an explicit objective of Phase 3, enhancing these relations—legally, financially, and in terms of needed technical infrastructure—should get careful attention.

Figure 1.1 China's National Innovation System

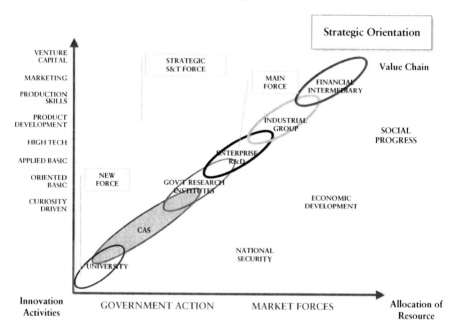

Source: Pan Jaiofeng, remarks at the China-Japan High Level Seminar on S&T Policy and Strategy, Sanya, Hainan, February 9, 2006.

The challenges facing the KIP in Phase 3 are interconnected. A major human resource challenge, for instance, is finding outstanding individuals who can lead the CAS institutes. The qualities required of such individuals might vary according to the understanding of innovation, which in turn is closely related to the clarity of the CAS's missions. These, in turn, cannot be understood independently of the broader NIS.

To secure a central role in the NIS, it will be necessary for the CAS, during Phase 3, to think in new ways about its institutional structure and the numerous and diverse functions it currently attempts to perform. This is clearly what the 10+1 formula is intended to accomplish. But whether the CAS can succeed in this endeavor remains to be seen, in light of the potential conflicts in performance standards and lines of accountability associated with its many functions and objectives.

For many CAS research staff, the Phase 3 objectives are best approached through a researcher-driven portfolio. On the one hand, this approach would accord with the Academy's basic research traditions and its interests in becoming a world-class graduate training facility. On the other hand, the desire to link the Academy's activities to the achievement of high-priority national objectives calls for a rather different management model, one that stresses more top-down direction and less investigator/researcher autonomy.

Attempting to serve national needs can be divided into two different managerial challenges. First, national needs can be met by serving commercial entities. Although the CAS has historically been weak in this respect, the commercial pressures it has faced over the past twenty years have produced considerable technology transfer experience and engendered a variety of transfer mechanisms, including contract research, the licensing of proprietary technologies, and the spinning off of companies from CAS institutes. An especially interesting approach has been the involvement of CAS units in high-technology zones established by local governments, such as the initiatives of the Institute of Computer Technology.[26] But many problems remain, and there are often mismatches between the relatively advanced technologies that the CAS develops and the willingness and ability of Chinese companies to adopt them. In some ways, the Academy's work fits more naturally with MNCs' technology interests. It is not surprising, therefore, that new commercial relations are developing between CAS institutes and MNCs.[27]

The second managerial challenge involves national demand for public goods that cannot be readily met through markets. For the CAS to help supply this demand requires that linkages be developed with other state bureaucratic systems (for instance, those dealing with public health, agriculture, defense, weather forecasting, and environmental protection), and technology transfer platforms that are distinct from those used in support of commercial transfers. The CAS has some experience in the provision of public goods—through its contributions to national defense technologies and natural resource surveys, for instance—but to do so effectively requires considerable familiarity with the procedures and expectations of different administrative agencies and their stakeholders. In some cases, the CAS's growing relationships with local governments may be quite useful for these purposes,[28] but they are no substitute for the deployment of substantial managerial resources and interagency coordination in support of national needs.[29] Some in the CAS consider too much involvement with local governments to be a diversion from the other, more central CAS missions that underpin its role as a *national* leader in S&T.

Pasteur's Quadrant and the CAS

It is against these issues of institutional purpose and identity that PQ is suggestive. As with other countries, China is keen to expand its presence in PQ. The new MLP stresses success in pushing the frontiers of knowledge and in linking subsequent, new understandings to commercial and social benefit. PQ is also consistent with "indigenous innovation" (*zizhu chuangxin*) emphasized in current policy discourse. Much of the policy rhetoric surrounding KIP also suggests that the CAS sees itself as especially well suited to a leading role in PQ, and the 10+1 formula for Phase 3 looks as if it was designed with PQ objectives in mind.

It is understandable that efforts to reinvent the CAS would take it in this direction. A future in PQ builds on certain traditions of the Academy and reflects the profile of its activities. The CAS devotes many of its resources to creating new knowledge, and in absolute terms this is increasing. Since the initiation of the KIP in 1998, expenditures on basic research have represented slightly more than one-third of total CAS R&D expenditures (rising from 30.8 percent in 1997 to 37.3 percent in 2006); meanwhile, the proportion of R&D manpower devoted to basic research rose from 36.3 percent in 1997 to 41.6 percent in 2004. The CAS's commitment to applied research, as measured by expenditures and manpower, fluctuated between 53 and 55 percent during the 1997–2006 period.[30] As indicated earlier, the outputs of CAS research can be measured in increasing patents and especially in an expanding number of scientific papers. Of note are the fields of greatest recognition. In 2003, for instance, chemistry and chemical engineering enjoyed the largest percentage of international citations, with physics second. Biology and materials science were third and fourth. Papers in "telecommunications and automatic control" did not have a significant presence in the *Science Citation Index* records, although they are better represented in Chinese domestic publications.[31]

Other sectors of the NIS are expanding their own knowledge creation and application roles. As Figure 1.2 shows, expenditures on CAS research constituted 3.7 percent of the national total in 2006 (down from 4.74 percent in 2004). In terms of expenditures, the CAS performed slightly more than 26 percent of the nation's basic research in 2006 (down from 28.4 percent in 2004) but only some 11 percent of its applied research and less than 1 percent of its development.

While the CAS would like to think of itself as the backbone of the NIS, and would cite its presence in PQ as evidence of its leadership role, it may have an increasingly difficult time supporting this claim. Recent data on expenditures and projects from China's national R&D programs across all sectors of the NIS suggest that challenges to the CAS position may be increasing.

In conception, the National Basic Research Program ("973") can be thought of as supportive of PQ objectives, in that it is intended both to push the frontiers of knowledge and to serve social needs. In design, it is organized around broad mission areas.[32] While the CAS certainly has been an important player in performing 973 research, the extent to which other sectors from the NIS have established a presence in this part of PQ is striking. In Table 1.2, for instance, we can see that the higher-education sector secured a larger percentage of 973 projects in the 2001–06 period than did the CAS, with institutes in the ministerial sector also securing a substantial number.

Figure 1.2 Ratio of R&D Expenditure, 2004

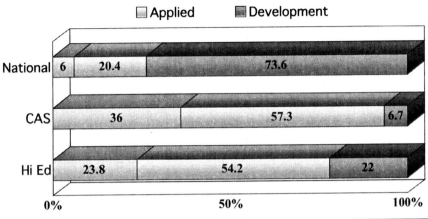

	R&D Expenditure (RMB 100 million)	Basic	Applied	Development
National	1966.61	118	401.19	1447.42
CAS	93.20	33.53	53.39	6.28
CAS as a Percentage of National	4.74%	28.42%	13.31%	0.43%

Source: Pan Jaiofeng, remarks at the China-Japan High Level Seminar on S&T Policy and Strategy, Sanya, Hainan, February 9. 2006.

Table 1.2 "973" Projects by Sector, 2001–2006

Sector	2001–2006	2001	2002	2003	2004	2005	2006
CAS	49	7	10	6	9	13	4
University	58	4	9	14	10	14	7
Ministry	30	6	5	3	5	9	2
Company	5	1	1	0	2	1	0
Institute	3	1	0	0	2	0	0
Local Government	14	1	1	4	3	3	2
Total	159	20	26	27	31	40	15

Source: Ministry of Science and Technology.

In Table 1.3, we see the distribution of 973 projects by mission orientation. Here, we can see the CAS's particular strengths in environmental sciences, but also surprisingly strong showings (relative to the CAS) from the higher-education sector in computer science and frontier research, energy, information science, and material sciences. These are areas in which the CAS would also be expected to do well.

The distribution of funds through the NSFC also points to the growth of R&D competence in institutions other than the CAS. While we have long known that the NSFC programs have helped the growth of university-based research, the extent to which the higher-education sector is able to capture the NSFC's resources, in comparison with the Academy, is notable. NSFC's "general programs" (*miansheng xiangmu*) are intended to encourage bottom-up investigator-initiated research, belonging for the most part in "Bohr's Quadrant," and are seen to be especially suitable for university-based researchers. NSFC's "leading" (*zhongdian*) and "major" (*zhongda*) programs, with their mission orientations, by contrast, seem to belong instead to PQ. Again, the CAS has been a significant player in competition for these programs, but as Figure 1.3 indicates, it does not have PQ to itself. But, as a sign that the KIP investments may be taking hold, its share of the *zhongda* project funding increased, rising from 28 percent in 2005 to 32.51 percent in 2007.

Figure 1.3 Funding of Key NSFC Projects by Sector, 2005

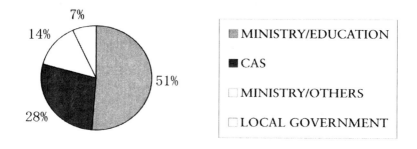

Source: NSFC.

China's National High Technology Program, "863," occupies a more ambiguous position with regard to PQ. While intending to encourage exploratory research, it has been oriented more explicitly toward applications than the 973 and NSFC programs. In this sense, we might think of 863 as having orientations to both PQ and Edison's Quadrant. Given the CAS's strong commitment to "strategic high technology," we would expect it to have been a prominent participant in 863.

Table 1.3 "973" Projects by Sector and Field, 2001–2006

Sector	Agriculture	Synthetic and Frontier Science	Energy	Environment	Health	Information	Materials
CAS	4	8	1	15	6	9	6
University	6	14	7	2	8	9	12
Ministry	7	3	2	4	13	1	–
Company	–	–	5	–	–	–	–
Institute	–	–	1	–	–	–	2
Local Government	2	6	–	–	4	–	2
Total	19	31	16	21	31	19	22

Source: Ministry of Science and Technology.

While the CAS has certainly been active, the growth of R&D capabilities in other sectors has clearly increased the general competition for 863 program funds. As Figures 1.4a and 1.4b illustrate, not only has the higher-education sector had a stronger presence in 863 than the CAS, but these data for 863 spending in 2005 plainly show the dramatic increase in R&D in Chinese industry.

Figure 1.4a Number of "863" Projects by Sector

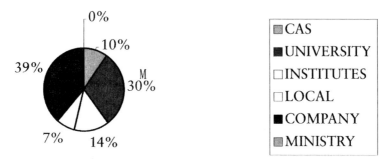

Source: Ministry of Science and Technology.

Figure 1.4b Number of "863" Projects by Funding Share

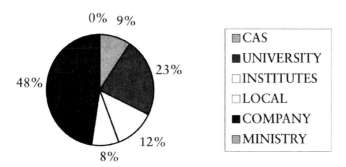

Source: Ministry of Science and Technology.

Conclusion

This analysis does not preclude CAS success in PQ, but it suggests that if the CAS is to meet its objective of becoming the backbone of the NIS through control of PQ, the Academy will have to overcome strong competition in PQ-related funding from national R&D programs. Thus, the question of the CAS's role in the NIS remains unresolved. With its strong traditions of basic research, contributions to strategic high technology, and service to national missions, the CAS has the ingredients to succeed in PQ. It could plausibly define its institutional identity, within the NIS, in these terms. But this assumes that the CAS can foster an internal environment of research creativity, and that the multiple functions and missions that it maintains do not work against one another and undercut the prospects for success in PQ.

In this connection, it may be useful to reflect on the CAS's funding sources. The Academy remains strongly dependent on government funding for its research activities. Indeed, it is notable that over the eight years since the KIP began, reliance on central government funding has increased from 50.6 percent of support in 1997 to 54.9 percent in 2004. In the same period, financial support for research projects "entrusted by enterprises" declined from 8.6 percent to 5.9 percent, and in spite of new CAS outreach activities, projects from local governments also declined from 6.5 percent to 4.5 percent.[33] These data suggest that in relative terms, and in light of the expansion of R&D activities in institutions of higher education and industry, the CAS share may have gone down. This points to the importance of the special KIP funding line that the CAS enjoys, and to the importance of significant government funding for other activities (outside the national program structures) to explain the expanded proportion of central government support in the Academy's overall research portfolio. Should there be more focus on concrete applications, as seems likely, it would draw CAS toward Edison's Quadrant and, conceivably, compromise the capabilities necessary for success in PQ.

The reinvention of the CAS through the KIP mechanism presents an intriguing case of large-scale reform in a major R&D institution. Much has been accomplished, but the challenges are also growing more daunting. Meeting them requires, above all else, that the Academy's identity and mission be widely understood and legitimized. Other sectors of the NIS have not always appreciated the special support that the CAS has received for the KIP initiative. The CAS is therefore under considerable pressure to live up to the expectations it has created in the minds of a political leadership that has generously supported the KIP.

In addition, as with other parts of the NIS, the rapid increases in financial support raise new questions about performance evaluation and financial accountability. Although the play of politics in Beijing differs from that of the Washington that Stokes describes, there are important similarities in the accountability of scientific institutions to a state that has supported research with increasing generosity. Unlike universities, which can justify their research expenditures in relation to the teaching and advanced training of students, and unlike corporations, which can justify them in relation to market competition, institutions such as the CAS face a more varied and therefore more difficult task.

Research and innovation in PQ is clearly becoming more important in national science policies throughout the world. Succeeding in PQ can be a key to solving institutional accountability problems.[34] Hence, it is reasonable for the CAS to try to do so. Nevertheless, its success is not guaranteed, in light of both the daunting internal challenges it confronts in effectively managing and coordinating such a large organization, and the other, ever more capable NIS institutions that seek to control the same arena.

Notes

[1] Parts of this paper are drawn from Richard P. Suttmeier, Cong Cao, and Denis Fred Simon, "China's Innovation Challenge and the Remaking of the Chinese Academy of Sciences," in *Innovations: Technology, Governance, Globalization* 1, no. 3 (Summer 2006), 78–97. A shortened version of the latter article appeared as "'Knowledge Innovation' and the Chinese Academy of Sciences," *Science* 312 (April 7, 2006), 58–59. Partial support for this work came from NSF grant OISE-0440422.

[2] While the discussion here focuses on its activities on the Chinese mainland, the CAS has built cooperative relations with universities in Hong Kong and maintains contacts with Taiwan, including Taiwanese firms invested in the mainland.

[3] Between 1998 and 2003, the KIP received a total of RMB 15.6 billion ($1.9 billion) from the Chinese government.

[4] Donald E. Stokes, *Pasteur's Quadrant: Basic Science and Technological Innovation* (Washington, DC: The Brookings Institution, 1997). Research belongs in Pasteur's Quadrant if it involves both the quest for fundamental understanding and consideration of its usefulness.

[5] See Richard J. Green and Will Lepkowski, "A Forgotten Model of Purposeful Science," *Issues in Science and Technology* 22 (Winter 2006), 69–73.

[6] The reduction in state funding, part of a package of reforms promoted by the State Science and Technology Commission (the predecessor of the Ministry of Science and Technology), was driven by a desire to force government research institutes to become more attentive to market forces and to the needs of users of the knowledge being generated. Not all members of the technical community, including CAS officials, see these reforms in a positive light. According to the dissenting view, China was not prepared for the reforms (resulting in much wasted effort), which left Chinese R&D seriously underfunded for more than a decade, and promoted an unhealthy dependency on foreign technology.

[7] The NIS concept was actively promoted in a report submitted to the then-SSTC by an international policy review team organized by the Canadian International Development Research Centre (IDRC). See IDRC and SSTC, *A Decade of Reform: Science and Technology Policy in China* (Ottawa: IDRC, 1997), especially chapter 5.

[8] The Academy's Medium- to Long-Term Plan, released in March 2006, set up a more ambitious goal of turning the CAS into an institution with a level of overall R&D capability that would put it among the top three institutions of its type in the world by 2020.

[9] These include fifteen in mathematics and physics, twelve in chemistry and chemical engineering, twenty in biological sciences, nineteen in earth sciences, and twenty-one in technological sciences. See Chinese Academy of Sciences, *CAS Statistical Data, 2005* (Beijing: Bureau of Comprehensive Planning, Chinese Academy of Sciences), 4.

[10] For earlier accounts of KIP reforms, see Hui Li, "Chinese Academy of Sciences: Reform Shatters Iron Rice Bowl," *Science* 279 (January 30, 1998), 649; Hui Li, "Chinese Academy of Sciences: Institutes Reinvent Themselves as Part of Well-Funded Reform," *Science* 283 (January 8, 1999), 150–53; Jeffrey Mervis, "Chinese Academy of Sciences: Neuroscience Institute Breaks New Ground," *Science* 283 (January 8, 1999), 150–51; "Chinese Academy of Sciences: CAS President Engineers Major Reform of Institutes," *Science* 286 (November 26, 1999), 1671–73; and Ding Yimin, "Chinese Academy of Sciences: In China, Publish or Perish Is Becoming the New Reality," *Science* 291 (February 21, 2001), 1477–79.

[11] Chinese Academy of Sciences, *CAS Statistical Data, 2005*, 6.

[12] Ibid., 7.

[13] Recruits into the 100 Talents Program are eligible to receive RMB 2 million ($250,000) in start-up funds.

[14] New graduate students do their basic coursework at the graduate gchool and then move to CAS institutes for their research and thesis preparation.

[15] Chen Zhu, "Historical Missions of Young Science Administrators: Rational Thinking on the Younger Leadership of China Research Institutions" (in Chinese), *Nature* 432, China Voices II (November 18, 2004), A24–A29.

[16] Chinese Academy of Sciences, *CAS Statistical Data, 2005*, 10–11.

[17] Setting aside the subjective issue of quality, the CAS believes that the sheer number of publications by its researchers is now double that of the Max Planck Society, its German analogue.

[18] The CAS has ten major companies under it, including Lenovo and the less well known but nevertheless large Di Ao Pharmaceutical Group based in Sichuan province. Many CAS institutes have spun off successful companies. The San Huan Corporation, for instance, is a spinoff from the Institute of Physics, which harnesses knowledge generated by research into rare earth metals to produce high-quality magnetic materials. Institute spinoffs now number about 490 in total. Overall, CAS enterprises employed approximately 60,000 people in 2003 and had an income of RMB 53.4 billion ($6.4 billion). Both the Academy and its institutes have divested themselves of ownership shares in companies and reduced the number of affiliated companies.

[19] The administrative arrangements include the establishment of three groups (*pian*) for basic research, strategic high technology, and sustainable development missions. The groups would help convert long-term strategic plans into annual plans, with a focus on identifying scientific opportunities that satisfy applied goals. They would also attempt to identify and cultivate human resources to support these plans. Experts from industry, the academic community, and government would advise the groups, but would have the authority to make decisions to support national missions that promote interdisciplinary, cross-institute cooperation.

[20] See, for example, David Cyranoski, "Biologists Lobby China Government for Funding Reform, *Nature* 430 (July 26, 2004), 495; Yi Rao, Bai Lu, and Chen-Lu Tsou, "Fundamental Transition from Rule-by-Man to Rule-by-Merit: What Will Be the Legacy of the Mid-to-Long Term Plan of Science and Technology?" (in Chinese), *Nature* 432, China Voices II (November 18, 2004),

A12–A17; Mu-ming Poo, "Big Science, Small Science" (in Chinese), *Nature* 432, China Voices II (November 18, 2004), A18–A23.

[21] This concern for ensuring a supply of young, new personnel reflects the persistence of a prereform mentality, its preoccupation with shortages, and the need for vertical integration to manage shortages. By extension, it also points to continuing imperfections in the market for highly skilled labor.

[22] In September 2005, CAS physicist Mao Guangjun committed suicide for reasons that included pressures he felt after his failure in performance evaluation. See Ding Yimin, "Scientists' Suicides Prompt Soul-Searching in China," *Science* 311 (February 17, 2006), 940–41.

[23] See Jia Hepeng, "China's Reverse Brain Drain Plan 'Risks Backfiring,'" *SciDevNet* (August 30, 2005), <http://www.scidev.net/News/index.cfm?fuseaction=readNews&itemid=2322&language=1>.

[24] In spite of the expansion of this policy advisory role, abuses in electing and rewarding academicians are attracting more public attention and have made the academician system increasingly controversial. See Cong Cao, *China's Scientific Elite* (London and New York: RoutledgeCurzon, 2004), especially chapter 9.

[25] CAS plans expand collaboration with universities during Phase 3. This will include fifty collaborative research projects, between twenty and thirty joint laboratories, joint graduate training programs, and the engagement of university professors as outside experts.

[26] Institute of Computer Technology initiatives include integrated circuit (IC) design centers in Suzhou and Ningbo, a mobile computing center in Shanghai, a software development center in Zhaoqing, and a center for intelligent electronic technology in Taizhou.

[27] Among the Phase 3 objectives is the establishment of some one hundred R&D centers with enterprises by 2010 and the development of new technology transfer platforms in science parks and high-technology zones.

[28] The Zhaoqing software development center of information and computer technology, for instance, is a venture that also involves MOST and the government of Brazil. It has as major objectives the development of low-cost information technology and the delivery of IT services to Chinese farmers.

[29] As part of its Phase 3 plans, the CAS has pledged to support a series of regional "scientific action plans."

[30] Chinese Academy of Sciences, *CAS Statistical Data, 2005*, 22, 27.

[31] Ibid., 34. Similarly, papers in agriculture and earth sciences have little presence in the international literature but do appear more significant in the domestic literature.

[32] Many of the criticisms leveled against 973 fault it on precisely this point. The strong mission orientation from above introduces excessive complexity in the formation of research teams and stifles original, investigator-initiated research.

[33] Chinese Academy of Sciences, *CAS Statisitical Data, 2005*, 28.

[34] For a recent discussion, see Green and Lepkowski, "A Forgotten Model of Purposeful Science," n. 5.

ITRI's Role in Developing the Access Network Industry in Taiwan

Kristy H. C. Sha, Tzung-Pao Lin, Chih-Young Hung, and Bao-shuh Paul Lin

The Industrial Technology Research Institute (ITRI) is Taiwan's largest government-sponsored applied research organization dedicated to setting up new industries and upgrading existing ones. Its Information and Communications Research Laboratories (ICLs) were assigned to conduct the national communication Technology Development Program (TDP) and to play the leading role in developing high-end communication technology. (The ICLs were previously known as the Computer and Communications Research Labs, CCLs, before ITRI's reorganization in March 2006.)

In the early 1990s, Taiwan was not competitive in terms of telecommunications technology. Major export products included low-end telephones, intercoms, analog modems, network interface cards, and so forth. The high-end telecommunication equipment required for building telecommunication infrastructure, such as telephone-exchange-system and transmission equipment, had to be imported from foreign telecommunication vendors and then integrated by domestic vendors to complete the localization jobs. Thus, it was impossible to have any control of the key technology. The advent of the broadband communication era offered an opportunity for the ICLs to prioritize their tasks concerning the sequential development of broadband communication technology.[1]

Essentially, telecommunication networks are classified as long-haul backbone networks, metro backbone networks, and loop-access (or last-mile access) networks. The loop-access equipment in this chapter is defined as the equipment from the central office (CO) site to home.

Given that the long-haul backbone and metro backbone equipment were dominated by international manufacturers, which local vendors in Taiwan could hardly penetrate, the loop-access technology was the key to providing broadband communication services and the best opportunity for local vendors. The ICLs decided to develop a series of critical technology or products with emphasis on the European Telecommunication Standard Institute (ETSI) digital loop carrier (DLC), digital subscriber line (xDSL), and fiber loop-access products (abbreviated as FTTx—fiber to the x; x can be home, curb, building, and so on). As a result, Taiwan successfully developed a flourishing broadband loop-access industry.[2]

The success of ETSI DLC development in 1994 was the first time that Taiwanese vendors dominated a key transmission technique. The execution of the xDSL TDP was an even bigger success than the ETSI DLC program. The worldwide market share of Taiwanese xDSL production was number one for four consecutive years (2004 to 2007), with 76 percent, 78 percent, 83 percent, and 74 percent of the global share, respectively. The production volume of small-office/home-office (SOHO) routers derived from xDSL technology also ranked number one globally in recent years.

In this chapter we summarize the ICLs' experiences in using TDP resources to partner with foreign and local vendors for the purpose of nurturing new industries. We explore the key success factors in the loop-access industry.

ITRI's Operations

ITRI was founded in 1973 to engage in applied research and industrial services. Evolving from a small research unit to a large, multidisciplinary research center with a heavy emphasis on leading-edge research and development (R&D), ITRI has been a major driving force of Taiwan's economic growth. It has successfully accumulated more patents than similar research and technology organizations (RTOs)—such as the Netherlands Organisation for Applied Scientific Research (TNO), the Australian Commonwealth Scientific and Research Organisation (CSIRO), and the National Research Council (NRC) of Canada—far surpassing them in the number of invention patents granted in the United States.

Lee and Chiang[3] explored ITRI's technology development strategies based on Rosenbloom and Burgelman's[4] framework and concluded that those strategies have progressed through sequential periods of technology importation, R&D project initiatives for value-added product features, and then-key-component and new-product development.

Hung and Hwang[5] drew on resource-based theory, organization evaluation, and technology commercialization to explain how ITRI exploits internal and external resources to speed the commercialization process. As for ITRI's technology implementation, the tools and methods used include the following: organizing of technology symposiums and training programs; technological resource registration (in books or online); thesis presentations; participation in international standard-formulation activities and exhibitions, client-commissioned contracted services; establishment of spin-off companies; joint research among industry, academic, and research sectors; testing certification; patent authorization (unidirectional); licensing exchange (bidirectional); intellectual property (IP) pooling; preliminary participation; technological strategic alliance; open laboratories; and incubation centers. Figure 2.1 depicts the ITRI generic business model.

ITRI creates value by transferring its R&D results to industry for the benefit of society. Taking a central position in technology development and commercialization, ITRI integrates government, academic, industrial, and foreign resources to make the most of industrial innovation. Hsu[6] illustrates the

"Adaptable Industrial Innovation Model" to explain the formation of ITRI's industrial innovation mechanism for developing new technologies, nurturing high-tech talents, and enabling new service industries.

Figure 2.1 The ITRI Generic Business Model

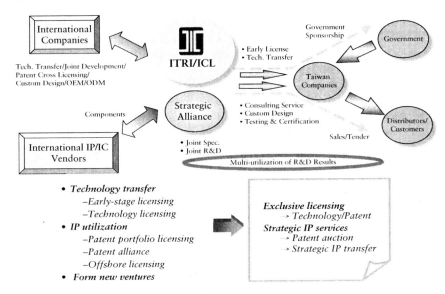

Analysis of Access Network Industrial Development in Taiwan

In this section we use the need-solution-differentiation-and-benefit (NSDB) model, a modification by ITRI of the need-approach-benefits-competition (NABC) model of SRI International. SRI proposed that the discipline of innovation—creating compelling customer value—starts from value proposition, a statement that includes the customer/market need, a compelling approach to meeting that need, and the benefits that customers will receive in comparison to the competition or other options. ITRI proposes that economic value can be created by providing a solution to market needs and by maximizing its benefits through differentiation with competitors.

Taiwanese communication enterprises are small to medium-sized, with R&D capabilities that are relatively weak, and thus the support from the TDP is influential. However, TDP resources are limited, so they have to be carefully allocated. We select the development strategies of the DLC, xDSL, and FTTx to show how the system works.

Digital Loop Carrier (DLC)

DLC can rapidly provide phone services to remote areas, and CO site transmission equipment is often used to improve national phone services. In the 1990s, the economy in China started to take off while its telecommunication infrastructure fell far behind (phone availability was 12 phone lines for every thousand people in China, compared to 415 lines per thousand people in Taiwan). The situation was similar in Southeast Asia. The phone availability was 40 phone lines for every thousand people in Thailand, 22 in the Philippines, and 11 in Indonesia. Obviously, the potential of DLC demand was quite promising in the 1990s.

At that time, major local communication manufacturers that concentrated on products with low technical barriers and low prices faced slim profits and the threat of steep competition from the newly industrializing countries. The fastest way to advance technical competence was adopting foreign technology and quickly disseminating to local industry. The ICLs' task was then targeted on building the know-how of those high-priced and high-tech CO site systems such as DLC. The first decision to be made was whether to follow the American or the European standard. Taiwan's shallow local-transmission experience had been fully focused on the American standard, yet the U.S. standard market scale for DLC was about one-third of the European standard market scale, with only three countries—Taiwan, the United States, and Canada—using it in the early 1990s. Regarding market size, the European standard was the better choice.

Advanced Fibre Communications (AFC), a small, California-based U.S. company, had the cutting edge DLC technology but was short on capital. Therefore, the ICLs chose AFC as the major source of technology. Both parties agreed that the ICLs retained the rights of reauthorization, utilization, modification, production, and selling. Although the intellectual property rights (IPRs) resulting from the project belong to AFC, this was an inviting deal, especially when all the big companies closely guarded their core technologies. Three Taiwanese companies—OPNET Technologies, Sun Moon Star, and Hitron—were attracted to the joint development.

The ICLs sent a team of engineers to AFC in the United States to codesign and transfer the core technology. ICLs' local staff and domestic manufacturers then worked to lower costs and construct the human-to-machine interface in Chinese, with testing, volume production planning, and technology receiving as the top-priority items.

Despite the presence of competitors such as AT&T, Nokia, NEC, Northern Telecom, Rockwell, Fujitsu, and Alcatel (now part of Alcatel-Lucent), the DLC technology that was transferred from AFC was marketable, particularly in the China and Southeast Asia markets. Its intrinsic advantage forced Lucent to drop out of the DLC market in mainland China.

By mid-1995 the ETSI DLC was produced in volume. The three manufacturers marketed own-brand products in China and Southeast Asia and were the first to export local transmission products. DLC products ranked number eight among communication products in 1996, with an annual growth rate of 171 percent. All partnering manufacturers together generated an annual revenue of about NT$1 billion (US$31.3 million). By 1998 the revenue accumulated by the three manufacturers came to approximately NT$3.4 billion (US$106.3 million).

Digital Subscriber Line (xDSL)

Since 1993 Internet traffic has grown explosively, as has large bandwidth demand. Telecommunication service providers expected to use existing phone lines for broadband communication services. The most critical technology is xDSL, and x can be asymmetric (ADSL), high data rate (HDSL), very high data rate (VDSL), and so on. HDSL is the early product of xDSL that is dedicated to the enterprise subscriber. In comparison with a traditional E1/T1 digital leased line, HDSL has a lower price, a faster installation, and a longer transmission range. The later-developed ADSL can provide Web access, data transmission, and SOHO services, enabling broadband data access and phone services simultaneously. The market potential is tremendous.

When most of the advanced countries developed the broadband copper-line communication technology, the communication industry in Taiwan was heading for a global original equipment manufacturing (OEM) base. As the local vendors got the key components from the global leading integrated circuit (IC) manufacturers to produce ADSL modems and routers and handed them over to downstream name-brand vendors to market them internationally, a business partnership in the supply chain was formulated. The ICLs' task then was to quickly disseminate the broadband copper-line communication technology to enable local vendors to win orders of ADSL modems from the leading global manufacturers, and to develop key components to swiftly enter the CO site equipment market.

However, domestic vendors had a meager understanding of system standards and specifications, too few R&D personnel, no available interoperability laboratory (IOL) test environment, and no familiarity with conducting end-product type approval. Besides, the key components were in the hands of foreign giants such as TI, Broadcom for the ADSL transceiver, and Lucent, Nokia, Alcatel, and Samsung for the digital subscriber line access multiplexer (DSLAM). Few Taiwanese vendors stood much of a chance.

The strategy adopted by the ICLs regarding the ADSL modem was to rapidly develop prototype products; help vendors to understand demand, standards, and specifications; and train R&D personnel. Meanwhile, Alcatel, the international leader in ADSL, favored making the ADSL modem in Taiwan and expected local manufacturers to partner with it. Seeing mutual benefits, the ICLs then signed a contract with Alcatel to set up the Alcatel ADSL IOL within the ICLs (the first in Asia and the third globally). The ICLs also cooperated with the government agency, the Directorate General of Telecommunications (DGT), which supervised the local telecommunication services, forming an industrial consortium consisting of major local equipment manufacturers, service providers, research institutes, and colleges—twenty-two members in total—to jointly formulate the HDSL local-type specification.

Key communication core components have always been a weakness for Taiwan. Since the ICLs already had development experience and know-how in DLC transmission products, when ITRI was promoting the core components of the ADSL transceiver, the ICLs adopted the model of simultaneously doing self-development and importing technology. The ICLs' acquisition of the key ADSL analog front-end IC technology, which was introduced by Valence Semiconductor Inc. (VSI), not only sped up development but also accumulated many intellectual property rights (IPRs) and trained the senior designers.

In 1994, ADC Telecommunications (which later merged with Motorola) and Level One Communications (LOC, later bought by Intel) asked the ICLs to develop the HDSL system with joint capital and common sharing of benefits. The condition was that ADC exchange the U.S. standard HDSL system for the European standard HDSL system developed by the ICLs. Meanwhile, LOC and ADC further authorized the ICLs to develop the key component Framer IC of the HDSL system. The model by then had changed from an entire foreign-technology transfer to a technology exchange and mutual licensing. In developing the DSLAM system in mid-1998, the ICLs were already able to initiate the requirements, and specify the key component, Queue Manager, and thus authorize Synopsys to develop it for the ICLs.

Among the vendors who participated in ICLs' technology of the ADSL modem, or who visited the ITRI Alcatel IOL, Askey, ZyXEL, Ambit Microsystems (which later merged with Foxconn), and TECOM have already become world-class manufacturers. The global market share of xDSL CPE from Taiwan by shipped quantity was number one in 2004 (76 percent), 2005 (78 percent), 2006 (83 percent), and 2007 (74 percent). The technology-transfer vendor CyberTAN (which also later merged with Foxconn) further developed SOHO router-stemming from the technology basis of xDSL CPE to meet the emerging demand for the digital home. The global market share of the SOHO router in 2004, 2005, 2006 and 2007 was number one, or 75 percent, 78 percent, 80 percent, and 80 percent by shipped quantity.

TrendChip Technologies, founded by the ICLs' technical team, has sold ADSL chips to twenty countries. The release schedule of the new-generation ADSL2+ solution package promises to keep up with the major international manufacturers. TrendChip Technologies reverses the situation of the ADSL chip market, which is dominated by foreign vendors, with a low-price strategy and fast response.

Fiber to the x (FTTx)

The fiber-to-the-home (FTTH) service is the focus of much next-generation network development worldwide. As bandwidth leaps from the current 8 Mbps

ADSL service to a 100 Mbps optical network service, wireless communications will also move to the 3G and 4G high-speed transmission era, supplying real-time transmission of voice and video. Many countries, especially in the Asia-Pacific region, have been developing FTTH technologies. The Japanese government is committed to investing 5 trillion yen ($45.5 billion) before 2010. With $80.4 billion, Korea aims to adopt FTTH for last-mile service with bandwidths from 50 Mbps to 100 Mbps. In Taiwan, the government expects that by 2010 there will be 4.2 million FTTx users.

Since Ethernet is uncomplicated, high-bandwidth, and low-cost, the application of Ethernet in access networks is widespread. A passive optical network (PON) does not require a power supply in the optical distribution network and uses the least number of optical fibers and transceivers. Therefore, the Ethernet passive optical network (EPON) is the most cost-effective way of delivering optical fiber communications.

Taiwanese manufacturers have the technical capabilities to develop CO site, remote-site, and subscriber terminal equipment. However, the optical fiber communication system or high-density CO site system is still a tough nut to crack quickly. The market of remote-site system equipment can be penetrated only by partnering with a leading foreign CO manufacturer. Meanwhile, the software and hardware components need to be strengthened in an effort to reduce the system cost and enhance the equipment competitiveness. Hence, the ICLs will assist manufacturers in entering the field so that local manufacturers can supply key components and market own-brand products.

Carrier-grade network equipment, such as CO systems, has always been primarily proprietary architecture. However, to reduce development time and cost, open architecture is becoming the trend. In 2004 the International Telecommunication Union Telecommunication Standardization Sector (ITU-T) established the Open Communications Architecture Forum (OCAF) Focus Group and a subdivision, Carrier Grade Open Environment (CGOE) Working Group, in the hope of standardizing the open-architecture specifications and interfaces. The aim is to encourage the development of low-cost standardized commercial off-the-shelf (COTS) components to be integrated into the communications system. The ICLs have been working with leading companies such as Intel, HP, and MontaVista to promote the CGOE architecture in Taiwan.

No company can build a very cost-effective system without partnering with subsystem manufacturers worldwide. Therefore, hardware manufacturers in Taiwan contract with major foreign manufacturers to provide certain cards or modules in telecommunication systems. Local vendors can gradually accumulate enough technical strength to jointly contract for critical FTTx projects, without ruling out the possibility of becoming the main bidder for domestic and foreign FTTx tenders. Figure 2.2 shows the FTTx access technology promotion strategy in Taiwan.

In light of the immense market potential in EPON and other carrier-grade network equipment, the ICLs have initiated the Broadband Carrier-Grade Open Platform Trial Project, with support from the National Science Council. Using

EPON as the platform, the project does R&D and performance analyses in high-availability (HA) middleware and creates a carrier-grade HA EPON system as the testing platform.

Figure 2.2 FTTx Access Technology Promotion Strategy in Taiwan

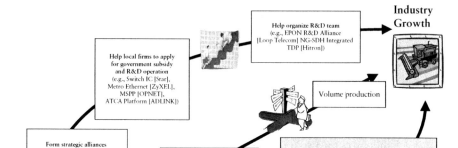

EPON is a fairly new technology, and the establishment of an EPON testing environment is very important to both carriers and equipment vendors. So the ICLs have established an EPON testing environment with Chunghwa Telecom, with support from the Ministry of Economic Affairs (MOEA) Department of Industrial Technology (DOIT). This environment will provide product testing for domestic EPON system manufacturers and for IC and component vendors. The EPON testing environment includes two major areas: EPON function and performance testing and EPON field trials.

We see that Taiwan followed the acquire-assimilate-improve path in the ETSI DLC and xDSL case. The major development momentum came from the advanced countries, but Taiwan skillfully exploited internal and external resources to nurture its comparative advantages, spur industrial upgrading and transformation, and incubate new industries. By now, Taiwanese vendors have changed their relationship with major foreign manufacturers from one of cooperation into one that is both cooperative and competitive.

Before the Taiwanese government sponsored the TDP, there was no broadband industry. Then Internet traffic exploded. From 1996 through 2005, through investment of over three hundred person-years and NT$690 million in technology development and dissemination, ITRI helped Taiwan establish broadband access and customer premise equipment, key component industries from scratch, and accelerated deployment of a broadband environment, making Taiwan one of the leading broadband countries in the world today.

The key factors lie in understanding various kinds of demand and constantly accumulating technical capabilities through many channels. Even after the competitive edge is enhanced, crisis consciousness will still exist, and considerable R&D resources must be allocated to sustain the edge. Domestic vendors need to be careful in assessing in-house technical capabilities. If it is sufficient, the self-development model is adopted; otherwise, the technology is acquired through the TDP in various ways.

Conclusion

Over more than twenty years, Taiwan has evolved into the world's most important base of OEM and original design manufacturing (ODM) by developing a host of products ranging from PCs, semiconductors, mobile phones, LCDs, and so on. The government has always played a critical role with its technological and industrial policies. These policies are tightly linked by the incubation of talented students, establishment of R&D organizations, science park planning, preferential tax benefits, market intelligence collection, a convenient fund-raising channel, and recruitment of overseas talent. Resources are integrated down to the industrial platform for carrying out R&D, add-on value, and expansion through the TDP. The results are that more than thirty items of information and communications technology products rank in the top three of the worldwide market.

The success of the local loop-access industry in Taiwan can be attributed to understanding domestic and international market demand, standard specifications, technical trends, and local industrial traits, as well as strengths and weaknesses, position in the global value-chain, and long-term planning and investment. Moreover, choosing suitable partners, effectively dividing work, employing different strategies in different stages—absorption (acquire-assimilate-improve) and joint collaboration and competition (cooperation-competition)—are the key factors for enhancing the industrial competitive edge.

To sustain success, more emphasis should be put on innovative business models, including patent portfolios, patent auctions, exclusive licensing, and strategic IP-transferring services, to maximize the value from accumulated intellectual assets.

Notes

[1] T. P. Lin and R. Tsai, "The Communication Electronics Technology Development Program and Future Prospect of the Communication Industry," *CCL Technical Journal*, no. 51 (1996): 26–33; and Lin and Tsai, "A Perspective on Broadband Communication System Technologies and Key Component Technologies," *CCL Technical Journal*, no. 61 (1997): 23–29.

[2] Lin and Tsai, 1996; Lin and Tsai, 1997; and T. P. Lin, "Digital Subscriber Loop Technology" (memorandum, ITRI's Technical Award).

[3] R. F. Lee and H. C. Chiang, "ITRI's Technological Development Strategy," in *Industrial Technology and the Industrial Technology Research Institute: A Visible Brain*, ed. C. Shih (Hsinchu, Taiwan: ITRI, 2005), 207–50.

[4] R. S. Rosenbloom and R. A. Burgelman, eds., *Research on Technological Innovation, Management, and Policy*, vol. 6 (Greenwich, CT: J.A.I. Press, 1997).

[5] S. C. Hung and Y. H. Hwang, "Resource Exploitation and Industrial Development," in *Industrial Technology and the Industrial Technology Research Institute: A Visible Brain*, ed. C. Shih (Hsinchu, Taiwan: ITRI, 2005), 161–203.

[6] C. W. Hsu, "Formation of Industrial Innovation Mechanism through the Research Institute," in *Industrial Technology and the Industrial Technology Research Institute: A Visible Brain*, ed. C. Shih (Hsinchu, Taiwan: ITRI, 2005), 105–58.

RESTRUCTURING CHINA'S RESEARCH INSTITUTES

Gary Jefferson, Bangwen Cheng, Jian Su, Paul Duo Deng, Haoyuan Qin, and Zhaohui Xuan[*]

China's research institutes are at the heart of the country's research establishment. In the mid-1990s, five to six thousand research institutes operated as independent accounting units (*duli hesuan danwei*) under the supervision of a wide array of government agencies. Nearly all received direct subsidies. In 1999 the government began a broad-based program to restructure them with the principal objective of forcing most to become independent, self-financing entities. With less dependence on government, reformers anticipated that their innovation activities would become increasingly responsive to the market. At the same time, the government retained substantial oversight and support of the core of the nation's basic research establishment, notably institutes within the Chinese Academy of Sciences (CAS) and the Chinese Academy of Social Sciences.

In this chapter, we estimate three performance functions. The first revenue received from institutes' technology models as a function of their science and technology (S&T) personnel and research equipment. The second models their production of patents, a critical output. Finally, we connect the two by estimating the contribution of patents to the generation of technology revenues.

Restructuring entails converting the institutes to one of several new organizational forms, mainly S&T companies in the private sector (henceforth called "enterprises"); not-for-profit research institutes; and nonprofit, nonresearch institutes. We use the three performance functions to estimate the impact of conversion on each of these new organizational forms. We also examine the characteristics of the institutes most likely to be converted to one or another of the three forms.

It is difficult to measure new knowledge, the most important product of these institutes. Unlike manufacturing firms (for example, automobile companies) in which production inputs and outputs largely occur within a single period, for research institutes the time between inputs (such as researchers and their equipment) and outputs (such as patents and technology revenues) is difficult to predict but is likely to stretch over many years. There is also the phenomenon

---

[*] The authors appreciate the cooperation of Liu Shumei and Gao Changlin, who made possible the research presented in this chapter. We also acknowledge the critical support of the National Science Foundation (award #SES-0519902).

of "selection bias."[1] If the analysis does not control for institute quality at the time of restructuring, institutes may later appear good or bad only because those candidates chosen for reform were initially good or bad. To control for this bias, our estimation method uses an instrumental variables approach.

This analysis is based on a balanced sample of 386 research institutes spanning the period 1995–2004. While this sample covers only 7.5 percent of the 5,187 institutes that participated in the 2004 Ministry of Science and Technology (MOST) survey, it represents an important cross-section of these organizations. The sample was randomly selected from a balanced panel that included five major sectors: agriculture, public service, and the three manufacturing sectors of chemicals, ferrous and nonferrous metals, and computing and electronic equipment.

Our key finding is that the restructuring has achieved some of its hoped-for efficiency gains. Institutes converted to S&T enterprises show substantial gains in technology revenues, controlling for their inputs of S&T personnel and equipment. Furthermore, their patents generate more revenue. While we do not find an efficiency increase in the *number* of patents produced, restructuring has led to *higher value* patents by those institutes converted to S&T enterprises. Only institutes converted to nonprofit, nonresearch status show an increase in efficiency in the number of patents produced.

The Restructuring Initiative

In 1999 the Chinese government started its restructuring program. Its first phase focused on 242 research and development (R&D) institutes, which were subordinated to 10 agencies under the State Economic and Trade Commission; these were mostly in the fields of technology and natural science. Of the 242 institutes, 131 were merged into enterprises outside the formal research institute sector, 40 became S&T enterprises under local authorities, and 29 were combined into 12 large S&T enterprises under a central authority. The remaining 42 were converted into a variety of other forms or liquidated.[2] As shown in Table 3.1, as of the end of 2001 the number of converted R&D institutes had grown to 891 (and by the end of 2004 to 1,214).

This restructuring program was associated with dramatic changes in China's entire research institute complex. As Table 3.1 indicates, in 1995 the MOST collected data from 5,828 institutes. By 1999, the year of the formal restructuring initiative, that number had eroded to 5,573. Then, in the aftermath of the restructuring, by the end of 2004 the number had declined to 3,973. Most of the decline involved institutes that merged with others or were acquired by enterprises outside the formal research establishment and were hence excluded from the annual survey conducted by the MOST.

Table 3.1 Summary Statistics—China's Research Institutes

	1995	1999	2001	2004
Number of institutes[a]	5,828	5,573	4,635	3,973
Number of restructured institutes[b]	0	132	891	1,214
Of which, converted to:				
• S&T enterprise	0	114	821	1,087
• nonprofit, nonresearch institute	0	18	70	127
Number converted to not-for-profit research institutes[c]	0	0	0	134
Number of merged/disappeared institutes[d]	0	73	147	371
Number of S&T personnel (1,000 person)	644.4	534.6	427.0	N/A
Scientists and engineers in S&T personnel (%)	59.03%	62.93%	64.80%	N/A
Patents/S&T personnel[e]	–	–	0.008	0.014

Sources: MOST, "S&T Technological Index of China, 2004" (Beijing: S&T Publishing, August 2005); MOST, "National Survey of S&T Institutes," 1995–2004.

[a] Above-county-level government research institutes and institutes of science information and literature, with an independent accounting system, but not including restructured research institutes. The number in 2004 is estimated.
[b] The cumulative annual number of the institutes that were successfully converted to enterprises or converted to nonprofit, nonresearch institutes.
[c] The number of institutes that remain as government not-for-profit research institutes.
[d] The estimated number of research institutes that disappeared as a result of mergers and acquisitions or shutdowns.
[e] The average number of patents owned per S&T personnel in all research institutes that have nonzero S&T personnel for that year.

During the late 1990s and into the next decade, R&D spending rose rapidly in the S&T sector. In 1995 spending on these organizations was 146.4 billion yuan, 42 percent of the national total. By 1999 it had risen to 206.4 billion, 38.4 percent of the national total, and in 2003 it soared again, to 399.0 billion yuan, 25.9 percent of the nation's total R&D spending. That the institutes' share of total national spending declined, even as their R&D spending rose by 173 percent over this eight-year period, largely reflects a 400 percent increase in the large- and medium-size enterprises sector.[3] During this period, both the numbers and the proportion of China's S&T workers in the research sector fell. While in 1995 the sector had 644,000 workers, 24.5 percent of the nation's total, by

1999 the proportion had fallen to 22.2 percent, and in 2003, with just 406,000 remaining workers in these establishments, it dropped to 12.4 percent. While nonscientists and nonengineers were leaving this sector, scientists and engineers as a proportion of the total workforce nevertheless rose. In 1995 the proportion was 59.0 percent; by 1999, it had risen to 62.8 percent, and in 2003 to 65.5 percent. Along with the deepening of skills within China's research institutes, the share of basic research in total R&D spending rose from 7.0 percent in 1995 to 11.8 percent in 2003. In that year, the research institutes accounted for a total of 53.5 percent of China's total spending on basic research—virtually unchanged from 1999—even as the share of research institute R&D spending in the national total declined to just 25.9 percent in 2003 as compared with 38.4 percent in 1999, the year the restructuring was initiated.[4] The restructuring, therefore, seems to have made the research institutes more specialized in skilled research personnel and more focused on basic research.

Table 3.1 shows the pattern of outcomes beginning in 1999. By 2004, 1,214 institutes, 23 percent of the total of 5,187, had been restructured. Nearly 90 percent of these had been converted to S&T enterprises, while most of the balance had been converted to either not-for-profit research institutes or nonprofit, nonresearch institutes. The MOST estimates that by 2004, approximately 371 research institutes had disappeared from the survey population due to mergers or liquidation.

Three basic facts stand out from this survey. First, increasing numbers are disappearing from the MOST's annual survey due to mergers or liquidation. Second, among the converted institutes, the vast majority are being converted to S&T enterprises. And finally, while the research institute sector is shrinking, it has become increasingly specialized in basic research.

Statistics

Summary statistics from the subsample of 386 research institutes are shown in Table 3.2. Over the ten-year period for which institute data are available (1995–2004), annual observations total 3,860. A much larger proportion of the institutes in this subsample—about 46 percent—was restructured, compared to just 23 percent for the total population. Still, there are enough unconverted research institutes—more than half the sample—to control for the conversion effect. As with the full population, the vast majority of those in our sample (76 percent) became S&T enterprises. Also, as with the full population, the total number of S&T personnel declined, while the proportion of scientists and engineers rose.

Table 3.2 Summary Statistics—Sample of Research Institutes

	1995	1999	2001	2004
Number of institutes,[a] of which:	386	386	386	386
• Restructured[b]	0	81	160	177
• Converted to an enterprise	0	75	136	149
• Converted to nonprofit, nonresearch institute	0	6	24	28
• Converted to nonprofit, research institute[c]	0	1	2	56
• Did not restructure[d]	386	304	224	153
Number of S&T personnel[e]	49,184	40,152	34,317	33,449
Scientists & engineers/S&T personnel[e]	58.45%	66.79%	67.49%	73.20%
Stock of technology assets/technical revenue[e]	–	3.14	4.53	4.48
Number of surviving technology revenue research institutes[f]	302	253	241	221
Patents/S&T personnel[g]	–	0.008	0.016	0.020

Source: MOST, "National Survey of S&T Institutes," 1995–2004.

[a] The sample of 386 research institutes spans 1995 to 2004 and includes five industries: agriculture, chemicals, ferrous and nonferrous metals, computing and electronic equipment, and public service. The sample included all the research institutes, both converted or nonconverted, for the metals and the computing and electronic equipment industries. For the other three industries, all converted institutes were included, but for the nonconverted, a systematic sampling method was used. The number of institutes included in the agriculture sample is 40; for the chemical industry, 32; and for public service, 45.
[b] The converted research institutes, including those converted to enterprises and those converted to nonprofit, nonresearch institutes. The figures are cumulative.
[c] The institutes that were converted to not-for-profit research institutes. They are still government research institutes. The figures are cumulative.
[d] The nonconverted research institutes, including those for which conversion has not been initiated and those for which conversion is incomplete. The number refers to the number of research institutes of that year (noncumulative).
[e] The average of the ratio between cumulative technology equipment investment and technology revenue for all the research institutes in the sample that have nonzero technology revenue in that year. Cumulative technology equipment investment is obtained by using the following formula, assuming a 15 percent annual depreciation rate: accumulative tech equipment = (tech equipment investment of current year) + (tech equipment investment with 1-year lag) + (tech equipment investment with 2-year lag)/1.15 + (tech equipment investment with 3-year lag)/(1.15*1.15).
[f] Institutes that have nonzero S&T activities.
[g] The average of number of patents owned per S&T personnel in all research institutes that have a nonzero S&T personnel for that year.

Table 3.3 shows average performance measures across all years for each of the five sectors examined in this chapter. It also shows the figures for 2004, and the comparison with "all years" shows the direction of change. In both periods, agriculture shows the lowest ratio of technology revenue relative to S&T personnel, as well as the lowest rate of profit per S&T worker. By comparison, ferrous and nonferrous metals register the highest level for each of these measures (except for 2004, when the chemicals sector had the highest profit per S&T worker). The chemical sector also had the highest incidence of patenting per S&T worker. Public service, followed by agriculture, shows the highest 2004 rate of government subsidy per S&T worker and also shows a large increase over time.

While these figures show substantial variation in all the measures, Table 3.3 shows that they generally rose over time. Finally, the ratio of technology revenue to total revenue declined in all five industries, which most likely exhibits an effort on the part of the converted organizations to find revenue sources other than technology sales.

Changes in Research Orientation

Table 3.4 shows the principal changes in S&T performance after restructuring, as measured by performance in 2004 in relation to 1998, the year before restructuring. One notable change is the large increase in patents granted, from 131 to 332, an increase of more than 2.5 times. The largest increase was in computing and electronic equipment. Among the unchanged institutes, patents increased notably only in the chemical industry.

Much of the overall increase was in invention patents, for which the review criteria are relatively rigorous, with an annual increase of 268 percent from 1998 to 2004. Invention patents in the 66 chemical industry institutes went from 15 to 83. In contrast, patenting abroad in this sample remained low.

Table 3.4 also shows a general increase in S&T papers, including those published abroad. The notable exception is the metals industry, where they fell sharply.

Research Methodology

It is a challenge to measure the performance of a research institute. This is more difficult than analyzing the performance of a conventional commercial enterprise for three reasons:

(1) *Different measurement units.* The economic value of outputs is difficult to measure. With multiple heterogeneous outputs, including patents, scientific papers, and new products, the lack of a common unit of output complicates a performance evaluation.

(2) *Ambiguous time structure.* Inputs and outputs share an ambiguous time structure. R&D inputs today may not produce value, such as a patent

Table 3.3 Summary Statistics by Sector

	Agriculture		Chemical		Metals		Electronic equipment		Public service	
	All years	2004	All years	2004	All years	2004	All years	2004	All years	2004
Number of organizations	–	98	–	66	–	50	–	71	–	90
Technology revenue/ S&T personnel	9.277 (23.38)	12.550 (23.08)	15.304 (32.20)	27.029 (62.13)	33.662 (82.26)	44.652 (67.22)	26.172 (55.55)	24.322 (54.60)	26.144 (78.75)	36.734 (72.24)
Profit/S&T personnel	8.547 (34.16)	17.081 (55.08)	36.891 (237.98)	103.915 (369.18)	54.579 (202.83)	69.160 (161.40)	17.121 (82.40)	27.341 (96.03)	13.124 (46.55)	35.460 (97.45)
Patent/S&T personnel	0.008 (0.03)	0.011 (0.03)	0.040 (0.13)	0.048 (0.14)	0.026 (0.06)	0.031 (0.06)	0.016 (0.07)	0.021 (0.09)	0.004 (0.02)	0.004 (0.01)
Government subsidy/ S&T personnel	42.443 (56.98)	56.935 (58.78)	18.613 (23.00)	19.960 (25.70)	21.291 (29.23)	22.551 (37.14)	19.360 (34.90)	20.518 (43.57)	39.081 (51.87)	76.574 (103.92)
Technology revenue/ Total revenue	0.088 (0.15)	0.075 (0.14)	0.160 (0.23)	0.159 (0.24)	0.179 (0.24)	0.178 (0.26)	0.224 (0.30)	0.155 (0.29)	0.210 (0.28)	0.189 (0.28)
R&D personnel/S&T personnel	0.338 (0.32)	0.351 (0.35)	0.173 (0.31)	0.136 (0.27)	0.237 (0.25)	0.209 (0.38)	0.122 (0.24)	0.144 (0.34)	0.247 (0.31)	0.308 (0.38)

Note: Numbers in parentheses are standard deviations.

Table 3.4 Change in Composition of S&T Output

Industry	Agriculture		Chemicals		Ferrous and nonferrous metals		Computing and electronic equipment		Public service	
Restructured	no	yes	no	yes	no	yes	no	yes	no	yes
No. of research outputs	40	58	32	34	9	41	27	44	45	56
					Ratio 2004/1998					
S&T papers	844/564	1,782/1,662	220/236	475/356	39/106	1,935/1,946	133/39	249/219	1,005/783	2,059/1,466
Published abroad	16/2	132/112	43/33	64/41	0/0	95/201	52/10	23/5	38/52	229/114
Patent applications	8/0	73/12	43/35	80/19	4/0	261/126	34/6	169/9	9/7	24/8
Patent grants	4/1	4/7	43/9	43/15	1/0	138/94	6/2	90/12	6/2	14/3
• Invention	4/1	2/5/7	41/6	42/9	1/0	64/51	5/0	17/3	4/1	6/0
• Abroad	0/0	0/1	0/0	0/1	0/0	2/0	0/0	1/0	0/0	0/0
Total patents owned	23/0	63/0	144/0	230/0	4/0	427/0	68/0	146/0	26/0	27/-
No. of projects	665/616	1,415/1,384	156/249	354/457	51/59	1,110/1,906	109/137	237/401	523/518	1,733/1,582
Avg. expenditure[a,b]	59.8/17.7	271.8/69.4	24.0/16.4	118.1/64.3	6.9/23.6	364.8/203.2	90.0/24.8	264.2/86.4	101.0/37.6	281.1/111.3
Technology sales[b]	0/0.045	17.9/9.6	9.4/4.2	30.3/37.7	3.0/0.07	430.4/271.1	33.8/16.3	27.3/43.4	16.0/5.3	57.5/28.2
• Abroad	0/0	0/0	0/0	0/0	0/0	2,388/8,029	12,825/0	0/0	0/0	513/0
Tech. purchases	0/80	320/1,420	0/0	0/206	0/0	38,368/37,830	0/243	1,950/294	0/1170	6,790/510
• Abroad	0/0	0/1,420	0/0	0/0	0/0	23,697/36,990	0/0	1,500/48	0/808	6,790/0

Notes:
[a] Avg. expenditure refers to "project expenditure-internal."
[b] Units = 1,000s.

or a new product or revenue, until well into the future—if ever. This uncertainty makes it difficult to specify an econometric equation with a time structure that relates inputs to outputs.
(3) *Technology spillovers.* Much of the value of research results in technology transfers and spillovers to other users.

Therefore, the social returns to innovation generally exceed the private returns to the research institute itself.

In this chapter, we do not attempt to measure technology spillovers, that is, the last of the measurement challenges identified above. However, we do address the measurement issues of different research outputs and the ambiguous time structure.

Measures of returns to R&D include revenue generated by the sale of technology and the number of patents. These are arguably the institutes' most important research outputs. We focus first on the resources required to generate them.

Our first task is to specify a general model that represents the relationship between R&D inputs (RI) and research outputs, notably patent grants (PG). The model is summarized by the patents production function:

$$PG_{it} = f^p(RI_{it})\varepsilon^p_{it} \qquad (3.1)$$

where f^p represents the functional relationship between research inputs and patent grants and ε^p represents the relevant error structure. In equation (3.1) and the other equations in this chapter, the subscript i represents the individual research institute ($i = 1 \ldots 386$), while the subscript t refers to a year ($t = 1995 \ldots 2004$). Later in this chapter, we estimate this function using a functional form that expands on equation (3.1).

We also formulate a general function for the other, nonpatent, research outputs (RO), and a production function for the creation of technology outputs other than patents:

$$RO_{it} = f^n(RI_{it})\varepsilon^n_{it} \qquad (3.2)$$

Because we expect RO—as constituted by patents and other research outputs—to generate revenue, we also formulate the following function on the relationship between RO and technology revenue (TR).

$$TR_{it} = f^o(PG_{it}, RO_{it})\eta^o_{it} \qquad (3.3)$$

Substituting equations (3.1) and (3.2) into (3.3), we obtain the following reduced-form TR function, which describes the relationship between RI and TR:

$$TR_{it} = f^r(RI_{it})\varepsilon^r_{it} \qquad (3.4)$$

The *RI*s consist of S&T personnel (*ST*) and equipment (*EQ*). To allow for differences in levels of the productive efficiency between *RI*, *RO*, and *TR* across different sectors, we include in the estimation equations a set of dummies for four of the five industry sectors (*IND*), with agriculture serving as the reference industry. We also include dummies for each year (*YR*) covered by the data, thereby allowing for year-to-year changes in productivity and prices. We estimate equation (3.3), the *TR* function, later in this chapter.[5]

Another issue is how to identify the impact of conversion on an individual institute's performance. Three different approaches are potentially available:

(1) *Test for increases in the magnitude and statistical significance of the post-restructuring-year dummies.* An increase in the magnitude of the year dummy—either for *TR* or for *PG*—may indicate that restructuring has enhanced revenue or patent production.
(2) *Test for structural change across subsamples.* We can split the sample into two subsamples—one for the earlier years, before or during which restructuring occurred, and the other following the restructuring. We can test for a structural shift in efficiency.
(3) *Estimate the direct impact of restructuring using data that identifies the year and nature of the restructuring.* Because our data include dates and the nature of the restructuring (converted to an S&T enterprise, a nonprofit research institute) we can directly test the impact, if any, of the change.

As a direct test, the last method is preferred. Below, we embed in equation (3.4) a set of institute conversion dummy variables (RCi_{kt}, in which $k = 1, 2, 3 \ldots$), that represent the form of conversion in each year (t) and for each institute (i).[6]

$$TR_{it} = f'(RI_{it}, RC_{ikt})\varepsilon^r_{it} \qquad (3.5)$$

Motivations for Restructuring Particular Institutes[7]

The factors we see as influencing the choice of one or another of the five restructuring types are:

(1) *Human capital intensity.* The proportion of scientists and engineers relative to total S&T personnel.
(2) *Capital intensity.* The proportion of assets relative to total revenue.
(3) *Autonomy.* The proportion of total revenue originating from nongovernmental sources.
(4) *Fixed effects.* Controls for geographical region (coast, middle, west) and the level of the supervisor (dummy if central government).
(5) *Changes.* We also test for changes associated with particular years.

The results may be summarized as follows:

- *Conversion to an S&T enterprise.* The greater the human capital intensity and proportion of revenues coming from nongovernmental resources, the more likely an institute is to become an S&T enterprise. The greater the physical capital intensity, the less likely it is to change to that status. Also, those research institutes in the three industrial sectors (relative to those in the agriculture sector) are more likely to be converted to enterprises, while public service institutes are less likely to be similarly converted. Institutes in the middle provinces are also less likely to be converted, while those whose supervisory authorities are at the center are more likely to be converted. Finally, conversions to enterprises appear to be concentrated in 1999 and 2000, rather than in the later years, for which the estimates of the dummy variables are significantly negative.
- *Conversion to a not-for-profit research institute.* Two factors stand out here. First, such conversions are more likely to take place among public sector institutes. Second, those in the western provinces are more likely to become not-for-profits. Controlling for these two factors, the incidence of conversion is relatively uniform over the period 1999–2004.
- *Conversion to a nonprofit, nonresearch institute.* Relatively high human capital is associated with a greater probability of conversion to this status, while dependence on nongovernmental revenue diminishes the probability of such conversion. Also, public service institutes are less likely to be converted to nonprofit, nonresearch status, while those reporting directly to the center are more likely to receive such a designation. Finally, such designations are most likely to occur in the coastal areas and became frequent only later in the restructuring period, from 2002 through 2004.

These three designations are general and convey only limited information. For one, S&T enterprises are for-profit; the other two categories are not. Two of the converted organizational forms may continue to be primarily engaged in research. The third—nonprofit, nonresearch institutes—is intended to carry out such activities as policy analysis and service delivery.

The Technology Revenue Function

We also estimated what happened to revenues derived from technology. Here, the main input variables are total S&T workers and equipment for the different industries and years.[8] The main result is that conversion to the status of S&T enterprise significantly helps revenues, while conversion to either of the not-for-profit statuses does not.

Table 3.5 Impact of Conversion on Research Input Performance (dependent variable = *ln* [technology revenue])

	1999–2004		
	OLS (ordinary least squares)		IV (instrumental variables)
	(1)	(2)	(3)
ln (S&T personnel)	0.861 (10.77)	0.846 (10.60)	0.836 (10.50)
ln (Equipment)	0.245 (9.19)	0.221 (8.06)	0.223 (8.24)
Converted to S&T enterprises	–	0.684 (3.84)	1.109 (4.89)
Converted to nonprofit, research institute	–	−0.690 (−2.26)	−0.713 (1.79)
Converted to not-for-profit, nonresearch organizations	–	0.277 (1.25)	0.126 (0.28)
• Chemicals	−0.166 (1.09)	−0.142 (0.93)	−0.059 (0.39)
• Metals	−0.029 (0.18)	−0.177 (1.10)	−0.223 (1.38)
• Electronic equipment	−0.705 (4.67)	−0.795 (5.25)	−0.759 (5.03)
• Public service	−0.402 (1.01)	−0.016 (0.09)	0.159 (0.91)
Middle region	0.153 (0.00)	0.102 (2.06)	0.201 (1.99)
Western region	−0.151 (1.49)	−0.200 (1.96)	−0.194 (1.92)
2000	−0.368 (1.56)	−0.354 (1.46)	−0.470 (1.99)
2001	−0.377 (1.59)	−0.352 (1.49)	−0.510 (2.14)
2002	−0.300 (1.28)	−0.308 (1.32)	−0.500 (2.07)
2003	−0.195 (0.83)	−0.197 (0.85)	−0.394 (1.62)
2004	−0.368 (2.31)	−0.356 (1.52)	−0.568 (2.24)
Constant	1.053 (2.94)	0.824 (2.23)	0.826 (2.29)
Number of observations	1,866	1,866	1,866
F (13, 1,866)	41.40	35.54	–
Prob > F	0.000	0.000	0.000
R squared	0.225	0.235	0.238
Adj. R squared	0.220	0.229	0.231

The results, shown in column 1 of Table 3.5, are that both inputs—S&T personnel and equipment—are statistically significant determinants of *TR*. Column 1 also shows that the computing and electronic equipment sector is significantly less productive in translating research inputs into *TR* relative to the reference sector, agriculture. We see that the institutes exhibit no measurable improvement in performance over the years up to 2003.

In addition, there may be a selection bias—in that the most promising revenue generators were selected for change—and Table 3.5 shows the results of controlling for this effect.

Production and the Returns on Patenting

We estimate the production of patents. Our model has the following inputs: S&T personnel, the original value of equipment, conversion dummies, industry, and year of conversion. The results are that patent production is highest in the chemicals and metals sectors and lowest in the public service sector.[9]

We also examined the impact of conversion on the generation of *TR*. The results suggest constant returns on patent production—that is, a 10 percent increase in patents owned yields a 10 percent increase in *TR*.

Conclusion and Issues for Further Investigation

We find evidence that the conversion of research institutes to S&T enterprises in China, which by 2004 represented 76 percent of those in our sample, is raising research productivity. We also find evidence, although rather weak, of a productivity improvement in institutes converted to nonprofit, nonresearch status. We find no evidence of an improvement in the performance of those converted to not-for-profit research status.

This analysis is preliminary in several respects. First, comprising a mere 7.5 percent of China's research institutes, the sample is small. Sample size is particularly problematic because of the heterogeneity of the sample and the variety of research questions that we wish to explore. The sample spans five sectors, pre- and postconversion periods, and at least four different conversion types. One reason that the results for conversions to S&T enterprises may be substantially more robust than those for the other types is that their relatively large number generates more efficient estimates than for the other types. Another issue is selection bias. While we have tried to correct for it, we know that many institutes dropped out of the population that the MOST annually surveys. With these omitted, we cannot assess key features—including the characteristics of exits, the performance of exiting organizations relative to that of surviving ones, and the effect of restructuring programs—on the propensity of these institutes to leave.

Three issues seriously complicate evaluating research organizations in China or, indeed, anywhere else: (1) the many measures of research output and productivity, (2) identifying the time lapse between R&D inputs and outputs, and (3) accounting for the spillover benefits from innovations that are not directly captured by *TR* or other measures that the organizations report. China's research institutes and their conversion possibilities are a heterogeneous set with many inputs and goals. Future research strategies might focus on individual subsamples of different converted forms (for example, S&T enterprises or not-for-profit research institutes) or investigate specific sectors (for example, separating the manufacturing industries from those with high concentrations of foreign direct investment). On the time gap issue, most of the explanatory power in our estimates is likely to result from cross-sectional relationships, in which individual organizations show the persistence of inputs and outputs over time. This condition suggests using time-series analysis to attempt to identify the time gap between various technology inputs and outputs, such as patents and revenue.

The third issue is estimating the impact of research, particularly in industrial companies. Access to extensive data on S&T activity in large- and medium-size industrial companies enables a search for such impacts. For example, by looking at clusters of R&D expenditures in institutes in specific industries and locations (for example, electronic equipment in Chengdu), one can try to see if specific technologies or geographic locations affect the performance of nearby industrial enterprises.

While these results are preliminary, this chapter sets forth a useful strategy for identifying the impact that China's restructuring program has had on its research institutes, and lists some robust results. Likewise, the chapter presents evidence that the program has begun to have promising effects on at least some of the institutes in the restructured subsample on which we have focused.

Notes

[1] See Gary Jefferson and Su Jian, "Privatization and Restructuring in China: Evidence from Shareholding Ownership, 1995–2001," *Journal of Comparative Economics* 34 (2006), 146–66.

[2] Ministry of Science and Technology (MOST), *China Science and Technology Indicators* (Beijing, Scientific and Technical Documents Publishing House, 2000), 45–46.

[3] National Bureau of Statistics (NBS) and MOST, *China Statistical Yearbook on Science and Technology, 2004* (Beijing: China Statistics Press, 2004), 6.

[4] NBS/MOST, *China Statistical Yearbook on Science and Technology, 2006* (Beijing: China Statistics Press, 2006), 6, 26–27.

[5] The time structure problem is discussed in a detailed appendix (A), available at <http://sprie.stanford.edu/publications/greater_chinas_quest_for_innovation>.

[6] While there are five conversion options, two types—not converted and to-be-but-not-yet-converted—are combined into one category and used as the reference intercept, thereby leaving three intercept dummies.

A potential problem here is selection bias. For example, if the best-performing institutes are converted to enterprises, the coefficient on the institute conversion dummy will be upwardly biased. The method of dealing with this problem is presented in a detailed appendix (B), available at <http://sprie.stanford.edu/publications/greater_chinas_quest_for_innovation>.

[7] The details of this model and model results are presented in a detailed appendix (C), available at <http://sprie.stanford.edu/publications/greater_chinas_quest_for_innovation>.

[8] The details of this model and model results are presented in a detailed appendix (D), available at <http://sprie.stanford.edu/publications/greater_chinas_quest_for_innovation>.

[9] The details of this model and model results are presented in a detailed appendix (E), available at <http://sprie.stanford.edu/publications/greater_chinas_quest_for_innovation>.

R&D BY MULTINATIONAL COMPANIES

The Political Economy of Technological Development: China's Information Technology Sector

Douglas B. Fuller

China's national innovation system (NIS) comprises uneven institutional terrain in which a certain type of firm stands out as a driver of technological development. We call this the hybrid foreign-invested enterprise (hybrid FIE); it combines financial sources from the advanced industrial world with a strategic commitment to China, and in doing so contributes much more to the nation's technological development than any other type of technology enterprise operating in China's information technology (IT) industry.

Two variables, source of finance and operational strategy (OS), explain the differing levels of technological upgrading among China's IT firms. Source of finance is a binary variable, with a firm being linked to either domestic or foreign financial sources. A firm's OS is also a binary variable; firms possess either a strategic commitment to basing core corporate activities in China, which we dub a China-based strategy, or a foreign-based strategy without such a commitment to China. Ethnic ties to the local economy explain the propensity of some firms to adopt a China-based OS.

The main finding presented in this chapter is that there is substantial differentiation among firm types in their contributions to China's technological development. Hybrid FIEs, combining China-based OSs with foreign financing, contribute the most, while firms with foreign financing and foreign-based strategies (regular FIEs) contribute moderately to technological upgrading. Firms with domestic financing contribute very little.

The larger implications of the findings are threefold. First, the behavior of NISs and individual technology firms cannot be evaluated outside the institutions of political economy, particularly the institutions of finance, within which they operate. This is particularly true in China, where the financial system remains wracked with problems. On a positive note, the findings also suggest that developing countries can ameliorate domestic institutional deficiencies by opening their economies to foreign institutions. In China, given the current deluge of criticism of the considerable foreign involvement in the economy and in the "reform and opening" policies more generally, there is strong evidence against such criticisms. And yet, the benefits of such openness to foreign investment must be placed in context. The third finding does this: co-ethnic transnational technology networks can serve to make FIEs commit to local development without resorting to Singapore-style strict state monitoring, which many states are incapable of pursuing.

The Causal Argument

The four types of firms found in China are favored domestic firms, neglected domestic firms, hybrid FIEs, and regular FIEs. They are distinguished by their politically determined sources of finance and their OSs.[1] The relationship of firms to the state determines their sources of finance—that is, whether they can access financial institutions. The sources of finance in turn determine each firm's incentives and capabilities to pursue technological activities. A firm's OS determines its motivation to upgrade in China as opposed to doing so elsewhere. Because of their different state-firm relations, FIEs rely on foreign finance and domestic firms do not. Hybrid FIEs differ from regular FIEs because the hybrids have a China-based OS. Domestic firms also have a China-based OS, as the literature on the nationality of multinational corporations (MNCs) would predict.[2]

Finance with hard budget constraints gives firms incentives to upgrade because it forces them to remain competitive to survive. Soft budget constraints encourage firms to become very lax, so their capabilities suffer.[3] A third financial situation is no source of finance. Firms without financing will not be able to invest in technological development.

As previously stated, a mix of interests and ideational factors cause these firms to perceive China either as the vital center of their operations (the China-based OS) or as just another location among many (the non-China-based OS). For technology firms, one core activity is corporate research and development (R&D). Thus, firms with a China-based OS will be more likely to pursue their core R&D activities in China.

This chapter (based primarily on 342 interviews with government and business participants in China's IT industry from 1998 through 2006) finds that the two types of FIEs are more likely to contribute to technological upgrading than the two types of domestic firms. Among the FIEs, the hybrids are more likely to contribute than the regular. The hybrids are the most successful upgraders because they have both disciplined finance (credit with relatively hard budget constraints) from foreign financial institutions and the motivation to upgrade in China due to their China-based OS. This access to foreign finance is the result of an arrangement between the Chinese state and FIEs to allow both types of FIEs to invest in China while limiting their access to the Chinese state financial system. The successful upgraders obtain technology from returnees, but so do most firms that have failed to upgrade.[4] Thus, technology flows from the developed world are necessary but are not sufficient to explain upgrading. The unsuccessful domestic upgraders lack finance (neglected domestic firms) or financial discipline (favored domestic firms) due to their particular relationships to the state. The favored domestic firms have a close relationship to the Chinese state, which provides them with finance and procurement without any effective monitoring,[5] and thus undermines both firm capabilities and incentives for upgrading. In the past, the favored firms were almost exclusively state-owned enterprises (SOEs), but now the mix of ownership types in the favored category has expanded, in part because reforms have changed many SOEs into shareholder companies.[6] According to our research, even firms that were clearly not SOEs leveraged the right connections to enjoy the state's lavish patronage. The neglected domestic firms were those that did not have access to the state financial system and other prerequisites of state patronage. They were almost exclusively private firms with a distant relationship to the state. Cut off from credit, they were unable to afford the costs of upgrading. The regular FIEs can upgrade due to their financial discipline and access to transnational technology networks, but they undertake less of it in China than the hybrids because they lack a China-based OS.

The FIEs' ability to access finance is as much a result of politics as is that of the domestic firms. The Chinese state has generally not allowed foreign firms to utilize its financial system.[7] At the same time, it has opened up China to investment in a manner that sets few barriers to or restrictions on the access of foreign firms to overseas capital.[8]

Distinguishing FIEs from domestic firms is done based on a firm's registration. Joint ventures are considered foreign or domestic depending on which type of firm holds a majority stake in the enterprise. For joint ventures in which neither side holds a majority, the actual balance of control is used to determine the firm's classification.[9] In any case, the registry of the firm (foreign versus domestic) is important only insofar as it indicates the source of financing. For those few foreign firms that have been able to enter the state's patronage and financing network, the soft budget constraints have resulted in the same disincentives and deteriorating capabilities in technological upgrading as with the state-supported domestic firms.

Explaining Operational Strategies

All the hybrid FIEs have ethnic Chinese management.[10] Some concepts about firm behavior that have been borrowed from the literature on the nationality of MNCs help to explain the propensity of firms with ethnic Chinese management to adopt a China-based OS. This literature, based on empirical observations of MNCs' behavior, argues that firms concentrate their core resources in their home bases. This research belies the idea of the truly multinational enterprise.[11] Even research that bemoans the demise of NISs due to the movement of R&D offshore shows that more than 87 percent of all corporate R&D among a large sample of the five hundred largest MNCs is conducted within the home borders. It also shows that more than 70 percent of overseas R&D was concentrated in three countries—Britain, Germany, and the United States.[12]

Explanations for this trend vary. Some argue that firms have developed competitive strengths that rely on the specific institutions of the home economy.[13]

An alternative, interest-based argument centers on the limited information available to firms about foreign economies. Firms generally know more about how to operate at home than they do abroad.

An ideational argument could be focused simply on patriotism. Firms are run by managers of the same nationality, and they do not simply make decisions based on cold, profit-maximizing principles if the decisions would adversely affect the home economy. Another ideational argument is that firms believe that their competitive advantages are linked to distinct features of the home economy even if that belief is not necessarily correct. A third could be that they have certain ideas about the just distribution of resources within their home economy that dictate what activities should be kept at home. Doremus and his colleagues[14] combine ideational and interest aspects in their arguments on the nationality of MNCs.

The connection between a firm's nationality and its OS does not imply that all ethnic Chinese firms will act as national Chinese firms. Indeed, there is evidence that some ethnic Chinese firms forcefully reject a China-based OS. The Taiwan Semiconductor Manufacturing Company (TSMC) has been quite hostile to China and is not any less of an ethnic Chinese firm because of this anti-China sentiment. Instead of assuming that all ethnic Chinese will embrace a China-based OS, the literature informs us of the ways that the sociocultural knowledge and ideas of ethnic Chinese are conducive to adopting a China-based OS.

By drawing on their sociocultural knowledge, ethnic Chinese firms have some of the same informational advantages in China that MNCs have at home, because ethnic Chinese firms have lower information barriers to understanding and operating in China. Socioculturally informed ideas influence their behavior as well. There is a nationalist/parochial component for some ethnic Chinese, just as there are nationalist/parochial motivations that encourage MNCs to favor their home base.[15] Finally, ideas influence interest. Ethnic Chinese managers (perhaps due to ethnonationalist pride) may believe there are benefits to being China-based even when these advantages do not actually exist.[16] MNCs exhibit the same attitude when they insist that certain activities must be done in the home country even when those supposed advantages are not realizable. Here, it is not necessary to decide whether it is only interest—or ideas shaping interest—that lead some ethnic Chinese firms to adopt a China-based OS. It is sufficient that either motive encourages them to adopt this strategy.

The firms that adopt China-based OSs attempt to maximize their utilization of China's resources because they believe this provides a competitive advantage. The hybrid firms do not maximize Chinese resources simply because they wish to help China. Nevertheless, such a strategy makes more sense for those attuned to the local culture. Furthermore, it would be foolish to ignore nationalism as part of the motivation to embark on a China-based strategy, because many ethnic Chinese say that they returned to help build a prosperous and strong China.[17]

The evidence in this chapter is that hybrids, in strategy and practice, utilize China as the main base of operations, and the regular FIEs do not. As for the hybrids, their OS has several characteristics that differentiate it from the OS of regular FIEs. Given evidence from elsewhere in the world that suggests the home-base bias of even large MNCs,[18] it is safe to assume that domestic Chinese firms treat China as the home base. Determining the OS of FIEs is a much more complicated task. Measuring the proportion of a firm's technology activities in China conflates the OS and the variable it affects, the technology strategy. Saying that firms that have more technological activities in China are more likely to upgrade begs the question of why they have more technological activities there.

To assess a firm's OS, a wider net must be cast. Firms must meet two criteria to have a China-based strategy. First, they must have a self-described China-based strategy. Second, their functional headquarters must be in China, meaning that both senior management and support staff are in China.

There are, in fact, two distinct subcategories of FIEs that often have China-based OSs: start-ups and established firms. For start-ups—typically technology enterprises that have been founded by mainland Chinese returning from abroad—the operational headquarters and a publicly articulated China-based strategy are required. Established firms are usually from the ethnic Chinese economies (ECEs) of Hong Kong, Taiwan, and Macao and have embraced a China-based strategy and moved some operations to the People's Republic. There is a bias if the returnee start-ups and the ECE firms are measured by the same standard. Because the new returnee firms have been founded at least with the idea of concentrating operations in China, they do not have large operations already built up outside China when they create their China strategy. In contrast, given that ECE firms already have established headquarters in their original home economy, it stands to reason that their headquarters are in one of the ECEs rather than in China. Thus, ECE firms are not required to have their headquarters in China to be considered as having a China-based OS. Instead, these firms are required to have a head count in Mainland China that is larger than any of their operations outside their original ECE home base.

There is also a difference between measuring the self-described China-based strategies of returnee firms and measuring those of ECE firms. Whereas the returnee start-ups must embrace a self-described China-based strategy, the ECE firms need only embrace a Greater China strategy—that is, they want to utilize intensely the resources of both their original home and Mainland China. While there are examples from Hong Kong of firms simply abandoning Hong Kong for the Mainland, most firms create "twin towers" (to borrow the phrase of Taiwan's Inventec), with one tower in China and one in the home economy. ECE firms that describe their strategies as such meet the standard of a self-described China-based strategy. Taiwanese firms also do not have to make this strategy a public one, because technology firms from Taiwan have been under great political pressure in recent years not to invest heavily in the Mainland. If these firms articulated such a strategy during the course of interviews conducted for this chapter, it was sufficient to meet the strategy criterion.

Non-ethnic-Chinese firms were held to the same criteria. MNCs had to meet only the standard that established ECE firms were required to meet. Foreign start-ups had to meet the more stringent requirements of returnee start-ups. The research did not uncover any non-ethnic-Chinese firms that embraced the China-based strategy.

Soft Budgets and Strategies of Favored Domestic Firms

Technological upgrading entails high risks and high costs. The risk is high because firms must invest human and financial resources in learning or developing new technologies, which is often difficult. Larger profits provide the incentive to upgrade.[19]

Why are favored domestic firms, which are showered with easy credit and procurement from the state, less likely to upgrade? One could plausibly argue that the softening of budget constraints would free them to take greater risks. However, the evidence from other studies,[20] and this one, suggests that even when such firms are willing to try upgrading under soft budget constraints, they do not conduct technology activities in an effective manner. This is a probabilistic judgment rather than an absolute one. They do not completely lack the capabilities to upgrade, but the likelihood of their upgrading is low because of the curse of state favor. There are three ways that the state undermines the ability of favored firms to upgrade: bureaucratic goals, insufficient resources, and disincentives to upgrade.

The disincentive of bureaucratic goals applies to only those favored firms that are SOEs. In SOEs, the bureaucrat-manager is given many goals other than profitability. Officials higher up the ladder set the goals by which the bureaucrats-as-managers will be judged.[21] Among those goals are preserving the SOE and increasing scale and perceived technological prowess. Preserving the firm mitigates the desire to undertake anything risky, such as learning new technical skills. Increasing its scale and perceived technological intensity induces the bureaucrat-as-manager to buy expensive capital equipment that makes the firm appear cutting-edge and increases its apparent scale and (at least if properly used) productivity. The new capital equipment also increases the fixed capital of the firm, another measure of scale.

These investments promote the survival of the bureaucrat's domain, the SOE, because the firm becomes too large or too important (in the eyes of national planners) to fail. China did go through a process, starting in the wake of the Fifteenth Party Congress in the autumn of 1997, of supporting the large SOEs while letting go of the small and medium-sized (the "grasp the large, release the small" policy, or *zhua da fang xiao*). The campaign served only to convince SOE managers that their firms should look big, though these incentives were implicit in the system earlier.

In sum, the bureaucratic goals induce a policy of capital investment without skill investment. Huang, Wu, and Jefferson and colleagues[22] have documented this tendency within Chinese SOEs. Wu points out that Chinese SOEs spend one dollar on assimilation for every ten on technology acquisition.[23] In contrast, the Japanese and Koreans spend ten dollars on assimilation for each dollar spent on acquisition. Assimilation is the process by which a firm learns how to use a technology that it has acquired. Jefferson and his colleagues[24] present evidence that SOEs invest less in skills even compared to other Chinese firms. They calculated that the marginal productivity of township and village enterprise (TVE) technicians was 250 percent higher than that of the technicians in SOEs. TVE technicians had four times the marginal productivity when new expenditures on productivity were incorporated into the calculus.

Firm incapacity is also an issue. Easy access to state finance and procurement encourages all sorts of inefficiencies in the favored firms that prevent them from doing much of anything well. They may attempt to do technological upgrading, but their general inefficiency in operations makes it difficult for them to do so effectively. Liu and Kornai and colleagues[25] note this problem of soft budget constraints undermining upgrading. This problem applies to state-favored and non-state-favored firms.

Disincentives to upgrading are a problem as well. State procurement plays a critical role. It encourages firms to forgo high-risk, high-reward upgrading for the less risky, if potentially smaller, profits to be made from feeding at the state's trough. These firms lack incentives to upgrade their skill set to be competitive in the marketplace. From the literature on China's industrial policy, we know that the Chinese state does not monitor firms effectively.[26] Without effective monitoring, the favored firms can simply enjoy the state's bounty without enhancing their performance in return. I will present evidence of how procurement undermines the incentives to upgrade in domestic firms. This problem can affect state-favored and non-state-favored firms alike, though the state firms already suffer from both the problems of bureaucratic goals and insufficient resources, so the lack of incentives is the least of their problems.

Just as the state has showered financial favor on some firms, it has neglected many other firms. This neglect creates a two-tiered structure of firms in China.[27] Domestic firms that are not favored by the state have difficulty in acquiring loans from the state banking system.[28] This scarcity of capital prevents them from embarking on technological upgrading because they do not have the means to do so despite whatever future rewards for technological advancement they might enjoy.

Summary of the Causal Argument

In short, the state's relations to firms determine their sources of financing. In turn, the sources of financing help determine the technological upgrading effort. The upgrading effort only partly explains the commitment to pursue technology activities in China (see Table 4.1). The combination of the sources of financing and the OS of the firm together determine the strategy, which in turn determines the upgrading outcomes. This chapter maintains that greater corporate technological efforts in China will on average lead to higher technological contributions.

Two Cases: Integrated Circuit Design and Evaluation of Information Technology R&D Activities

The following case of integrated circuit (IC) design provides concrete evidence that foreign firms outperform domestic firms. The case also demonstrates that hybrids are the firms contributing most to China's technological development.

Metrics of Evaluations

Several metrics were used to evaluate firms. First, firms were evaluated on technical criteria, such as the process technology through which they designed the ICs, as measured in microns, and the part of the design process they carried out themselves. Second, they were evaluated on whether they were engaging in actual design work. Much, perhaps most, of the IC design work performed in China is not design work at all but simply reverse engineering. Firms engaging only in reverse engineering were not counted as contributors to upgrading. Finally, firms were evaluated by their ability to commercialize products. Only products sold on the commercial market were considered proof of technological development. The commercialization standard was used in order to eliminate consideration of firms that relied solely on the beneficence of state patronage. Many of these firms are not able to create products at the proper cost to be competitive in the marketplace. Economists who stress the importance of technological development, such as Khan, and Pack and Westphal,[29] have assumed the criterion of economic efficiency in technological development. Firms that create products inefficiently do not meet this criterion. A total of fifty eight IC design firms were surveyed.

For process technology, the smaller the lithography, as measured in microns, the more sophisticated the design. Process design cannot be compared across technologies, such as bipolar versus complimentary metallic oxide silicon (CMOS), but within a given technology, microns are a good measure of design sophistication. Design can also be measured down the design chain. Generally speaking, the front-end activities are more sophisticated than the back-end activities.

Given the technical backwardness of the IC design industry in China, one should not set the bar for these metrics very high. Also, other metrics that are often used, such as gate-count controlling for intellectual property (IP) cores purchased or digital versus analog and mixed-signal design, are of much more value in differentiating average from sophisticated design rather than trying to differentiate crude from average design. Furthermore, in the case of gate counts, many firms in China are reluctant to reveal their gate counts to outsiders, fearing that this would reveal too much information about their products to competitors.

Table 4.1 Factors Determining Contribution to China's Technological Upgrading

Firm type	Finance	Technology inputs (follows the finance)	Capabilities and incentives to upgrade	Operational strategy (OS) (motivation)	Technology strategy (technology activities in China)	Contribution to upgrading in China
Neglected domestic firms	Little access to finance.	No access to finance, so no access to technology inputs.	Low capabilities: • No finance • No technology	China-based OS but no capabilities to undertake technology activities.	Do not pursue upgrading because of low feasibility and high risk.	Low
Favored domestic firms	State banks and procurement offer unmonitored and lavish financial support—that is, soft budget constraints.	Have access to finance, so have access to technology inputs.	Mixed capabilities: • Access to finance • Access to technology • Soft budgets undermine capabilities • Disincentives because of soft budgets	China-based OS but disincentives to upgrade, so motivation is irrelevant.	Generally do not attempt to do technological upgrading; the few attempts are poorly executed.	Low
Regular FIEs	Access to finance, but finance offers harder budget constraints relative to Chinese state banks/procurement.	Same as above.	High capabilities: • Access to finance • Access to technology High incentives: • Hard budget constraints plus technology-based competition	Non-China-based OS, so lack positive bias of interpreting market signals in favor of placing activities in China.	Try to enhance technological capabilities but not necessarily in China.	Variable (low to moderate)
Hybrid FIEs	Same as above.	Same as above.	Same as above.	China-based OS interprets market signals in a manner favoring activities in China.	Try to enhance technological capabilities in China.	High

Setting the bar high to show that the current design activities are closer to the international technology frontier than the design activities of the recent past (1999), a good standard is CMOS[30] process technology at .8 microns. The very few design teams that were not engaged in reverse engineering nine years ago were using .8-micron and larger process technologies. Since the mainstream design activity of five years ago was reverse engineering, moving to actual IC design at .8 microns still represented movement toward the technological frontier. Likewise, layout skills were fairly common even seven years ago, but most of the other design skills were lacking. Thus, firms that did more than layout in the IC design flow were considered to be contributing to technological upgrading.

To set up a harder test for the view that foreign firms are contributing more to technological development than domestic firms, the bar for foreign firms is set higher. Foreign firms must have process technology of at least .35 microns, several generations in advance of .8 microns. Furthermore, firms that appear to be, at best, marginally upgrading technologically are counted if they are domestic firms and are disqualified if they are foreign firms. These measures make the test more robust.

There were data on CMOS process technology for forty-four firms. Two firms used only bipolar or bi-CMOS technology, while four others declined to reveal their process technology. Table 4.2 presents the foreign and domestic firms and their ranges of process technologies. Interviews with other firms in the IC value chain and in the venture capital industry provide evidence that these reported figures often exaggerate the technical prowess of the reporting firm, especially among domestic firms.

Table 4.2 Reported Process Technologies of Foreign and Domestic Firms

	FIEs (28 firms)	Domestic (16 firms)
Average range of common design metrics	0.5 (2 firms) to 0.18 (13 firms)	0.8 (1 firm) to 0.35 (6 firms)
High	0.8 (1 firm)	1.5 (1 firm)
Low	0.13 (5 firms)	0.13 (1 firm)

Note: This table includes data from interviews with firms through September 2004. The process-technology profiles of firms interviewed before 2004 were updated during August and September 2004.

Reverse Engineering

Reverse engineering means taking apart products to learn how they are put together. Reverse engineering and engineering activities are hard to separate and use much of the same skill set.[31] Reverse engineering in the IC industry means peering into another firm's chip circuitry and copying it, but copying the circuitry alone, unfortunately, does not really lead to a more profound understanding of how the circuitry works in order to build an improved product. While it is often taken as a first step to create better designs, reverse engineering by itself will not create design skills.

An appropriate analogy would be photocopying a book in a language we do not understand. We have replicated the book but we would not be able to write a single sentence in that language. This analogy may be somewhat extreme, but chip design and chip reverse engineering are separate and distinct activities, as a number of advanced chip designers with knowledge of China's reverse-engineering tendencies pointed out. One foreign chip designer, who had worked with several of the foreign-invested design start-ups in China and was actively involved in recruiting designers, reported that he soon discovered that engineers with experience in the local firms had only reverse-engineering experience. This had two implications: they were not experienced or trained in design work, and they had bad habits, in that they would rather fall back on copying competitors' chips than create their own designs. Consequently, the designer avoided hiring any more designers with experience in the local firms and hired only fresh university graduates who had not been tainted by years of reverse engineering.

Daryl Hatano, vice president of public policy at the Semiconductor Industry Association, explains the difference: "Reverse-engineering a chip to design an original and better product is allowed under the layout design laws. However, in these [Chinese] cases the chips were essentially photocopies of the U.S. design, which we know because they included the U.S. company's part number etched in a submask level and unused circuits that the U.S. firm had placed on the chip to reserve space for future product development."[32]

Given the illegal nature of this activity, trying to determine which firms were conducting reverse engineering as the mainstay of their activities required finding reliable sources outside the firms that were studied, in order to assess firm practices. A few firms were quite open about the fact that they were only reverse engineering even though I never broached this sensitive topic directly in interviews with design operations. For most firms, interviews with upstream suppliers—such as suppliers of electronic design automation (EDA) tools, the many venture capitalists who concentrate on the technology sector, and returnee designers who were attempting to recruit local talent—provided extensive information on the extent of reverse engineering in the firms being studied. I also had the opportunity to take three electrical engineering professors from elite microelectronics programs in the United States to various companies to help assess the capabilities of those companies. These short visits were not conclusive on their own, but they served to point out certain companies to investigate further.

Overall Assessment of Integrated Circuit Design in China

The overall evaluations of which firms were contributing to technological upgrading are summarized in Table 4.3. These firms employed approximately half the estimated design workforce in China.[33] Of the six domestic firms deemed contributors, at least two were making only marginal contributions, because reverse engineering constituted a large part of their activities. Only two of the thirteen domestic firms that were not contributing were also not proven to be reverse engineering. One was essentially reselling an IP core it had received for free from the Ministry of Information Industry (MII). The other had received technology transfer from an MNC a decade ago and had been producing the same chip for the last ten years. In contrast, only two of the thirty-three foreign design houses were reverse engineering. The regular foreign firms were few in number, but the fact that fewer foreign firms were interested in establishing design operations in China was in itself telling and conforms with the predicted behavior of the non-China-based OS of the regular FIEs. I interviewed five other FIEs in the IC industry that had not set up any design operations in China even though they had offices in China. With the vast majority of IC design work still conducted in Japan, the European Union, and the United States, these firms were underrepresented in China's design operations. This underrepresentation suggests a relative lack of interest or ability to run technology activities in China, which precisely fits this chapter's prediction that regular FIEs are less willing or less able to pursue upgrading in host countries such as China.

Table 4.3 Contributors to Upgrading by Firm Type

Firm types	Contributors to upgrading	Total IC designers in contributing firms	Non-contributors	Reverse-engineering firms	Total firms
Hybrid FIEs	21	1,204	6	2	27
Regular FIEs	3	118	4*	0	7
Domestic firms	6	488	13	11	19

Source: Company interviews.

Note: *Includes one firm that subsequently sold its Chinese back-end design operations.

Evaluation of R&D Activities

This section describes two distinct types of data that I used to evaluate R&D activities in China. First, I analyzed the number of U.S. utility patents issued to China-based researchers in IT firms. Second, I employed a modified version of the Amsden-Tschang framework for evaluating R&D[34] to assess that undertaken by various types of firms in China.[35] The modified version of this framework breaks down R&D activities into four stages, from the most technically sophisticated to the least: basic research (stage 1), applied research (stage 2), detailed design (stage 3), and final development (stage 4).

Patent Data

One overall measure—albeit a crude one, given how discretely R&D can be divided in this era of cheap global communications—is the number of U.S. utility patents earned by corporate researchers based in China. U.S. utility patents are a relatively good standard of technology patents because the United States is still the world's largest and one of the most sophisticated technology markets and, unlike China,[36] has an effective legal mechanism to protect the property rights that patents confer. For this section, data were gathered on the U.S. utility patents originating from China of MNCs, large Chinese firms, and Taiwanese firms active in the IT industry. In addition, U.S. utility patent data were collected for interviewed or otherwise prominent smaller Chinese firms, although none of these smaller firms—with the exception of SMIC—had any U.S. utility patents. The time frame for this analysis was 1997 through 2004. When one looks at the U.S. utility patent trends in China (see Table 4.4) for the IT industry, several patterns emerge. First, the hybrids, followed by the FIEs, are the dominant holders of U.S. utility patents. Three of the top five patent holders are hybrids, and none is a domestic firm. Of the firms ranked in the top ten (there are twenty such firms, due to ties), twelve are FIEs, five are hybrids, and only three are domestic firms. Hybrids account for 464 of the 606 utility patents from China, whereas the FIEs account for 127, and the domestic firms account for only 15. Even when the largest hybrid patent holder, Hon Hai, is not counted, the hybrids still have many more patents (109) than domestic firms and almost as many as the regular FIEs.

One significant pattern that shows the commitment of the hybrid firms to a China-based strategy is the extent to which their portfolio collection is weighted toward China. Only two foreign firms, Microsoft and UMC, have more than 1 percent of their patent portfolio in China, whereas every hybrid firm has more than 1 percent from China. Furthermore, UMC is a Taiwanese firm, so it could be on the cusp of embracing a China-based OS. Several large firms have much larger portions from China. For example, Hon Hai and Inventec have significant patent portfolios from their Taiwanese home bases that they acquired before embracing a China-based strategy in the late 1990s, and have much larger portions from China, at 14 percent and 40 percent, respectively. Even Microsoft's anomalous position can be explained by the nature of the work done in China versus what is done in the United States. One of the projects Microsoft is conducting in China is hardware R&D for its Xbox game machines, and hardware R&D generates far more patents than software R&D. This fact was illustrated when IBM opened up its patent portfolio in software, which amounted to only 500 patents out of IBM's more than 40,000 global patents. Beyond Microsoft, the only FIE firm to have even 1/1,000 of its patents from China was Agere (see Table 4.4).

The patent data also confirm the prediction that MNCs will favor their national home base. The hybrid, a type of firm that embraces China as part of the national home base, has a significant presence in China. Moreover, the few patents generated by domestic Chinese firms also tend to be generated in China. TCL had the lowest number, with only one-third of its patents originating in China, due to its acquisition of foreign assets abroad. All the other domestic firms generated more than half their patents in China.

Table 4.4 Corporate U.S. Utility Patents from China, by Firm

Rank	Firm	2002–2004	1997–2001	Total China	Firm/Type	Total world	Percentage from China
1	Hon Hai	304	51	355	Hybrid	2,539	14
2	Inventec	47	31	78	Hybrid	197	40
3	Microsoft	37	3	40	FIE	3,467	1
4	UMC	12	17	29	FIE	2,519	1
5	Winbond	14	8	22	Hybrid	761	3
6	Intel	7	1	8	FIE	9,327	0.09
6	Nokia	6	2	8	FIE	3,301	0.2
7	Delta	6	1	7	Hybrid	293	2
7	Epson	6	1	7	FIE	5,310	0.1
7	Siemens	6	1	7	FIE	19,623	0.04
7	Philips	5	2	7	FIE	19,421	0.04
7	Motorola	4	3	7	FIE	16,389	0.04
8	Huawei	6	0	6	Domestic	7	86
9	Fujitsu	3	1	4	FIE	16,142	0.02
9	Lucent	2	2	4	FIE	7,572	0.05
9	Haier	4	0	4	Domestic	4	100
10	Agere	2	0	2	FIE	1,377	0.1
10	ZTE	2	0	2	Domestic	2	100
10	Matsushita	2	0	2	FIE	20,846	0.01
10	SMIC	2	0	2	Hybrid	3	67
11	TCL	1	0	1	Domestic	3	33
11	Hitachi	1	0	1	FIE	30,443	0.003
11	Tongfang	1	0	1	Domestic	1	100
11	BYD	1	0	1	Domestic	1	100
11	Seagate	0	1	1	FIE	2,626	0.04
	Total	481	125	606			

Unfortunately, the domestic firms simply did not generate many patents. Their lack of learning and innovation is precisely why the hybrids have played such a prominent role in China's upgrading. Beyond the fact that the domestic firms generated only fifteen patents, Haier's four patents were in white goods (refrigerators and washing machines), which only loosely fit into the IT category. More dismally, many of the so-called leading firms of China's *guojiadui* (the national team of firms championed by the state) were conspicuously absent from any contribution. Major PC and telecommunications firms, such as Lenovo, Founder, Bird, Konka, and Ziguang, were absent, as were consumer electronics giants Changhong and Hisense.

For the regular FIEs, conspicuous in their absence from the list of China-origin patents were thirteen major IT firms from the 2004 Organisation for Economic Co-operation and Development (OECD)[37] ranking of global IT giants: IBM, Hewlett-Packard (HP), NEC, Samsung, Toshiba, Nortel, Mitsubishi, Dell, Canon, EDS, Sanyo, Sun Microsystems, and Apple. Cisco, the twenty-first-largest IT firm and the largest networking-equipment firm in the world, also does not have any patents from China. In fact, seven of the top ten telecommunications equipment firms[38] did not have any patent activity in China. Also, many large, nonhybrid Taiwanese firms do not have a patent presence in China, including TSMC, Acer, Quanta, Compal, Asustek, and Tatung. The only nonhybrid Taiwanese firm with U.S. patents originating from China was UMC.

One final observation about the patent trends in China is that the patenting has accelerated. Corporations received 481 patents from 2002 through 2004, and only 125 before 2002. In 1997 corporations received only 9 utility patents by Chinese researchers.

Size and Stage of R&D Activities Conducted in China

Through my interviews, I have been able to confirm the size and activities of the R&D centers (Table 4.5). The R&D centers of the regular FIEs (listed as MNCs in the table) are all top twenty-five global IT firms or top ten global firms in software or telecommunications, except for two 1990s start-ups marked with asterisks in Table 4.5. I have also included Taiwanese firms that have eschewed the China-based strategy but are still operating some R&D in China. Finally, I have included the category "Other Taiwan," which denotes those large product manufacturers from Taiwan that employ a few design-for-manufacturing engineers in their local Chinese plants. I estimated that there were twenty plants with twenty such engineers each, although the actual plant total is probably much higher. This category is based on my own observations from interviews and plant visits, and on the works of Chen, and of Yang and Hsia.[39] These Taiwanese scholars observed that many Taiwanese product plants employed design-for-manufacturing engineers at the manufacturing site.

For the hybrids in Table 4.6, the first four firms are established firms; that is, each had a home base in one of the ECEs before embarking on a China-based

Table 4.5 Foreign R&D Centers: Personnel and Research Activities

Code	Nationality	Researchers	Products	Amsden-Tschang framework
MNC1	N. America	900	TC/software	3 and 4
MNC2	N. America	200	PC/software	4
MNC3	European Union	700	TC	3 and 4
MNC4	Japan	13	PC/software	4
MNC5	Japan	3,000	IT/software	4
MNC6	EU	114	TC/software	4
MNC7	N. America	300	TC/software	4
MNC8	EU	200	TC/software	3 and 4
MNC9	Japan	1,500	Consumer electronics	4
MNC10	N. America	150	IT/hard and soft	2, 3, and 4
MNC11	N. America	1,400	TC/hard and soft	3 and 4
MNC12	Japan	48	Consumer electronics	4
MNC13	EU	30	PC/software	4
MNC14	N. America	30	Software	4
MNC15	Japan	20	Machinery	4
MNC16	N. America	60	IT/software	3 and 4
MNC17	N. America	2,000	TC/software	3 and 4
MNC18	Korea	700	Consumer electronics	4
MNC19*	N. America	2,000	TC/hard and soft	3 and 4
MNC20*	N. America	400	TC/ind. design	3 and 4
MNC21	Korea	200	Consumer electronics	4
TW1	Taiwan	200	PC/software	4
TW2	Taiwan	100	TC/software	3 and 4
TW3	Taiwan	300	TC/software	4
TW4	Taiwan	75	TC/soft; mfg.	3 and 4
	Other Taiwan	400		4
Total	Regular FIEs	15,040		

Notes: TC: telecommunications; IT: general information technology; PC: personal computers.
* Not in the top twenty-five of global IT firms or the top ten of global firms in software or telecommunications.

Table 4.6 Hybrid R&D Centers: Personnel and Research Activities

Code	Nationality	Researchers	Products	Amsden-Tschang framework
Hybrid 1	Taiwan	3,000	Comp; TC/soft and hard	1–4
Hybrid 2	Taiwan	1,100	PC/soft and hard	2–4
Hybrid 3	Taiwan	400	Comp; Mfg.	1, 2, and 4
Hybrid 4	Hong Kong	120	Comp.	2–4
Hybrid 5	Overseas	50	TC/soft amd hard	2–4
Hybrid 6	Overseas	50	Consumer/soft and hard	2–4
Hybrid 7	Overseas	200	TC/soft and hard	2–4
Hybrid 8	Mixed	55	Comp.; TC/soft and hard	1–4
Total	Hybrids	4,975		

Note: TC: telecommunications; Comp.: components; PC: personal computers.

strategy. The other four were start-ups or, in the case of hybrid 8, a new business for an established firm. As shown by the totals, far more engineers were trained by the regular FIEs than by the hybrids. When we turn to examine the activities pursued, the hybrids come out ahead. Only one firm among the regular FIEs was pursuing new product generation (stage 2) research. Among the hybrids, all the firms were conducting new product generation research, and some were involved in basic research for which a doctorate degree is required.

When we consider the domestic firms, almost none qualifies as a contributor to technological development. TCL is a marginal contributor and has only about one hundred engineers engaged in product development in China. The firm is, at best, doing detailed design, according to the reports of suppliers. If the standards are lowered further, Huawei (despite violating the standard of commercialization by selling the majority of its products to the state) could be included, but Huawei's reported tens of thousands of design engineers are at best engaged in detailed design based on other firms' products.[40]

The Implications of Hybrid-Driven Development

There are three main implications of hybrid-driven development. The first is the need to place NISs in the context of their larger political economy. Second, the hybrid-driven model challenges the long-held belief of revisionist political economists that foreigners must be kept at bay or at least under strict state monitoring to foster development. The third implication is that in today's increasingly globalized world, developing nations have options beyond sticking to their respective (and often malfunctioning) domestic sets of institutions.

I would argue for analyzing NISs in their wider political economy context by introducing politically determined sources of finance as an important causal variable when looking at these systems. This is important because much of the literature assumes that the effectiveness of the institutions found in a given country and studies of particular national systems often are simply laundry lists of the extant science and technology institutions.[41] By showing the greater effectiveness of hybrids in generating technology, due in large part to their different relationship with local institutions, the evidence from the two cases indicates that China's NIS is not a level playing field, equally accessible to all actors, where the local institutions actually function to create innovation (or in the developing-country context, technological learning). The two cases show that the politics of financing have a differential impact on the behavior of firms operating within the same NIS. In effect, China's uneven institutional terrain provides an empirical case and an institutional explanation as to why a given system may be ineffective and inefficient. Being both a developing economy and a postsocialist transitional economy, China's NIS can be understood only in its wider and evolving institutional context.

While many previous studies[42] have looked at the indigenous learning process in developing countries and others have examined the process specifically in China,[43] these works do not anticipate the sharp dichotomy in the contributions of various types of firms to technological upgrading in developing countries such as China, because they do not attempt or do not go far enough to place the institutions they examine in relation to the local or national political economy. Many scholars of Chinese technology have suggested an improving picture of firm performance, as well as improvement in the performance of the wider institutions of the economy and innovation system,[44] even while recognizing that problems remain, particularly the problem of the soft budget constraint of China's SOEs.[45] Few studies have attempted to use empirical work to differentiate the technological performances among Chinese firms, despite an extensive literature on the difference in overall performance between SOEs and other firms.[46] For at least one high-technology product area, the overall judgments of China's progress or lack of progress are not contingent solely on the economic or innovation system, but instead depend on how firms relate to that system.

The leading role of the hybrid FIEs in developing technical skills in China also suggests that in the increasingly globalized economy the traditional deemphasis of foreign firms as part of the integral process of economic and technological development[47] (as opposed to being a source of technology) should be questioned. However, any embrace of foreign firms must be tempered by the fact that only a particular kind of FIE appears to play a critical positive role in development. In China, the hybrid FIEs behaved differently from the typical FIE because they embraced a China-based OS. Without the technological diaspora of educated Chinese and prosperous ECEs located close to China, it is doubtful whether such hybrid firms would emerge. In short, co-ethnics in wealthy centers of technology were the key to changing the behavior of a subset of foreign firms. One could imagine a similar scenario for some foreign firms in India with returning technology experts from the West. Other nations, ranging from Thailand to Armenia, arguably have nascent co-ethnic technology networks they can utilize.[48]

Another implication of the findings is that developing countries are not irrevocably stuck with their domestic set of institutions. If domestic institutions perform reasonably well, then the developing country should probably not pursue the path of radical openness. For the many developing nations where the set of domestic economic institutions has serious deficiencies, China's opening up of access to better-performing (more efficient) foreign financial institutions shows the developing world a new route for institutional improvement. Instead of being limited by the domestic institutions, openness brings new possibilities of institutional configurations along with the influx of technology and capital.

How sustainable are these divergent patterns? They have rested on a foundation of access to state capital by favored domestic firms and on the many restrictions on private firms going abroad to seek capital. As the foreign financial presence grows in China, even in the absence of an overhaul of domestic regulations, neglected domestic firms may have more opportunities to seek foreign capital. The growing private sector in China also suggests that there may soon be many domestic alternatives to state finance. Furthermore, the state banking sector's inefficient practices may be unsustainable, so the coterie of favored firms may sooner or later be given a cold shoulder by a reforming financial sector in China. Over the medium term, the domestic arrangements that have made the hybrid-driven path possible are under stress. Nevertheless, allowing foreign firms to play a large role in technological upgrading has proven an effective strategy, given the current institutional shortcomings in China's economy.

Notes

[1] Robert Wade, "Globalization and Its Limits: Reports of the Death of the National Economy Are Greatly Exaggerated," in *National Diversity and Global Capitalism*, ed. S. Berger and R. Dore (Ithaca, NY: Cornell University Press, 1996); Paul Hirst and Grahame Thompson, *Globalization in Question* (Cambridge: Polity Press, 2001); and Yao-su Hu, "Global or Stateless Corporations Are National Firms with International Operations," *California Management Review* 34, no. 2 (1992): 107–26.

[2] Yasheng Huang, *Selling China: Foreign Direct Investment during the Reform Era* (New York: Cambridge University Press, 2003); Edward S. Steinfeld, *Forging Reform in China* (New York: Cambridge University Press, 1998); and Neil Gregory, Stoyan Tenev, and Dileep Wagle, *China's Emerging Private Enterprises: Prospects for the New Century* (Washington, D.C.: International Finance Corporation, 2000).

[3] Janos Kornai, Eric Maskin, and Gerard Roland, "Understanding the Soft Budget Constraint," *Journal of Economic Literature* 41 (2003): 1095–1136.

⁴ Xielin Liu's surveys indicate that domestic firms were actually favored over foreign firms as the destination for returning Chinese technologists for most of the 1990s. See his *Ershiyi Shiji de Zhongguo Jishu Chuangxin Xitong* [The twenty-first century's Chinese system of innovation] (Beijing: Peking University Press, 2001), 204.

⁵ Huang, *Selling China*; Dwight Perkins, "Industrial and Financial Policy in China and Vietnam: A New Model or a Replay of the East Asian Experience?" in *Rethinking the East Asian Miracle*, ed. J. Stiglitz and S. Yusuf (New York: Oxford University Press, 2001); and Thomas G. Moore, *China in the World Market: Chinese Industry and International Sources of Reform in the Post-Mao Era* (New York: Cambridge University Press, 2002).

⁶ Gilles Guiheux, "The Incomplete Crystallisation of the Private Sector," *China Perspectives* 42 (2002): 35.

⁷ Nicholas Lardy, *Integrating China into the Global Economy* (Washington, D.C.: Brookings Institution, 2002), 114–19.

⁸ David Zweig, *Internationalizing China: Domestic Interests and Global Linkages* (Ithaca, NY: Cornell University Press, 2002), chapters 2 and 3.

⁹ There was only one fifty-fifty joint venture in the study, and the firm involved was clearly controlled by the Chinese partner.

¹⁰ I use the term *management* because part of the ownership of these firms is from foreign financial institutions, which may or may not be ethnic Chinese. Nevertheless, even in the case of start-ups where the majority of shareholders (the venture capitalists) are often not ethnic Chinese, the teams of managers in charge of running these firms are ethnic Chinese and have a significant equity stake. The managers sold investment stakes of their firms on the basis of their China-based strategy, so there is no reason to think that the non-ethnic-Chinese foreign finance will try to change the strategic orientation toward China of these hybrids. I uncovered no instances of this and I heard reports of just the opposite, foreign venture firms attracted to invest in hybrid firms precisely because these firms were "China plays." See Ann Grimes, "Venture Firms Seek Start-ups That Outsource," *Wall Street Journal*, April 2, 2004, B1.

¹¹ Paul N. Doremus, William W. Keller, Louis W. Pauly, and Simon Reich, *The Myth of the Global Corporation* (Princeton, NJ: Princeton University Press, 1998); Wade, "Globalization and Its Limits"; Hu, "Global or Stateless Corporations"; and Hirst and Thompson, *Globalization in Question*.

¹² Pari Patel and Keith Pavitt, "National System under Strain: The Internationalization of Corporate R&D" (Science and Technology Policy Research, University of Sussex [SPRU] Electronic Working Paper Series, no. 22, 1998), 1–27. Patel and Pavitt's work is concerned about the integrity of NISs, but their main example of the strains on national systems comes from Europe. Europe is exceptional in that the EU itself is arguably becoming the home market. Thus, one would expect individual European countries to have large R&D bases outside their national borders, such as Belgium, which has two-thirds of its R&D outside its national borders. In contrast to European

countries, the United States and Japan had well over 90 percent of their R&D in their home economy, and the change from the early 1980s was not dramatic. The United States has increased foreign R&D by 2.2 percent, and Japan has actually decreased global R&D by 7 percent. The main measure that Patel and Pavitt used was the origin of U.S. utility patents by the Global 500. When the actual R&D budget is examined, the United States had slightly more than the average for MNCs, at 11.9 percent, and Japan had much lower than average, at 2.1 percent. However, it is unclear whether the R&D budget is a better measure than the patents. In any case, under neither metric are the two largest national economies in the world very global in their core R&D functions.

[13] Peter Hall and David Soskice, eds., *Varieties of Capitalism* (New York: Cambridge University Press, 2001). The varieties of capitalism literature of which Hall and Soskice are proponents are not very applicable to the case of hybrid FIEs in China, because this literature argues that the microeconomic foundations of domestic corporations utilizing the strengths of the home economy are a product of a long historical coevolution. The hybrids and China's institutions simply have not had the time to coevolve in this manner.

[14] Doremus et al., *The Myth of the Global Corporation*, 16–17.

[15] For some countries, one could argue that immigrants also adopt a nationalist bias toward their adopted country, but because China is not an immigrant country, the chance of this happening among non-ethnic-Chinese foreigners is quite low. But for ethnic Chinese, it is a different matter. There is an ideology in China regarding ethnic-Chinese foreigners, so-called "overseas Chinese" (*huaqiao*), as fellow Chinese rather than as foreigners. Sun Yat-sen, the founding father of modern China, revered on both sides of the Taiwan Straits, was a U.S. citizen. Sun first heard the news of the unfolding Chinese revolution he had supposedly started while on a train in the middle of the United States.

[16] One of the most interesting examples of imagining that economic interest is crucial to explaining social behavior, even in the face of countervailing objective economic interests, is Yoshiko Herrera's *Imagined Economies: The Sources of Russian Regionalism* (New York: Cambridge University Press, 2005).

[17] I did not bring up the nationalist motivation in interviews, due to the sensitive nature of talking about such topics with a foreigner, but some of the interview subjects confided that this factor was part of the motivation. However, it is equally obvious that not all ethnic Chinese feel a strong desire to help the People's Republic of China. Thirty-eight percent of ethnic Chinese in Taiwan supported Chen Shui-bian, the anti-China candidate, in the 2000 presidential election in Taiwan, and just over 50 percent supported Chen in 2004. Taiwanese business people I encountered in China often were Taiwanese nationalists who were strongly anti-China. Still, for Hong Kong and Taiwan natives, the cultural barriers in China are obviously lower than they are for non-ethnic-Chinese foreigners, regardless of which conception of nationality these ethnic Chinese from outside the Mainland may ascribe to. Adam Segal also found nationalism to be a motivation for local Chinese entrepreneurs; see

his *Digital Dragon: High-Technology Enterprises in China* (Ithaca, NY: Cornell University Press, 2003), 166–68.

[18] Hu, "Global or Stateless Corporations"; Wade, "Globalization and Its Limits"; and Hirst and Thompson, *Globalization in Question.*

[19] Mushtaq H. Khan, "Rents, Efficiency and Growth," in *Rents, Rent-seeking and Economic Development*, ed. M. H. Khan and K. S. Jomo (Cambridge: Cambridge University Press, 2000), 40–53.

[20] Janos Kornai, *Highways and Byways: Studies on Reform and Post-Communist Transition* (Cambridge, MA: MIT Press, 1995); Kornai et al., "Understanding the Soft Budget Constraint"; and Liu, *The Twenty-first Century's Chinese System of Innovation.*

[21] Kornai, *Highways and Byways*; and Huang, *Selling China*, 137–140.

[22] Huang, *Selling China: Foreign direct investment during the Reform era*; Qiang Wu, "Ri, Han Jishu Yinjin Zhengce ji Zhongguo de Fazhan Zhanlue" [Technology import policies of Korea and Japan, and China's development strategy] *Jingji Guanli* (*Economic Management*) issue 9 (1996): 12–15; and Gary H. Jefferson, Thomas G. Rawski, and Yuxin Zheng, "Innovation and Reform in China's Industrial Enterprises," in *Enterprise Reform in China*, ed. G. H. Jefferson and I. Singh (New York: Oxford University Press, 1999).

[23] Wu, "Technology import policies of Korea and Japan."

[24] See Jefferson et al., "Innovation and Reform."

[25] Liu, *The twenty-first century's Chinese system of innovation*, 39 and 43; Kornai et al., "Understanding the Soft Budget Constraint," 1105.

[26] Moore, *China in the World Market*; Huang, *Selling China*; and Perkins, "Industrial and Financial Policy in China and Vietnam."

[27] Huang, *Selling China*, chapters 2–5.

[28] Huang, *Selling China*, chapter 4.

[29] Khan, "Rents, Efficiency and Growth"; and H. Pack and L. E. Westphal, "Industrial Strategy and Technological Change: Theory versus Reality," *Journal of Development Economics* 22, no. 1 (1986): 87–128.

[30] CMOS technology is the standard used here because it is the most common technology of the firms interviewed, as well as the most common process technology globally. Forty-four of the firms used CMOS technology. I did not include one additional CMOS firm, because I first interviewed that firm in 2005.

[31] I acknowledge the contribution of a group of scholars at the Institute of Developing Economies (IDE) in Chiba, Japan, to this distinction between reverse engineering in other industries and IC design. One scholar of the automotive sector in Japan and China pointed out how reverse engineering was critical to building the design skill set in China's motorcycle industry. This traditional, benign view of reverse engineering is also expressed in *Industrial Technology Development in Malaysia: Industry and Firm Studies*, ed. K. S. Jomo, Greg Felker, and Rajah Rasiah (New York: Routledge, 1999), 6.

[32] This quotation by Daryl Hatano appears in Thomas R. Howell, Brent L. Bartlett, William A. Noellert, and Rachel Howe, *China's Emerging*

Semiconductor Industry: The Impact of China's Preferential Value-Added Tax on Current Investment Trends (Washington, D.C.: Dewey Ballantine LLP, 2003).

[33] Chris Hsieh, *China Semiconductors* (Taipei: ING, 2004).

[34] Alice Amsden and F. Ted Tschang, "A New Approach to Assessing the Technological Complexity of Different Categories of R&D (with Examples from Singapore)," *Research Policy* 32, no. 4 (April 2003): 553–72.

[35] The main modification made to the Amsden-Tschang framework was the removal of the stage of "pure science."

[36] One mark of the ineffectiveness of domestic Chinese patents as opposed to foreign ones is the absence of a connection between Chinese patents and new product creation. Yifei Sun found that there was no statistically significant relationship between domestic Chinese patents and the ability to create new products. However, Sun found a significant positive correlation between foreign technology inputs and the ability to create new products. In other words, foreign technology inputs was much more worthwhile in new product creation than holding Chinese patents. See Yifei Sun, "Sources of Innovation in China's Manufacturing Sector: Imported or Developed In-house?" *Environment and Planning A* 34 (2002): 1059–72.

[37] *OECD Information Technology Outlook 2004* (Paris: OECD, 2004).

[38] According to the OECD (2004), in 2003, the top ten telecommunications equipment firms were Nokia, Motorola, Cisco, Alcatel, Ericsson, Nortel, Lucent, L-3 Communications, Avaya, and Qualcomm.

[39] Shin-horng Chen, "Taiwanese IT Firms' Offshore R&D in China and the Connection with the Global Innovation Network," *Research Policy* 33, no. 2 (2004): 337–49; and You-ren Yang and Chu-joe Hsia, "Local Clustering and Organizational Governance of Trans-Border Production Networks: A Case Study of Taiwanese IT Companies in the Greater Suzhou Area, China" (paper presented at the Association of Asian Studies annual meeting, March 31–April 3, 2005, in Chicago).

[40] Dimitri Kessler, "The Pride and Humility of Tech Transfer: Confrontation and Cooperation in Mainland China and Taiwan" (paper presented at the Association of Asian Studies annual meeting, March 31–April 3, 2005, in Chicago). In addition to Huawei's dubious claims to commercial viability, a number of its competitors that were interviewed claimed that the firm has egregiously violated their IP rights and that it stands out in this regard even considering China's lax IP-enforcement environment.

[41] See Wei Xie and Guisheng Wu, "Differences between Learning Processes in Small Tigers and Large Dragons: Learning Processes of Two Color TV (CTV) Firms within China," *Research Policy* 32 (2003): 1463–79; Xielin Liu and Steven White, "Comparing Innovation Systems: A Framework and Application to China's Transitional Context," *Research Policy* 30 (2001): 1091–14; Shulin Gu, *China's Industrial Technology: Market Reform and Organizational Change* (New York: Routledge, 1999).

⁴² Alice Amsden, *Asia's Next Giant: South Korea and Late Industrialization* (New York: Oxford University Press, 1989); Michael Hobday, *Innovation in East Asia: The Challenge to Japan* (Brookfield, MA: Edward Elgar, 1995); John A. Mathews and Dong-sung Cho, *Tiger Technology: The Creation of a Semiconductor Industry in East Asia* (Cambridge: Cambridge University Press, 2000); Linsu Kim, *Imitation to Innovation: The Dynamics of Korea's Technological Learning* (Boston, MA: Harvard Business School, 1997); and Keun Lee and Chaisung Lim, "Technological Regimes, Catching-up and Leapfrogging: Findings from the Korean Industries," *Research Policy* 30 (2001): 459–83.

⁴³ See Xie and Wu, "Differences between Learning Processes in Small Tigers and Large Dragons;" Liu and White, "Comparing Innovation Systems;" and Gu, *China's Industrial Technology.*

⁴⁴ Gu, *China's Industrial Technology*; Xie and Wu, "Differences between Learning Processes"; Liu and White, "Comparing Innovation Systems;" and Qiwen Lu, *China's Leap into the Information Age* (Oxford: Oxford University Press, 2000).

⁴⁵ Liu and White, "Comparing Innovation Systems," 1108; Liu, *The twenty-first century's Chinese system of innovation*, 43.

⁴⁶ See the literature review in Huang, *Selling China*, chapters 1 and 2.

⁴⁷ See, for example, Amsden, *Asia's Next Giant*; Hobday, *Innovation in East Asia*; Mathews and Cho, *Tiger Technology*; and Kim, *Imitation to Innovation*. A major exception is Lee and Lim, "Technological Regimes."

⁴⁸ Douglas B. Fuller, "Creating Ladders Out of Chains: China's Technological Development in a World of Global Production" (PhD dissertation, MIT, Cambridge, MA, 2005).

MULTINATIONAL R&D IN CHINA: MYTHS AND REALITIES

Lan Xue and Zheng Liang[*]

Although conducting research and development (R&D) abroad is usually one of the last components of an internationalization strategy, many companies have at last begun to embrace it. This trend has generated some uneasiness in the developed countries. A *Business Week* article in 2005 posed a question that perhaps was in many minds: "First came manufacturing. Now companies are farming out R&D to cut costs and get new products to market faster. Are they going too far?"[1] At the same time, the *World Investment Report 2005* pointed out the growing importance of developing countries as recipients of foreign direct investment (FDI) in knowledge-intensive activities. For example, while in 1994 developing countries accounted for 7.6 percent of overseas R&D expenditures by U.S. multinational corporations (MNCs), that percentage had grown to 13.5 by 2002.[2] There are also concerns in developing countries about potential talent loss for local companies and the "crowding out" of local markets.

This phenomenon raises a number of questions: Why would MNCs do something of arguable concern in both types of societies? What are their motivations? Are such R&D activities geared mostly toward local markets, or are they focused mainly on global R&D networks? In what industries do MNCs set up R&D centers in developing countries, and are they different from those in developed countries? What are the implications for the national innovation system (NIS) of the local country? This chapter aims to improve our understanding of MNCs' R&D activities in developing countries, with a focus on their R&D centers in Beijing, where over half of those in China are located. This follows up on our 1999 study, in which we surveyed over thirty major MNC R&D centers around China. In this chapter we focus on two related issues. The first is the scope, organization, activities, and linkages with local institutions of major MNCs' R&D centers in Beijing. The second is the goal of better understanding their motivations.

---

[*] The study on which this chapter is based was funded by the Beijing Municipal Science and Technology Commission, International Science and Technology Cooperation Program, Ministry of Science and Technology (MOST), No. 2005DFA00020, and Tsinghua University Fund for Promotion of Humanities and Social Sciences, No. 2007WKYB017. We would also like to acknowledge the support of the Foreign Investment Administration, the Ministry of Commerce, the Department of Science and Technology, the Ministry of Education (MOE), and the Department of Policy, Regulation and Reform.

A Brief Literature Review

The study of the globalization of R&D has been a hot topic since the 1980s. Most of the literature concentrates on motivations, objectives, and locations.[3] Japan and Western Europe remain the main host locations of MNCs' overseas R&D activities, but there is a clear trend toward more of them being outside the triad, especially in the newly industrialized economies (NIEs) of East and Southeast Asia.[4]

Local adaptation is the dominant type.[5] With more production overseas and increased product complexity, R&D facilities near production sites are needed to give technical support to localized manufacturing. Usually, the larger the host market, the greater the need for local adaptation of goods and services. Technology sourcing is an increasingly important reason to do R&D in countries with centers of excellence.[6] A well-functioning NIS with strong public research institutions, science parks, an adequate system of intellectual property rights (IPR) protection, and government incentives is a key determinant.[7]

The recent expansion of R&D outside the triad suggests that a new set of drivers—costs and the availability of research personnel—have become increasingly important.[8] These lead MNCs to locate R&D in developing countries with low-cost and ample scientific personnel.[9]

Knowledge management has become a key topic in the research on MNCs.[10] Although foreign affiliates doing R&D have become a part of the NIS, the transfer of R&D resources between a parent and its affiliates does not automatically lead to a diffusion of knowledge in the host economy. Cheung and Lin[11] suggest that FDI can benefit the host country by means of such channels as reverse engineering, skilled-labor turnovers, demonstration effects, and supplier-customer relationships. Linkages between MNCs and domestic business entities are vital,[12] as well as interactions with local universities and public research institutes.[13] However, there is evidence that there are few spillovers, if any.[14]

Patents are widely seen as the most important output measure of corporate R&D. Its internationalization means that inventors and the ownership of such inventions cross national borders more frequently.[15] As Tomoko[16] estimated, of a total of 12,920 patent grants by the United States Patent and Trademark Office (USPTO) to 137 Japanese MNCs, 7.4 percent were based on overseas inventions. Some view IPR and patents as tools to promote innovation and encourage the dissemination of information, while others hold that they are simply tools to capture market share.[17] Some scholars show that the international protection of IPR entails high costs for the economies of developing countries and does not really help them.[18]

Since the second half of the 1990s, technology-intensive MNCs have located many R&D activities in China.[19] Armbrecht[20] emphasizes that while cost savings matter, MNCs do this primarily for strategic reasons: to tap the vast pool of talent and ideas and to stay abreast of competitors in increasingly sophisticated markets. Walsh and Ernst[21] identify three types of MNCs'

R&D in China: satellite, contract, and (more) equal partnership. Satellite and contract laboratories still dominate,[22] but there are examples of more equal partnerships, especially in the development of China's alternative standards in mobile telecommunications, open-source software, and digital consumer electronics.[23]

Research Methodology

MNCs' R&D activities include those in manufacturing and sales facilities, R&D cooperation with universities and research institutions, and the setting up of R&D centers. In this chapter we focus on the autonomous R&D center, defined as a stand-alone facility with 50 percent or more control by a foreign parent company. This type of center typically has its own budget and managers, separate from sales and manufacturing facilities of the parent company's operations. In addition, we examine other kinds of R&D facilities, especially those in manufacturing and operating corporations without their own budgets.

Given the absence of official data, we chose the *Business Week 1000* (2004) for our investigation. We also added Korea's corporations in the Fortune Global 500 list (2003) because they have rapidly added R&D operations in China.

According to the new classification system based on the Global Industry Classification Standard (GICS),[24] these companies are in 10 main categories: energy, materials, industry, consumer discretionary, consumer staples, health care, financials, information technology, (IT), telecommunication services, and utilities. (We excluded financials and utilities and all subdivided industries belonging to services except telecommunication as unlikely to do R&D in China.) Adding 12 Korean companies brought the total number of corporations to 483.

With further investigation, the set was narrowed to 335, and we received responses from 289. Table 5.1 shows their industrial distribution.

Initial Findings: The Status Quo of MNCs' R&D Investment in China

Through interviews, we obtained information on about 215 R&D subsidiaries in China. Of those subsidiaries, 107 were autonomous, 59 were inner R&D units, 6 were ongoing R&D organizations, and 43 were other R&D organizations (including some that had been coestablished with universities, science and research institutes and enterprises, and some laboratories also with R&D functions).

Table 5.2 shows that among all autonomous centers, the IT industry accounted for 61 percent; it also accounted for 50 percent among total R&D organizations. Other industries with a relatively large percentage among total organizations were industrial equipment and components (12.7 percent), biotechnology and drugs (9.0 percent), and automobiles (7.2 percent). Household electronics (5.4 percent), chemicals (4.2 percent), and food and beverages (4.2 percent) had low proportions. The distribution of R&D units was different, with industrial equipment and components having the largest proportion (27.1 percent), and biotechnology and other IT products having the second- and third-largest proportion (15.3 and 13.6 percent, respectively). Industry characteristics evidently have a large influence on R&D investment by MNCs, and MNCs' R&D investment in China was focused mainly on software, telecommunication, and semiconductors.

Table 5.1 Industrial Distribution of the Firms

Industry code in GICS		Industry name	Total firms	Have operations in China	Responded
10 Energy	101010	Energy equipment, and services	8	1	1
	101020	Oil and gas	29	17	14
15 Materials	151010	Chemicals	22	19	15
	151020	Construction materials	4	1	1
	151030	Containers and packaging	4	4	4
	151040	Metals and mining	12	9	9
	151050	Papers and forest products	6	5	4
20 Industry	201010	Aerospace and defense	12	7	6
	201020	Building products	3	2	2
	201030	Construction and engineering	3	0	0
	201040	Electrical equipment	16	15	12
	201050	Industrial conglomerates	17	17	17
	201060	Machinery	19	18	16
	20201020	Data processing services*	3	1	1
25 Consumer discretionary	251010	Auto components	6	6	6
	251020	Automobiles	15	12	12
	252010	Household durables	18	12	9
	252020	Leisure equipment and products	5	5	5
	252030	Textile, apparel, and luxury goods	17	7	5
	254010	Media	28	12	9

30 Consumer staples	302010	Beverages	11	9	
	302020	Food products	14	10	
	302030	Tobacco	6	4	2
	303010	Household products	7	5	5
	303020	Personal products	11	9	8
35 Health care	351010	Health care equipment and supplies	18	12	9
	352010	Biotechnology	9	6	4
	352020	Pharmaceuticals	48	34	31
45 Information technology	451010	Internet software and services	11	5	5
	451030	Software	19	17	13
	452010	Communications equipment	10	9	8
	452020	Computers and peripherals	11	10	9
	452030	Electronic equipment and instruments	6	5	5
	452040	Office electronics	3	3	3
	452050	Semiconductor equipment and products	17	14	12
50 Telecommunication services	501010	Diversified telecommunication services	22	9	7
	501020	Wireless telecommunication services	11	3	3
Others			2	1	1
Total			**483**	**335**	**289**

Note: *Data processing services (20201020) is one of the subdivided industries under the category of commercial services and supplies (202010), which has similar characteristics to the IT industry.

The number of autonomous R&D centers increased from 32 to 107 between 1999 and 2004, and their industry domains became more diversified, with increases in biotechnology and drugs and in commodity chemicals. Household electronics, other IT products, chemicals, and food and beverages emerged. This is in sharp contrast to the United States, where industrial, chemical, and biomedicine lead other industries.[25]

Table 5.2 The Industrial Distribution of R&D Organizations from *BusinessWeek 1000* (2004)

Industry name	Total organizations		Autonomous R&D centers		R&D units	
	Number	Percent	Number	Percent	Number	Percent
Software	30	18.1	26	24.3	4	6.8
Telecommunication	25	15.1	20	18.7	5	8.5
Semiconductors	15	9.0	15	14.0		0.0
Industrial equipment and components	21	12.7	5	4.7	16	27.1
Automobiles	12	7.2	7	6.5	5	8.5
Commodity chemicals	8	4.8	7	6.5	1	1.7
Biotechnology and drugs	15	9.0	6	5.6	9	15.3
Household electronics	9	5.4	6	5.6	3	5.1
Other IT products	13	7.8	5	4.7	8	13.6
Chemicals	7	4.2	4	3.7	3	5.1
Food and beverages	7	4.2	3	2.8	4	6.8
Industrial conglomerates	2	1.2	2	1.9		0.0
Others	2	1.2	1	0.9	1	1.7
Total	166	100	107	100	59	100

In terms of parent countries, MNCs of North America set up the most (47 percent), Japan and Europe were next (both 21 percent), and Korea was fourth (10 percent), with hardly any from other countries. However, Korea has recently become a more important parent country with the rise of players such as Samsung. There was also little difference between the country distribution of autonomous R&D centers and total organizations, suggesting that the parent nation does not influence the form of R&D organizations of MNCs in China.

Between 1999 and 2004, the percentage of MNCs from North America, England, France, Germany, and Northern Europe decreased, while the percentage of MNCs from Japan and other countries increased. The number of autonomous R&D centers from Japan rose from five in 1999 to twenty-three in 2004, the largest increase among all countries. This indicates that Japanese companies, conservative in technology transfer to and R&D investment in China, have quickened their R&D investing pace. This is directly related to the depression of Japan's economy and the high speed of development of China starting in the late 1990s.

Table 5.3 The Country Distribution of R&D Organizations Reported by *BusinessWeek 1000* (2004)

Parent countries	Total organizations		Autonomous R&D centers		R&D units	
	Number	Percent	Number	Percent	Number	Percent
North America[1]	78	47.0	52	48.6	26	44.1
Japan	35	21.1	23	21.5	12	20.3
Korea	17	10.2	9	8.4	8	13.6
Europe	35	21.1	22	20.6	13	22.0
UK, France, and Germany	23	13.9	11	10.3	12	20.3
Northern Europe[2]	6	3.6	6	5.6	–	0.0
Other European countries[3]	6	3.6	5	4.7	1	1.7
Other countries	1	0.6	1	0.9	–	0.0
Total	166	100	107	100	59	100

Notes: [1] "North America" includes the United States and Canada; [2] "Northern Europe" includes Denmark, Sweden, Norway, Finland, and Iceland; and [3] "Other European countries" excludes the UK, France, Germany, and Northern European countries.

Table 5.4 The Change of Autonomous R&D Centers Reported by *BusinessWeek 1000* (Distributed by Parent Countries)

Parent countries	1999	2004
North America	19	52
Japan	5	23
UK, France, and Germany	5	11
Northern Europe	3	6
Other European countries	0	5
Other countries	0	1
Total	32	98

Note: To compare the data of the 1999 investigation, we have excluded the data on Korea's companies.

R&D organizations of MNCs are mainly in Beijing (39 percent) and Shanghai (29 percent), followed by Guangdong Province (9 percent), Jiangsu Province (8 percent), and Tianjin (5 percent). Over 80 percent of autonomous R&D centers were in Beijing and Shanghai, but less than 50 percent of R&D units. In contrast, the proportion of R&D units in Jiangsu, Guangdong, and

Tianjin (at 15 percent, 12 percent, and 10 percent, respectively) is much higher than the proportion of autonomous centers within them. This suggests that R&D organizations outside Beijing and Shanghai are attached mostly to manufacturing activities and support localization activities.

Table 5.5 Regional Distribution of R&D Organizations Reported by *BusinessWeek 1000* (2004)

Regions	Total organizations		Autonomous R&D centers		R&D units	
	Number	Percent	Number	Percent	Number	Percent
Beijing	65	39.2	51	47.7	14	23.7
Shanghai	49	29.5	35	32.7	14	23.7
Tianjin	8	4.8	2	1.9	6	10.2
Guangdong	15	9.0	8	7.5	7	11.9
Jiangsu	14	8.4	5	4.7	9	15.3
Other regions	15	9.0	6	5.6	9	15.3
Total	166	100	107	100	59	100

Beijing and Shanghai had the largest increases in autonomous R&D centers (by thirty-two and twenty-three, respectively) but with some dispersion in recent years, and there was a large increase in Guangdong and Jiangsu. Some MNCs have set up second or even third R&D branches in "second-line" cities such as Nanjing, Hangzhou, Suzhou, Xi'an, and Chengdu.

Beijing

From October 2004 through February 2005, we sent questionnaires to many companies, from which thirty-six organizations returned valid answers. Twenty-eight of the thirty-six were listed in the *BusinessWeek 1000*, and eight others were added in the second phase. Twenty-seven of these were autonomous R&D centers. Nearly 80 percent were from the IT industry (including software, telecommunication, semiconductors, and computer and other electronic equipment), while 14 percent were from biotechnology and drugs, and less than 10 percent were from other industries. This is similar to research by Yu, Wang, and Shi.[26]

Table 5.6 The Quantitative Change of the Autonomous R&D Centers Settled by *BusinessWeek 1000* (Distributed by Located Regions)

Year	Beijing	Shanghai	Tianjin	Guangdong	Jiangsu	Other regions	Total
1999	19	12	1	1	0	0	33
2004	51	35	2	8	5	6	107

Table 5.7 Industrial Distribution of R&D Organizations of MNCs in Beijing

Industry name	Total organizations		Respondent organizations	
	Number	Percent	Number	Percent
Software	28	35.9	11	30.6
Telecommunication	17	21.8	8	22.2
Semiconductors	11	14.1	6	16.7
Computer and other electronic equipment	7	9.0	3	8.3
Biotechnology and drugs	6	7.7	5	13.9
Industrial equipment and components	3	3.8	2	5.6
Commodity chemicals	3	3.8	1	2.8
Household electronics	2	2.6	0	0.0
Others	1	1.3	0	0.0
Total	78	100	36	100

Note: The organizations included in this table are those to which we sent questionnaires.

Among the thirty-six organizations, eighteen were from the United States and Canada, eight from Japan, seven from Europe, two from Korea, and one from Taiwan. The distribution of their parent countries was generally the same as that of all China.

Beginning in 2000, there was an increase of about five new organizations every year; those that were set up after 2000 account for 60 percent of the total. Most of those established before 1998 were inner R&D units attached to manufacturing companies, while later ones are mainly autonomous.

Reviewing the size of investments helps us to better understand these R&D activities. However, because this topic is sensitive, only sixteen of the thirty-six firms reported their initial investment amount, twelve told their accumulated amount (up until September 2004), and nine gave us both. (All these were autonomous R&D organizations.)

Table 5.8 Country Distribution of R&D Organizations Settled by MNCs in Beijing

Parent countries	Total organizations		Respondent organizations	
	Number	Percent	Number	Percent
North America	33	42.3	18	50.0
Japan	17	21.8	8	22.2
Korea	7	9.0	2	5.6
UK, France, and Germany	6	7.7	3	8.3
Northern Europe	4	5.1	3	8.3
Other European countries	3	3.8	1	2.8
Chinese Hong Kong	2	2.6	–	0.0
Chinese Taiwan	5	6.4	1	2.8
Other countries and areas	1	1.3	0	0.0
Total	78	100	36	100

Note: The organizations included in this table are those to which we sent questionnaires.

Table 5.9 The Establishing Time of R&D Organizations Settled by MNCs in Beijing

Years	Before 1998	1998	1999	2000	2001	2002	2003	2004	No reply	Total
Samples	7	4	2	4	5	1	6	3	4	36
Percent	19.4	11.1	5.6	11.1	13.9	2.8	16.7	8.3	11.1	100

Of the sixteen firms, half had invested less than RMB10 million ($1.4 million); 37 percent less than RMB5 million ($700,000); and 12 percent between RMB5 million ($700,000) and RMB 10 million ($1.4 million). Overall, there was a great difference among organizations, with most of them at less than RMB10 million ($1.4 million) and very few at RMB50 million ($7 million) or more.

Table 5.10 The Initial Investing Scale of R&D Organizations Settled by MNCs in Beijing

Initial investing scale	< 5	5–10	10–50	50–100	> 100	Number of respondents
Samples	6	2	6	1	1	16
Percent	37.5	12.5	37.5	6.3	6.3	100

Note: One unit equals RMB1 million.

Among the thirty-five organizations that responded, fourteen were global R&D centers, four were Asia-Pacific region centers, ten were China-focused, and seven were supported parent manufacturing branches in China. (One organization did not answer this interview question.) This shows their relatively high standing in the global R&D networks.

Twenty-eight companies answered the question about number of workers (as of September 2004). In total, they had 3,001 R&D employees, 107 on average, with the largest at 900 and the smallest at 5. The median was 60. There was much potential to expand, and since the year 2000 many companies have done this. For example, from 1999 to 2002 the R&D staff of Motorola doubled and that of Lucent Technologies (China) and Bell Labs (Beijing) grew by 300 to 628.[27] Since 2003 R&D subsidiaries of companies such as Microsoft and LG Electronics have greatly expanded.

Table 5.11 Scale of R&D Employees of R&D Organizations Settled by MNCs in Beijing

Scale of R&D employees	< 10	10–50	50–100	100–500	> 500	Number of respondents
Samples	2	11	6	7	2	28
Percent	7.1	39.3	21.4	25	7.1	100

We gathered the following data about managers:

- Principals of over half the organizations come from the MNC parent countries or other countries, ten from the China Mainland, and two from Chinese Taiwan.
- Managers from Mainland China predominate. Taking the Intel China Research Center (ICRC) as an example, the head is a foreign citizen from China who was educated in the United States, joined Intel, and returned to take charge of the ICRC. Of the thirteen senior managers of the ICRC, four are from the United States and nine are from Mainland China.
- The R&D "backbones" of most organizations all come from China (74 percent) or are mostly Chinese (16 percent), with only 10 percent having few Chinese R&D backbones. This is consistent with our finding that MNCs view high-quality Chinese human resources as important.

We asked companies to rank the following items: technology service and support (product testing); improvement of products, equipment, or techniques; exploration of new products or techniques; applied research; and fundamental research. We found that applied research ranked first, followed by exploration of new products and fundamental research.

Table 5.12 Human Resources of MNCs' R&D Subsidiaries in Beijing

	Parent country or other countries	Mainland China	Chinese Taiwan	Valid sample
Principals				
Number	15	10	2	27
Percent	55	37	7	100
Senior managers	Parent country or other countries	Both China and other countries	All Chinese	Valid sample
Number	5	10	15	30
Percent	16	33	50	100
R&D backbone	Few Chinese	Most Chinese	All Chinese	Valid sample
Number	3	5	23	31
Percent	9	16	74	100

Notes: "Principals" refers to general managers of R&D organizations, deans of academe, or ministers of R&D ministries. "Senior managers" refers to vice general managers, vice deans, and vice ministers. "R&D backbone" refers to R&D talents who play a vital role in R&D projects, such as experienced researchers, senior researchers, and chief researchers.

We asked respondents to also rank factors that attract R&D investments. Acquiring talent was first, while economic opportunities and the level of science ranked second and third. The fourth-ranked factor was protection of knowledge—the factor that causes MNCs the most worry. Policies that attract R&D, infrastructure, the efficiency of government, and being a native were not ranked as important.

Among sources of knowledge, "R&D headquarters of parent company" was ranked the most important, "literature" and "other overseas R&D subsidiaries" were second and third, and "Chinese university or research institute" was fourth. Contact with R&D centers of other MNCs, local production companies, foreign academic institutions, and suppliers and clients of parent companies were relatively less important.

Respondents were asked about the uses of R&D outcomes. The most common response was R&D headquarters. Less important—and about equal—were uses by production companies in China, R&D centers in other countries, Chinese affiliates, and production centers in other parts of the world. Interactions with the R&D headquarters of the parent was considered the main resource in acquiring knowledge and applying it.

As for cooperation with local entities, among the thirty-six organizations, nineteen had agreements with universities, research institutes, or enterprises; ten did not have such agreements; and seven did not respond to the question. It seems that the main cooperating organizations for MNC R&D units were universities or research institutes, but not enterprises. This may be ascribed to the low level of the local industrial R&D and the research orientation of MNCs' R&D.

We also asked about joint R&D centers with the "211 project"[28] for universities of the Ministry of Education. Among the fifty-one R&D centers that gave valid responses, thirty-three were coestablished with MNCs listed in the Fortune 500, and eighteen with MNCs not listed in Fortune 500. Of the thirty-three Fortune 500 MNCs, twenty-eight had independent R&D centers. Large MNCs belonging to the Fortune 500 are, not surprisingly, the main drivers of foreign R&D in China, and their cooperation with universities is an important part of their R&D strategy in China. However, only three out of the eighteen non–Fortune 500 firms had independent R&D centers, and for them, joint R&D centers are important. This finding is similar to that of the United Nations Conference of Trade and Development (UNCTAD).[29] In developing countries such as China, MNCs choose to set up joint R&D centers with universities or research institutions at the initial stage, due to the weak R&D capabilities of domestic enterprises.

The activities of joint R&D are distributed widely, from "exploring basic research" to "new products, new technology exploration" and "technical services and support." The connection between academe and companies, "teaching and technology consulting," is important, as are "technical services and support."

MNCs' R&D Outputs in China: Patent Data[30]

The relationship between R&D and patents can be seen as a cycle: R&D induces patents, which in turn require further development in order to reach the market.[31] Patents are an indicator of R&D success.[32]

We asked each R&D affiliate for the number of patent applications, patent grants, copyrights, software copyrights, and published dissertations its parent company and the Chinese subsidiary, respectively, acquired in 2003. To our surprise, only two R&D affiliates told us the number of parent applications and grants, and only one told us the number of software copyrights of either the parent company or the Chinese subsidiary. (We would like to learn about the patent activities of the nonrespondents.)

We turned to the online national patent searching database[33] for information on patents granted by the State Intellectual Property Office of China (SIPO) to the thirty-six R&D affiliates, and found there were almost none. Based on interviews, it seems that these R&D centers are of three types regarding IP: the first kind, in effect, acts as a subcontractor of the parent and does not have the IP of the whole "product;" many R&D centers are of this type. The second kind works on only part of the R&D chain. For example, France Télécom's R&D center in Beijing is a bridge between basic research and commercialized technology, and therefore much of its output is not suitable for patenting.

The third kind has considerable research output but does not apply directly for patents in China because of legal issues or because of the company's IP management policy. Microsoft Research Asia (MRA) is an example.

An expanded search found that, from 1986 to 2005, 63,293 patents were granted to these thirty-six MNCs. Among them, 30,028 went to Japanese MNCs, 9,104 to U.S. MNCs, and 15,653 to Korean MNCs. Twelve MNCs were each granted more than 1,000 patents over the twenty years, 60,560 altogether, making up 96 percent of the patents. Six MNCs—Samsung, Matsushita Electric, Mitsubishi, Hitachi, Toshiba, and Siemens—had over 5,000 grants each. In total, they had 46,776 (74 percent). It is evident that large MNCs are the main foreign patenters.

In the SIPO database, "inventions" represented 92 percent of all foreign patents; "utility" models were 1 percent and "industrial designs" 7 percent. This pattern is very different from the domestic pattern. From 1995 to 2005, 20 percent of domestic grants were inventions, 45 percent were utility, and 35 percent were industrial designs.[34] Such a difference is attributable to the fact that foreign companies have technologies that are more advanced than those of their Chinese counterparts. In some MNCs, the share of inventions was particularly high, such as in France Télécom, Novozymes, DoCoMo (all 100 percent), Monsanto (99 percent), Intel (99 percent), Hitachi (98 percent), NEC (98 percent), Qualcomm (98 percent), and Yokogawa (97 percent). Most were biotech or telecommunication services companies. There was a smaller share of inventions in companies such as Schlumberger (18 percent), Computer Associates International (19 percent), Petrolor (20 percent), and Hyundai (56 percent). Figure 5.1 shows the annual invention grants to the thirty-six MNCs from 1986 through 2005. In 2005, 16,934 inventions were granted—nearly fifty times as many as in 1986. Foreign patenting began to get active in 1995. After 2000, invention grants increased annually by about 50 percent.

Of the 58,092 granted inventions, 55,605 have prior rights.[35] Among the 2,487 inventions with no prior rights, only 924 originated from Chinese subsidiaries of these MNCs. This ratio is trivial compared to the patenting of Japanese MNCs in the United States, 7.4 percent, and shows the difference in the nature of foreign R&D activities conducted in the two countries.

Expanding the sample to 135 MNCs (using data from the *China Market Statistical Yearbook* 1997–2005),[36] we found that total revenue, total profit, foreign revenue, and foreign profit showed a high correlation between patenting and profits. This strongly supports Sun's[37] finding that foreign patents are driven largely by demand factors. Ernst[38] concluded that MNCs' patenting leads to excellent market performance. The relationship between patenting and market performance can be seen as a cycle: patenting supports market share, which in turn inspires more patents for further expansion.

Figure 5.1 Overall Invention Grants (1986–2005)

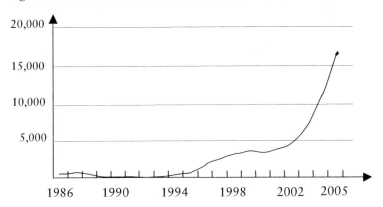

Implications

Low R&D costs, especially wages, are the most important factor for MNCs. In IT industries, especially software, nearly every manager reported this. Facing rising wages in Beijing and Shanghai, many companies are considering setting up other R&D branches in "second-line cities," such as Nanjing, Xi'an, and Chengdu. More abundant—and educated—talent and a higher level of technology cause MNCs to establish key R&D centers in countries such as India, Singapore, Taiwan, and mainland China. These often have close connections with their parent companies but not close connections locally. They are thus called global R&D centers.

We also found that one-third of the R&D organizations in Beijing fall into the "knowledge exploiting"[39] category, supporting overseas R&D subsidiaries or exploring products adapted to markets in China; about one-third are in the "knowledge-exploring" category, aiming at obtaining and using knowledge globally; and the final one-third comprise a combination of the first two types. We noticed that the Chinese domestic market is becoming more important in these investments, with some MNCs closely following such technologies in China as 3G, NGN, IPv6, and digital TV. When business opportunities come up, these MNCs can plunge into pertinent R&D fields quickly.

Finally, we found that a high percentage of these R&D centers are enclaves that interact mostly within the companies' own networks. Few of them apply for patents in China; their research outputs are incorporated into final products or into materials for further exploiting. Although many have frequent exchanges with Chinese universities or research institutes, and have even cofounded joint research centers with them, knowledge exchanges in most cases are not very deep. If the foreign R&D centers really want to be native innovative entities, as they allege, they should become more deeply embedded in China's innovation system.

Notes

[1] P. Engardio and B. Einhorn, "Outsourcing Innovation," *Business Week* (March 21, 2005): 51–57.

[2] United Nations Conference on Trade and Development (UNCTAD), *World Investment Report 2005: Transnational Corporations and the Internationalization of R&D* (New York: United Nations, 2005): 129.

[3] D. Archibugi and S. Iammarino, "The Policy Implications of the Globalisation of Innovation," *Research Policy* 28 (1999): 317–36; R. Boutellier, O. Gassmann, and M. von Zedtwitz, *Managing Global Innovation* (Berlin: Springer, 1999); J. A. Cantwell and O.E.M. Janne, "Technological Globalization and Innovative Centres: The Role of Corporate Technological Leadership and Locational Hierarchy," *Research Policy* 28, no. 2–3 (1999): 119–44; W. Kuemmerle, "Foreign Direct Investment in Industrial Research in the Pharmaceutical and Electronics Industries: Results from a Survey of Multinational Firms," *Research Policy* 28 (1999): 179–93; R. Florida, "The Globalization of R&D: Results of a Survey of Foreign-Affiliated R&D Laboratories in the USA," *Research Policy*, 26 (1997): 85–103; R. Florida, "The Globalization of R&D: Results in the Home and Foreign Plants of Multinationals," *Journal of Industrial Economics* 45, no. 3 (1997): 341–58; F. Meyer-Krahmer and G. Reger, "New Perspectives on the Innovation Strategies of Multinational Enterprises: Lessons for Technology Policy in Europe," *Research Policy* 28 (1999): 751–76; the Organisation for Economic Co-operation and Development (OECD), *Globalization of Industrial R&D: Policy Issues* (Paris: OECD, 1999); R. D. Pearce, "Overseas R&D and the Strategic Evolution of MNEs: Evidence from Laboratories in the UK, *Research Policy* 28 (1999): 23–41; D. H. Dalton and M. G. Serapio, "Globalizing Industrial Research and Development: An Examination of Foreign Direct Investments in R&D in the United States," *Research Technology Management* 28 (1999): 303–16.

[4] M. von Zedtwitz, "Managing Foreign R&D Laboratories in China," *R&D Management* 34, no. 4 (2004): 439–52; and UNCTAD, *World Investment Report 2005*, xxvi.

[5] E. B. Roberts, "Benchmarking Global Strategic Management of Technology," *Research Technology Management* 44, no. 2 (2001): 25–36; J. Edler, F. Meyer-Krahmer, and G. Reger, "Changes in the Strategic Management of Technology: Results of a Global Benchmark Survey," *R&D Management* 32, no. 2 (2002): 149–64; OECD and Belgian Science Policy, "Internationalisation of R&D: Trends, Issues and Implications for S&T Policies: A Review of the Literature" (background report for the Forum on the Internationalisation of R&D, Brussels, March 29–30, 2005); and B. Ambos, "Foreign Direct Investment in Industrial Research and Development: A Study of German MNCs," *Research Policy* 34, no. 4 (2005), 395–410.

[6] Cantwell and Janne, "Technological Globalization and Innovative Centres," 1999; W. Kuemmerle, "The Drivers of Foreign Direct Investment into Research and Development: An Empirical Investigation," *Journal of*

International Business Studies 30, no. 1 (1999): 1–24; P. Patel and M. Vega, "Patterns of Internationalisation of Corporate Technology: Location vs. Home Country Advantages," *Research Policy* 28, no. 2–3 (1999): 145–55; Roberts, "Benchmarking," 2001; C. Le Bas and C. Sierra, "Location versus Home Country Advantages in R&D Activities: Some Further Results on Multinationals' Location Strategies," *Research Policy* 31, no. 4 (2002): 589–609; T. Iwasa and H. Odagiri, "Overseas R&D, Knowledge Sourcing, and Patenting: An Empirical Study of Japanese R&D Investment in the U.S.," *Research Policy* 33, no. 5 (July 2004): 807–28; and Ambos, "Foreign Direct Investment," 2005.

[7] M. von Zedtwitz and O. Gassmann, "Market versus Technology Drive in R&D Internationalization: Four Different Patterns of Managing Research and Development," *Research Policy* 31, no. 4 (2002): 569–88; P. Reddy, *Globalization of Corporate R&D: Implications for Innovation Systems in Host Countries* (London and New York: Routledge, 2000); M. H. Toh, "R&D Activities of TNCs in Singapore: Its Role in Building the National Innovation System and Economic Development" (paper presented at the ASEAN-UNCTAD Annual Seminar on Key Issues of FDI: Attracting Quality FDI, Bangkok, May 5–6, 2005).

[8] N. Kumar and A. Aggarwal, "Liberalization, Outward Orientation and In-house R&D Activity of Multinational and Local Firms: A Quantitative Exploration for Indian Manufacturing," *Research Policy* 34 (2005): 441–60; S. H. Chen, "Taiwanese IT Firms' Offshore R&D in China and the Connection with the Global Innovation Network," *Research Policy* 33 (2004): 337–49.

[9] Economist Intelligence Unit (EIU), *Scattering the Seeds of Invention: The Globalisation of Research and Development* (London: EIU, 2004); and Deutsche Industrie- und Handelskammertag (DIHK), *R&D Offshoring: Examination of Germany's Attractiveness as a Place to Do Research* (Berlin: DIHK, 2005).

[10] I. Zander, "The Formation of International Innovation Networks in the Multinational Corporation: An Evolutionary Perspective," *Industrial and Corporate Change* 11, no. 2 (2002): 327; A. Phene and P. Almeida, "How Do Firms Evolve? The Patterns of Technological Evolution of Semiconductor Subsidiaries," *International Business Review* 12 (2003): 349–67; S.-C. J. Fang, L. Lin, L.Y.C. Hsiao, C.-M. Huang, and S.-R. Fang, "The Relationship of Foreign R&D Units in Taiwan and the Taiwanese Knowledge-Flow System," *Technovation* 22, no. 6 (2002): 371–83; F. J. Contractor and W. Ra, "How Knowledge Attributes Influence Alliance Governance Choices: A Theory Development Note," *Journal of International Management* 8 (2002): 11–27; P. J. Buckley and M. J. Carter, "Process and Structure in Knowledge Management Practices of British and U.S. Multinational Enterprises," *Journal of International Management* 8 (2002): 29–48; N. J. Foss and T. Pedersen, "Transferring Knowledge in MNCs: The Role of Sources of Subsidiary Knowledge and Organizational Context," *Journal of International Management* 8 (2002): 49–67; C.-J. Chen and B.-W. Lin, "The Effects of Environment, Knowledge Attributes, Organizational Climate, and Firm Characteristics on Knowledge Sourcing Decisions," *R&D Management* 34, no. 2 (2004): 137–46.

[11] K. Cheung and P. Lin, "Spillover Effects of FDI on Innovation in China: Evidence from the Provincial Data," *China Economic Review* 15 (2004): 25–44.

[12] D. Mowery, J. Oxley, and B. Silverman, "Technological Overlap and Interfirm Cooperation: Implications for the Resource-Based View of the Firm," *Research Policy* 27, no. 5 (1998): 507–24; G. D. Santangelo, "Inter-European Regional Dispersion of Corporate Research Activity in Information and Communications Technology: The Case of German, Italian, and UK Regions," *International Journal of the Economics of Business* 7, no. 3 (2000): 275–95; UNCTAD, *World Investment Report 2001: Transfer of Technology*, UNCTAD Series on Issues in International Investment Agreements (New York: United Nations, 2001).

[13] J. E. Jankowski, "Measurement and Growth of R&D within the Service Economy," *Journal of Technology Transfer* 26, no. 4 (2001): 323–36; Engardio and Einhorn, "Outsourcing Innovation," 2005.

[14] von Zedtwitz, "Managing Foreign R&D Laboratories," 2004; Kumar and Aggarwal, "Liberalization," 2005.

[15] D. Guellec and B. van Pottelsberghe de la Potterie, "The Internationalization of Technology Analyzed with Patent Data," *Research Policy* 30 (2001): 1253–66.

[16] Iwasa and Odagiri, "Overseas R&D," 2004.

[17] S. Kortum and J. Lerner, "Stronger Protection or Technological Revolution: What Is Behind the Recent Surge in Patenting?" *Carnegie-Rochester Conference Series on Public Policy* 48 (1998): 247–304; B. H. Hall and R. H. Ziedonis, "The Patent Paradox Revisited: An Empirical Study of Patenting in the U.S. Semiconductor Industry, 1979–1995," *The Rand Journal of Economics* 32, (2001): 101–28; C. Shapiro, "Navigating the Patent Thicket: Cross Licenses, Patent Pools, and Standard-Setting (Cambridge, MA: MIT Press, 2001); S. Macdonald, "When Means Become Ends: Considering the Impact of Patent Strategy on Innovation," *Information Economics and Policy* 16 (2004): 135–58.

[18] A. Panagariya, "The Regionalism Debate: An Overview," *The World Economy* 22, no. 4 (1999): 477–511; K. Maskus, *Intellectual Property Rights in the Global Economy*. (Washington, D.C.: Institute for International Economics, 2000): 157–69; Commission on Intellectual Property Rights (CIPR), *Integrating Intellectual Property Rights and Development Policy* (London: CIPR, 2002).

[19] Y. Wu, "Investigation and Research of R&D Organizations Established by Overseas Companies," *China Soft Science Magazine* 1 (2000): 64–66; X. Jiang, "R&D Activities of Organizations Invested by MNCs in China," *Science & Technology Review* 9 (2000): 27–28; L. Xue and S. Wang, "Globalization of R&D by MNCs in China: An Empirical Analysis" (working paper, U.S.-China Conference on Technical Innovation, Beijing, April 24–25, 2000); L. Xue, "Globalization of Technology and Opportunities, Challenges and Strategies of China," *Science of Science and Management of S&T* 21, no. 9 (2000): 4–8; L. Xue and Q. Shen, "Globalization of S&T and Its Political Meaning to China's Science and Technology Developments," *The Journal of World Economy* 10

(2001): 57–62; L. Xue, S. Wang, and Q. Shen, "Analysis of Influential Factors of the Establishment of R&D Organizations of MNCs in China," *Journal of Science Research Management* 22, no. 4 (2001): 132–42; L. Xue, Q. Shen, and S. Wang, "R&D Investment Layout of MNCs in China under International Strategy—Based on Empirical Analysis of the Distribution Differences of Independent R&D Organizations of MNCs in China," *Management World* 3 (2002): 33–42; X. Liu and J. Zhao, "Analysis of R&D Activities of 19 R&D Organizations of MNCs in Beijing," *Forum on Science and Technology in China* 4 (2003): 65–69.

[20] R. F. M. Armbrecht, "Siting Industrial R&D in China: Notes from Pioneers" (slide presentation, Arlington, Va.: Industrial Research Institute, 2003).

[21] K. Walsh, *Foreign High-Tech R&D in China: Risks, Rewards, and Implications for U.S.-China Relations* (Washington, D.C.: Henry L. Stimson Center, 2003), <www.stimson.org/techtransfer/pdf/FinalReport.pdf>; and H. Ernst, "Patent Applications and Subsequent Changes of Performance: Evidence from Time-Series Cross-Section Analyses on the Firm Level," *Research Policy* 30 (2001): 143–57.

[22] von Zedtwitz, "Managing Foreign R&D Laboratories," 2004; O. Gassmann and Z. Han, "Motivations and Barriers of Foreign R&D Activities in China," *R&D Management* 34, no. 4 (2004): 423–38; J. Li and J. Zhong, "Explaining the Growth of International R&D Alliances in China," *Managerial and Decision Economics* 24, nos. 2–3 (2003): 101–15.

[23] D. Ernst and B. Naughton, "Building Capabilities within Global Networks: China's Upgrading and Innovation in the IT Sector" (Honolulu, Hawaii: East-West Center, 2005), photocopy.

[24] As an industrial classification standard, the GICS was first issued by Morgan Stanley Capital International (MSCI) and Standard & Poor's in 1999. The GICS is currently the most authorized industrial classification standard and has been used by *BusinessWeek 1000* since 1999. In the initial edition of the GICS, all the industries were divided into seven main categories: energy, material, capital equipment, consumer goods, services, finance, and others. And in the GICS edition issued in 2004, the main categories were increased to ten. Consumer goods was divided into consumer discretionary and consumer staples; health care was separated from consumer goods as an autonomous main category; telecommunication services and utilities were separated from services; IT was separated from other categories and became a main category; the former capital goods now concentrate on industrial fields; and services was dispersed into all the main categories. These changes reflect the current of industrial evolution in recent years, especially the rise of information and digital technology, and the integration of service and other industries.

[25] See, for example, Dalton and Serapio, "Globalizing Industrial Research and Development," 1999.

[26] S. Yu, X. Wang, and W. Shi, "On Country Choice of R&D Investment of MNCs," *Management World* 1 (2004): 46–54.

27 Liu and Zhao, "Analysis of R&D Activities," 2003.

28 All the principals of these organizations are Chinese; their researchers are mainly teachers and students in universities. Respondents and investigation targets are the principals or related personnel of these organizations.

29 UNCTAD, *World Investment Report 2005*, 2005.

30 Chinese patents are classified into three types: invention, utility model, and industrial design. "Invention" refers to technical proposals to products, methods, or both. "Utility model" refers to technical proposals on shape or structure of a product, or both. "Industrial design" refers to aesthetics and industry-applicable designs for the shape, design, or color of a product, or a combination of these. Inventions represent major technological creations, while utility models and industrial designs are more incremental in nature.

31 B. H. Hall, Z. Griliches, and J. A. Hausman, "Patents and R&D: Is There a Lag?," *International Economic Review* 27, no. 2 (1986): 265–83; E. Duguet and I. Kabla, "Appropriation Strategy and the Motivations to Use the Patent System: An Econometric Analysis at the Firm Level in French Manufacturing," *Annales d'Économie et de Statistique* 49–50 (1998): 289–327; B. Crépon, E. Duguet, and J. Mairesse, "Research, Innovation, and Productivity: An Econometric Analysis at the Firm Level. *Economics of Innovation and New Technology* 7, no. 2 (1998): 115–58; E. Brouwer and A. Kleinknecht, "Innovative Output and a Firm Propensity to Patent: An Exploration of CIS Micro Data," *Research Policy* 28, no. 6 (1999): 615–24.

32 F. M. Scherer, "Firm Size, Market Structure, Opportunity, and the Output of Patented Inventions," *American Economic Review* 55, no. 5 (1965): 1097–1125.

33 See <http://www.sipo.gov.cn/sipo2008>.

34 See <http://www.sipo.gov.cn/sipo2008>.

35 Prior-rights information on patents includes priority and equivalent applications in other countries before the China granting.

36 National Bureau of Statistics of China, *China Market Statistical Yearbook* (*CMY*) 1997–2005 (Beijing: China Statistics Publishing House, 1998–2006).

37 Y. Sun, "Determinants of Foreign Patents in China," *World Patent Information* 25 (2003): 27–37.

38 Ernst, "Patent Applications," 2001.

39 See Kuemmerle, "Foreign Direct Investment in Industrial Research," 1999; and Kuemmerle, "Drivers of Foreign Direct Investment" 1999.

Linking Local and National Innovation Networks: Multinational Companies' R&D Centers in Taiwan

Yuan-chieh Chang, Chintay Shih, and Yi-Ling Wei

Research and development (R&D) has long been internationalized in the triad of North America, Europe, and Japan. However, the expansion of foreign-affiliated R&D centers in Asia has generated increasing interest among policymakers, industrial R&D executives, and scholars.[1] The reported share of R&D in developing countries by U.S. companies increased from 7.6 percent in 1994 to 13.5 percent in 2002. The share of U.S. companies' foreign R&D in Asia went from 3.4 percent in 1994 to 10 percent in 2002, about 74 percent of U.S. R&D in all developing countries that year.[2] The main destinations were China, India, Singapore, Hong Kong, Malaysia, South Korea, and Taiwan.

The number of studies of multinational corporation (MNC) R&D in developing countries, such as China[3] and India,[4] has increased because of those countries' massive scientific and engineering human resources and big local markets. However, little research has been done on foreign R&D activities where human resources and local markets are small, such as in Taiwan. This chapter attempts to bridge this gap. It addresses the following issues:

- Investment motives behind MNC R&D centers in Taiwan
- The types of R&D activities (basic research, applied research, prototype development) conducted by foreign R&D centers in Taiwan
- Intrafirm R&D networks, domestic interfirm R&D networks, and domestic research-based networks that MNC R&D centers established
- Policies to attract MNC R&D investment in Taiwan and to create a mutually beneficial relationship between MNCs and local innovation actors

This chapter considers the existence and the direction of technology transfers among R&D centers in their home country, in Taiwan, and at other sites. Ronstadt suggests three categories of foreign R&D operations: (1) technology transfer units (TTUs), which focus on technical services to the subsidiaries; (2) indigenous technology units (ITUs), where new products are developed abroad to service host countries; and (3) regional or global technology units (RTUs/GTUs), where foreign R&D centers develop products for simultaneous launch in regional and global markets.[5]

This chapter examines the function, coordination, and control of these R&D centers. It also looks at how these centers tap into national innovation systems (NISs). Two kinds of innovation networks are investigated: interfirm networks and research-based networks. The types of interfirm networks are subcontract production, such as original equipment manufacturing (OEM) and original design manufacturing (ODM); technology licensing; contract research; and collaborative R&D.

Background

In Taiwan, according to the *Indicators of Science and Technology* of the National Science Council (NSC), foreign-funded R&D has been minimal. It ranged from NT$35 million to NT$110 million—less than 0.06 percent of total national R&D expenditure and less than 0.09 percent of business expenditure on R&D (BERD)—in 1999 through 2004 (Table 6.1).[6] However, the NSC figures seriously underestimate foreign R&D activities. One measure of such activity is foreign patenting. According to the Taiwan Intellectual Property Office of the Ministry of Economic Affairs (MOEA), 16,093 patents were granted to foreigners, and 32 percent of that total was granted in 2004, as reported by the NSC.[7] Notably, statistics from the Taiwan government-launched Program to Encourage Multinational Corporations to Set Up Innovative R&D Centers in Taiwan (hereafter MNC R&D Centers Program) show that the NSC underestimated its figures on foreign R&D expenditure in Taiwan.

Table 6.1 Foreign-Funded R&D as a Proportion of National R&D in Taiwan

Year	Foreign-funded R&D (A)	Total national R&D (B)		Business R&D expenditure (C)	
		Amount	Share (A/B*100%)	Amount	Share (A/C*100%)
1999	110	190,520	0.058	125,712	0.088
2000	74	197,631	0.037	128,386	0.058
2001	35	204,974	0.017	132,950	0.026
2002	38	224,428	0.017	141,695	0.027
2003	60	240,820	0.025	151,550	0.040
2004	60	260,851	0.023	168,079	0.036

Source: National Science Council (NSC), 2005.
Note: Unit: NY$ million.

Under the MNC R&D Centers Program, more than twenty-five such R&D centers were established from 2001 to June 2005. These MNC R&D centers received subsidies of NT$265 million in 2003, NT$551 million in 2004, and NT$578 million in 2005. In contrast, as can be seen in Table 6.1, for the period 2002–2004, the NSC reports a total of onyl NT$158 million in foreign R&D expenditure.

The top R&D centers were set up by American, European, and Japanese parent companies, which set up thirteen, five, and three centers, respectively. These R&D centers are mainly in the information and communications technology (ICT) fields: semiconductors (seven centers), information and computer products (six centers), and telecommunications (four centers) (Table 2). More than 85 percent chose to locate in Taipei and Hsinchu. Only four MNC R&D facilities had more than one hundred R&D personnel. Dell, NEC, IBM, Sony, and GSK had the largest planned R&D budgets, with a combined total of more than NT$8.3 billion; this was more than 60 percent of the total foreign-planned R&D investment. In summary, MNC R&D centers are U.S.-dominated, are in the ICT field, and are concentrated in Taipei and Hsinchu.

Literature Review

The literature focuses on three areas of the internationalization of MNC R&D: (1) motives, (2) types of R&D activities, and (3) network strategies to tap into Taiwan's innovation system.

Motives

The motives for MNCs establishing R&D centers abroad are domestic market adaptation, the abundance of low-cost R&D personnel, and the scale of national technological effort.[8] Granstrand classified demand-oriented and supply-oriented driving forces, which accounted for centrifugal and centripetal R&D by MNCs.[9] The demand-oriented forces include: (1) a substitute transferring technology from parent to subsidiary, (2) integration with local production, (3) local ambitions among subsidiaries, (4) government regulations, and (5) the need for proximity to local customers. Supply-oriented forces include: (1) foreign mergers and acquisitions (M&A), (2) access to foreign science and technology (S&T), and (3) access to low-cost R&D personnel. The conditions of economic liberation and weak protection of intellectual property rights (IPRs) in the developing countries have not prevented MNCs from making R&D investments in both India[10] and China.[11] In chapter 5 of this book, Xue and Liang report on motivations for setting up such centers in China, with a focus on Beijing.

Table 6.2 Leading MNC R&D Centers in Taiwan

Company and country of origin	R&D centers	Industry	Location	Year est'd	R&D personnel/ total employees
Dell (U.S.)	Design Center	Computer	Taipei	2003	63/75
NEC (Japan)	Innovative Product Joint Development Center	Computer	Taipei	2004	30/398
IBM (U.S.)	xSeries R&D Center	Server	Taipei	2004	215/1,560
Sony (Japan)	Design & Engineering Center Taiwan	Information product	Taipei	2000	8/219
GSK (UK)	R&D Operation Center	Biotechnology	Taipei	2004	11/350
HP (U.S.)	Product Development Center	Computer	Taipei	2002	89/800
Synopsys (U.S.)	VDSM EDA R&D Center	Integrated circuits (ICs)/ EDA provider	Taipei	2004	48/150
Motorola (U.S.)	Product Development Center	Telecommunications	Taipei	2004	44/190
Microsoft (U.S.)	Technology Center	Software	Taipei	2003	10/275
Ericsson (Sweden)	Innovation Center	Telecommunications	Taipei	2003	84/430
IBM (U.S.)	M e-Business R&D Center	Telecommunications	Taipei	2003	115/1,405
Aixtron (Germany)	Manufacturing-Oriented Research Lab	Optoelectronics equipment supplier	Hsinchu	2002	NA/23
Intel (U.S.)	Taiwan Innovation Center	IC network platform	Taipei	2003	16/16
DuPont (U.S.)	Taiwan Technical Center	Advanced material	Taoyuan	2004	28/545
Pericom (U.S.)	Advanced Mixed-Signal R&D Center	IC and analog IC design	Taipei	2003	60/238
IBM (U.S.)	Bioinformatics R&D Center	Biotechnology	Taipei	2003	115/1,405
Becker (Germany)	Avionics Certification Skill & Key Technology Development Center	Aviation service	Taipei	2002	196/280
Sony (Japan)	LSI R&D Center	IC SoC technology	Taipei	2004	4/430
Broadcom (U.S., Singapore)	Network SoC R&D center	IC design	Hsinchu	2003	50/120
AKT (U.S.)	Asia-Pacific R&D Center	Semiconductor equipment supplier	Hsinchu	2004	NA/60
Alcatel (France, Italy)	ICT Application R&D Center	Telecommunications	Taipei	2004	NA/300

Source: Adapted from Abstracts of Operational Proposals, Department of Industrial Technology, Ministry of Economic Affairs, Taiwan.

Types of R&D Activities

Serapio and Dalton report that foreign R&D facilities in the United States involve R&D activities ranging from technological scanning, product application research, product customization, and product design and development, to basic research.[12] For developing countries, R&D activities can be organized into eight categories: (1) blue sky and basic research (longer than two years), (2) medium-term product/process research (one to two years), (3) medium-term applied research (one to two years), (4) short-term innovation and prototype development (less than one year), (5) significant adaptation and improvement to existing technologies, (6) implementation and operation of new equipment, (7) technology and engineering support for manufacture, and (8) production-related skill and capabilities.

Network Strategies to Tap into Taiwan's Innovation System

MNC R&D centers in Taiwan could adopt three network strategies to tap into Taiwan's innovation system. First, MNC R&D centers could establish *intrafirm* R&D networks to coordinate and integrate headquarter-subsidiary R&D activities within MNCs themselves. Second, MNC R&D centers could build *interfirm* R&D networks with Taiwanese suppliers, customers, and partners. Third, MNC R&D centers could leverage local S&T development by *cooperating with Taiwan's research base*.

Intrafirm R&D Networks within MNCs

Kuemmerle proposes two types of global R&D: the *home-based exploiting* R&D center and the *home-based augmenting* R&D center.[13] The first type seeks mainly to exploit firm-specific capabilities, and the main knowledge flow is from home R&D sites to foreign ones. The exemplars are foreign direct investment (FDI) in manufacturing and marketing. The second type seeks mainly to access unique resources and to capture knowledge that is created locally. The major knowledge flow in this type is from home-based, augmenting, foreign R&D centers to home-based R&D centers.

Domestic Interfirm Networks

Various countries have various areas of technological competencies. MNCs set up R&D facilities to absorb spillovers of a host nation firm's R&D activities.[14] One way to do this is to tap into domestic interfirm R&D networks. The other is to tap into or create domestic research networks with universities and R&D institutions. Foreign R&D centers tend to locate in such clusters as Silicon Valley and the Research Triangle Park in the United States,[15] and Beijing and Shanghai in China.[16]

However, interfirm R&D networks can operate by means of subcontract production systems, such as OEM and ODM, technology licensing, contract research, codevelopment, collaborative R&D, and R&D consortia.[17] These distinct types reflect various degrees of mutual commitment, trust, risk, and benefits among partners. The subcontracting, technology licensing, and R&D outsourcing type tend to be short-term, arm's-length, transactional-based activities, while the codevelopment, collaborative R&D, and R&D consortia, tend to involve longer-term and relational-based activities.[18]

Domestic Research-Based Networks

One of the benefits of doing basic research or establishing research links with universities and R&D institutions is that it enhances firms' absorptive capacity.[19] This helps firms to evaluate new scientific information and to use it in an innovative way. Faulkner and Senker argue that the best mechanism for knowledge transfer from research bases to firms is to establish R&D cooperative mechanisms, both formal and informal.[20] Geisler and Rubenstein find that university-research-based links involve various organizational commitments—some short-term, some long-term.[21]

Methods and Survey Scope

The survey of MNC R&D Centers in Taiwan (2004), covers centers that are subsidized by the R&D Centers Program (Table 6.2). About twenty-five stand-alone R&D centers, each with more than 50 percent control of shares by a foreign parent company, have been established since 2001. We exclude R&D sections of foreign firms and also R&D performed by third-party organizations that are financed by foreign companies. Some companies, including DuPont, NEC, and IBM, had established R&D facilities before opening stand-alone R&D centers. These MNC R&D centers comprise more than 75 percent of total foreign R&D investment in Taiwan.

Data Collection

The authors were invited as program evaluators and, therefore, were able to interview in more than twenty centers from April to December 2005. We designed a questionnaire to collect information on intrafirm, interfirm, and firm-research networks. The questionnaires were sent mainly to chief technology officers and vice presidents of R&D, as well as to managers in charge of these R&D centers. A total of twenty-five questionnaires were returned.

The study used the motives for internationalization proposed by Granstrand and others,[22] but we added another supply-led motive: government promotion and R&D subsidy. The survey asked the respondents to weigh the importance of motives on a Likert scale, by selecting (1) extremely unimportant, (2) very unimportant, (3) unimportant, (4) fair, (5) important, (6) very important, or (7) extremely important.

We asked about a wide spectrum of R&D activities (as outlined by the United Nations Conference on Trade and Development, UNCTAD),[23] as is appropriate for developing countries. The questionnaire covered the eight categories noted earlier in this chapter.

Results and Motives

Of the seven motives for MNC R&D centers in Taiwan (Table 6.3), center managers regard government promotion and incentives as a "very important (6)" to "extremely important (7)" factor (6.5 on average) in attracting them. They rate four other motives as "important (5)" to "very important (6)": tapping into Taiwan university and R&D networks (5.58 on average), accessing low-cost local knowledge and spillovers (5.5 on average), responding to the local market (5.5 on average), and employing local talent (5.08 on average). In contrast, supporting local production (4.42 on average) and acquiring local technology (4.25 on average) are rated between "fair (4)" and "important (5)." These results suggest that R&D centers in Taiwan have been stimulated mainly by government policy.

Due to strong clustering effects, semiconductors, computers, and telecommunications hardware are in the Hsinchu Science-based Industrial Park (HSIP), and MNC R&D centers have been attracted there or to Taipei to tap into local universities and R&D institutes. Supporting local production is less significant because MNC brand companies mainly subcontract their production to local manufacturers. Finally, acquiring local technology is the least significant motive.

However, motives vary across sectors. Managers in telecommunications give greater weight to six of the seven suggested motives than to those in the semiconductor industry and other sectors. The ranking of motives in telecommunications is consistent with the overall ranking. Managers in semiconductors place high-quality talent in this field second. Managers in miscellaneous industries, such as biotechnology, advanced material, and aerospace, rate "responding to local market" as one of the most important motives.

Types of R&D Activities

Overall, the MNC R&D centers reported a medium- to short-term (one- to two-year) outlook for product/process research, applied research, and prototype development (Table 6.4). Only one MNC R&D center reported undertaking more than a two-year outlook in semiconductor technology. Very few centers reported the provision of engineering support and production skills.

Table 6.3 Motives for MNC R&D Centers in Taiwan Given by Senior R&D/Technical Executives

Motives	Importance by industry[a]			Total (N = 12) (rank)
	Semiconductor (N = 3) (rank)	Telecom (N = 4) (rank)	Miscellaneous[b] (N = 5) (rank)	
Government promotion and incentives	6.67 (1)	7 (1)	6 (1)	6.5 (1)
Tapping into Taiwan university and R&D networks	5.67 (2)	6.25 (2)	5 (3)	5.58 (2)
Accessing low-cost local R&D resources	5.33 (4)	6.25 (2)	5 (3)	5.5 (3)
Responding to local market	5.33 (4)	6 (4)	5.2 (2)	5.5 (3)
Deploying local high-quality human resources	5.67 (2)	5.25 (5)	4.6 (6)	5.08 (5)
Supporting local production	3 (7)	5 (6)	4.8 (5)	4.42 (6)
Acquiring local technology	4.33 (6)	5 (6)	3.6 (7)	4.33 (7)

Source: MNC R&D Centers in Taiwan Survey, 2004.

Notes: [a] 1 = extremely unimportant, 2 = very unimportant, 3 = unimportant, 4 = fair, 5 = important, 6 = very important, 7 = extremely important.
[b] "Miscellaneous" includes R&D centers involved in biotechnology, advanced material, aerospace, and computers.

Nine out of twelve firms reported that the most important channel of intrafirm technology transfer is from multinational home-based and other R&D sites to Taiwan (Table 6.5). However, there were a few exceptions in which technology was transferred from the Taiwanese centers back to home or to other R&D sites. These were in the semiconductor and biotechnology fields. So the MNC R&D centers in Taiwan are mainly home-based exploiting, with a few of them home-based augmenting.

On coordination and control among headquarters, other R&D sites, and Taiwan's R&D centers, two centers functioned as international procurement offices to domestic OEM or ODM (Table 6.6). Eight out of twelve firms reported their R&D centers as providing only technical support and services. Eleven out of twelve centers conducted local R&D activities in Taiwan. Six out of twelve

Table 6.4 Types of R&D Undertaken by MNC R&D Centers in Taiwan, by Sector

Types of R&D	Number of firms reported, by sector			
	Semiconductor (N = 3)	Telecom. (N = 4)	Misc.[a] (N = 5)	Total (N = 12)
Basic research (> 2 yrs)	1	0	0	1 (8%)
Medium-term product / process research (1–2 yrs)	2	3	3	8 (67%)
Medium-term applied research (1–2 yrs)	0	3	3	6 (50%)
Short-term innovation and prototype development (< 1 yr)	1	3	1	5 (42%)
Improving existing technology	2	1	2	5 (42%)
Operating new equipment	0	0	0	0 (0%)
Engineering support	0	1	2	3 (25%)
Production skills	0	1	1	2 (17%)

Note: [a] "Misc." includes R&D centers involved in biotechnology, advanced material, aerospace, and computers.

Table 6.5 Intrafirm R&D Networks of MNC R&D Centers in Taiwan (2004)

Sector	Number of firms reporting the most important channels of intrafirm R&D network			
	Home-based exploiting		Home-based augmenting	
	Home →Taiwan	Other R&D sites →Taiwan	Taiwan →home	Taiwan →other R&D sites
Semiconductor (N = 3)	2	0	0	1
Telecommunication (N = 4)	3	1	0	0
Misc.[a] (N = 5)	2	1	1	1
Total (N = 12)	7	2	1	2

Note: [a] "Misc." includes R&D centers involved in biotechnology, advanced material, aerospace, and computers.

R&D centers played the role of regional hubs to coordinate and integrate the company's other R&D labs in the Asia-Pacific region. Finally, eight out of twelve R&D centers functioned as global R&D hubs to coordinate or integrate the company's worldwide R&D sites. These results suggest that the R&D centers in Taiwan play an increasing role as both regional (Asia-Pacific) and global R&D hubs in the semiconductor and ICT industries.

Table 6.6 Coordination and Control of MNC R&D Centers in Taiwan

R&D coordination and control	Number of centers reported			
	Semiconductor (N = 3)	Telecom. (N = 4)	Misc.[a] (N = 5)	Total (N = 12)
Coordination				
International procurement office	0	1	1	2
Technical support and services	1	4	3	8
Local R&D	3	4	4	11
Asia-Pacific regional R&D integration	1	3	2	6
Global R&D integration	2	3	3	8
Control of initiating R&D projects				
Parent company decided	2	1	0	3
Parent company and Taiwan R&D center jointly decided	1	4	1	6
Regional R&D center and Taiwan R&D center jointly decided	2	2	3	7
Taiwan R&D center decided, with permission from parent company	2	2	1	5
Taiwan R&D center decided, with permission from regional R&D center	0	2	0	2
Taiwan R&D center independently decided	0	0	2	2

Note: [a] "Misc." includes R&D centers involved in biotechnology, advanced material, aerospace, and computers.

The results also show that the major regional R&D centers and the Taiwan centers engaged in joint decisions (seven centers, 58 percent), followed by joint decisions between the parent company and the Taiwan center (six centers, 50 percent), and top-down decisions made by the parent company (three centers, 25 percent) (Table 6.6). Relatively few Taiwan R&D centers (two centers, 17 percent) could independently decide their R&D agendas. This suggests that the regional polycentric type and the parent-centralized type are the main coordination mechanisms.[24]

Links to National Innovation Systems

Finally, we examine how the centers in Taiwan tap into NISs. We investigated two kinds of innovation networks: interfirm networks and firm-research-based networks. To tap into domestic interfirm knowledge nets, our MNC R&D centers, on average, established collaborative R&D mainly with three domestic firms, doing contract research with 2.33 firms, OEM subcontracting with 1.67 firms, ODM subcontracting with 1.58 firms, and technology licensing with 1.17 firms.

Table 6.7 Interfirm Networks of MNC R&D Centers in Taiwan (2002–2004)

Types of interfirm networks	Number of firms cooperating (mean)			
	Semiconductor (N = 3)	Telecom. (%) (N = 4) No. of partners	Misc.[a] (N = 5)	Total (N = 12)
OEM	15 (5)	3 (0.75)	2 (0.4)	20 (1.67)
ODM	12 (4)	4 (1)	3 (0.6)	19 (1.58)
Technology licensing	13 (4.3)	0 (0)	1 (0.2)	14 (1.17)
Contract research	14 (4.67)	2 (0.5)	12 (2.4)	28 (2.33)
Collaborative R&D	9 (3)	17 (4.25)	10 (2)	36 (3)

Note: [a] "Misc." includes R&D centers involved in biotechnology, advanced material, aerospace, and computers.

Tapping into domestic-firm knowledge varies significantly. Centers in the semiconductor industry established the densest links with domestic firms by using various mechanisms. With the "pure foundry" business model and more than two hundred inegrated circuit (IC) design houses in Taiwan, it is not surprising that OEM and ODM are the major interfirm links for MNC R&D centers in semiconductor firms. R&D centers in the telecommunications industry mainly establish collaborative R&D links with domestic firms. For the centers in miscellaneous industries, contract research is the major domestic interfirm connection.

The results suggest that these R&D centers use collaborative R&D links with domestic firms as the major mechanism for acquiring their knowledge. They do this mainly by means of OEM and ODM relationships. MNCs are learning the cost structure of OEM and ODM operations. It might be shrinking the profit margin of Taiwanese firms. However, reverse technology transfer may also occur as Taiwanese firms learn from the foreign firms.

One link is between R&D centers and universities and research institutes. On average, R&D centers established 1.42 internship/training projects, 0.5 contract research projects, 2.5 collaborative R&D projects, and 0.75 licensing agreements with local universities and research institutes (Table 6.8). Most frequent is collaborative R&D, followed by internship/training projects, and licensing agreements. Centers cooperated most with the National Chiao Tung University, National Tsing Hua University, National Taiwan University, National Central University, and National Chung-Hsing University. Eleven out of twenty centers had research links with various research labs of the Industrial Technology Research Institute (ITRI). One biotechnology-related R&D center had links with Academic Sinica, a center of excellence in biotechnology. However, this link category is relatively weak, except for collaborative R&D. This particular biotechnology R&D center clearly chooses prestigious universities and centers of excellence, such as ITRI and Academic Sinica.

Table 6.8 Research-based Networks of MNC R&D Centers in Taiwan (2002–2004)

Types of research-based networks	No. of institutions cooperating/ No. of projects cooperating			
	Semiconductor (N = 3)	Telecom. (N = 4)	Misc.[a] (N = 5)	Total (N = 12)
Training/internship projects	4/4	2/2	11/11	17/17
Contract research	2/2	2/2	2/2	6/6
Collaborative R&D	3/3	7/7	12/20	22/30
Licensing agreements	1/1	2/2	6/6	9/9

Note: [a] "Misc." includes R&D centers involved in biotechnology, advanced material, aerospace, and computers.

We propose two types of R&D networks via which MNC R&D centers can tap into the Taiwan innovation system: *industrial R&D networks* and *research-based networks*. Industrial R&D networks consist of the R&D links that an MNC establishes with domestic suppliers, customers, and partners. Research-based networks consist of the R&D links that an MNC establishes with domestic universities and research institutes, such as National Tsing Hua University, Academic Sinica, and ITRI. Based on the strength of MNC R&D

centers' industrial networks and research-based networks, we have developed a matrix to locate MNC R&D centers (see Figure 6.1).

We can arrange MNC R&D centers into four quadrants. MNC R&D centers located in the lower-left quadrant are called *independent R&D centers*, which tap few or no industrial R&D networks and few or no research-based networks. Examples are Sony DECT, Sony LSI, Motorola, Pericom, Synopsys, Alcatel, and Broadcom. MNC R&D centers located in the lower-right quadrant are called *industrial-deepening R&D centers*, which have many cooperative links with local firms but few with domestic research bases. The computer- and telecommunications-related R&D centers, such as IBM-M, Ericsson, Hewlett-Packard (HP), NEC, and Dell, are mainly located in the lower-right quadrant. MNC R&D centers located in the upper-left quadrant are *research-based, deepening R&D centers*, which have many links with domestic research bases such as universities and research institutes, but few links with domestic firms. Examples of firms in this upper-left quadrant include AKT, Aixtron, and Microsoft. Finally, MNC R&D centers located in the upper-right quadrant are called *integrated-networks R&D centers*, which have strong links with domestic firms, universities, and research institutes. Becker, GSK, Intel, and DuPont are located in this quadrant.

Figure 6.1 Taxonomy of Foreign R&D Centers: Networks Tapping into the Domestic Innovation System

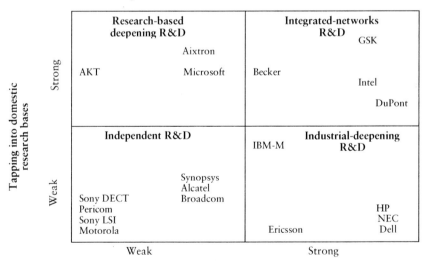

Discussion

These findings are in line with studies by Odagiri, Yasuda, and Kumar.[25] These scholars have concluded that, in developing countries, overseas R&D involves mainly local adaptation. However, we argue that, due to a small local market and an OEM-dominant business model, MNC R&D affiliates in Taiwan acted as a lead market (for example, in ICT products) and as a test bed for the regional and global market, rather than as production support and as securers of low-cost talent. Moreover, given ICT's industrial strength in these sectors, it is not surprising that over 90 percent of MNC R&D centers are in ICT. This is consistent with Kumar's finding that industrial specialization attracts foreign R&D activities to absorb spillovers of national innovations.[26]

The most popular sites for foreign R&D facilities in the United States are Silicon Valley for computers and computer equipment, computer software, semiconductors, and consumer electronics; Los Angeles for automotive design and styling; and Boston for electronics and new materials.[27] Based on a database (1995–2001) of 378 foreign R&D facilities in China, about 50 percent of R&D centers are in Beijing.[28] These studies suggest that the geographical R&D sites are concentrated in established industrial clusters. This is also true of Taiwan.

However, two issues headline debates about the effects of MNC R&D centers on their hosts. One is fierce competition for local talent, including senior R&D personnel. The other is a boomerang effect of international technology transfer; for example, the shortage of postgraduate engineers and scientists in Taiwan has intensified since the program began. There is competition between MNCs and domestic firms because MNCs can recruit the cream of domestic talent. This tension might be eased through expanded postgraduate training programs. As for the boomerang effect, domestic firms hold that the proximity of foreign centers in Taiwan may cause knowledge leakage from domestic firms to MNCs while reducing the competitive advantages of domestic OEM firms. They need to understand that knowledge inevitably moves both ways. It is crucial to create a mutually beneficial, reciprocal relationship between domestic firms and MNC partners, rather than to behave opportunistically.

There are two major benefits to Taiwan from MNC R&D centers: improved managerial skills and practices of global R&D operations,[29] and increased innovation of Taiwanese firms and research organizations as they connect to the global R&D networks of MNCs.

Conclusion

MNC R&D centers in Taiwan were motivated mainly by supply-driven factors. They undertook R&D activities that mainly exploited home-based technology, with more or less the functions of regional and global R&D integration, and a network strategy to tap into domestic innovation systems. At the beginning, the centers were mainly technology recipients. Despite the parent companies' exerting tight control, the centers gradually evolved as coordinators of regional R&D activities in Asia, Greater China, and global markets. We argue that the main motivations for setting them up have been government policy and tapping into local research networks. These motivations are quite different from those in developing countries with big pools of talent and massive (potential) markets. To attract foreign R&D facilities, small countries such as Taiwan should focus on specialized areas of S&T (such as industrial clusters and centers of research excellence) and emphasize the domestic market with an intelligent user strategy, such as in ICT products and services. Then MNCs might want to learn from Taiwan and treat these R&D centers as places for lead market testing. The industries that provide the best examples are mobile phones and computers.

Potential strategies include: (1) strengthening domestic innovation systems by providing abundant R&D human resources, creating strong niche research bases, and enforcing strong IPR protection; (2) promoting R&D-related FDI in Taiwan through government promotion, R&D incentives, and the establishment of science parks; and (3) enhancing the benefits of domestic firms by deepening MNC-local collaborative R&D networks. These strategies might improve indigenous R&D, innovative capabilities, and the entrepreneurial competence of local firms.

Notes

[1] See P. Reddy, *The Globalization of Corporate R&D: Implications for Innovation Capability in Developing Countries* (London: Routledge, 2000); L. Xue and S. Wang, "Globalization of R&D by Multinational Corporations in China: An Empirical Analysis" (EAP Report Memorandum #01-06, National Science Foundation (NSC), Tokyo Regional Office, 2001); K. Walsh, *Foreign High-Tech R&D in China: Risks, Rewards and Implications for U.S.-China Relations* (Washington, D.C.: The Henry L. Stimson Center, 2003); S.-H. Chen "Taiwanese IT Firms' Offshore R&D in China and Connection with the Global Innovation Network," *Research Policy* 33 (2004): 337–49; M. von Zedtwitz, "Managing Foreign R&D Laboratories in China," *R&D Management* 34, no. 4 (2004): 439–52; and the United Nations Conference on Trade and Development (UNCTAD), *World Investment Report: Transnational Corporations and the Internationalization of R&D* (New York: United Nations, 2005).

[2] UNCTAD, *World Investment Report: Transnational Corporations*, 2005.

[3] Xue and Wang, "Globalization of R&D"; Walsh, *Foreign High-Tech R&D in China*; von Zedtwitz, "Managing Foreign R&D Laboratories in China"; and J. Li and D. Yue, "Managing Global Research and Development in China: Patterns of R&D Configuration and Evolution," *Technology Analysis and Strategic Management* 17, no. 3 (2005): 317–37.

[4] Reddy, *The Globalization of Corporate R&D*.

[5] R. Ronstadt, "International R&D: The Establishment and the Evolution of R&D Abroad by Seven Multinationals," *Journal of International Business Studies* 9, no. 1 (1978): 7–24.

[6] National Science Council (NSC), *Indicators of Science and Technology* Taipei, Taiwan: NSC, 2005), 123–24.

[7] NSC, *Indicators*, 229–30.

[8] N. Kumar, "Determinants of Location of Overseas R&D Activity of Multinational Enterprises: The Case of U.S. and Japanese Corporations," *Research Policy* 30 (2001): 159–74; Xue and Wang, "Globalization of R&D"; Walsh, *Foreign High-Tech R&D in China*; von Zedtwitz, "Managing Foreign R&D Laboratories in China"; and Li and Yue, "Managing Global Research and Development in China."

[9] O. Granstrand, L. Hakansson, and S. Sjolander, "Internationalization of R&D: A Survey of Some Recent Research," *Research Policy* 22 (1993): 413–30.

[10] Kumar, "Determinants of Location."

[11] Li and Yue, "Managing Global Research and Development in China," 334.

[12] M. Serapio and D. Dalton, "Foreign R&D Facilities in the United States," *Research Technology Management* 36, no. 6 (Nov./Dec. 1993): 33–39.

[13] W. Kuemmerle, "Building Effective R&D Capabilities Abroad," *Harvard Business Review* (March–April 1997): 61–70.

[14] Kumar, "Determinants of Location," 165.

[15] Serapio and Dalton, "Foreign R&D Facilities," 35–36.

[16] Von Zedtwitz, "Managing Foreign R&D Laboratories in China," 444.

[17] A. Hagedoorn and J. Schakenraad, "Inter-firm Partnership and Cooperative Strategies in Core Technologies," in *New Exploration in the Economics of Technical Change*, ed. C. Freeman and L. Soete (London: Pinter, 1990), 221–25; M. Dodgson, *Technological Collaboration in Industry: Strategy, Policy and Internationalization* (London: Routledge, 1993), 1–12; and R. Coombs, K. Richards, P. Saviotti, and V. Walsh, eds., *Technological Collaboration: The Dynamics of Cooperation in Industrial Innovation* (Cheltenham: Edward Elgar, 1996), 1–9.

[18] O. E. Williamson, *The Economic Institutions of Capitalism* (New York: Free Press, 1985): 1–15.

[19] A. Arora and A. Gambardella, "Evaluating Technological Information and Utilizing It," *Journal of Economic Behavior and Organization* 24 (1994): 91–114; W. M. Cohen and D. Levinthal, "Absorptive Capabilities: A New Perspective on Learning and Innovation," *Administrative Science Quarterly* 35 (1990): 128–52; and K. Pavitt, "What Makes Basic Research Economically Useful?" *Research Policy* 20 (1991): 109–19.

[20] W. Faulkner and J. Senker, *Knowledge Frontiers: Public Sector Research and Industrial Innovation in Biotechnology* (Oxford: Oxford University Press, 1995), 124–26.

[21] E. Geisler and A. Rubenstein, "University-Industry Relationship: A Review of Major Issues," in *Cooperative Research and Development: The Industry-*

University-Government, ed. A. Link and G. Tassey (Boston and London: Kluwer Academic, 1989), 1–13.

[22] Granstrand, Hakansson, and Sjolander, "Internationalization of R&D," 414.

[23] UNCTAD, *World Investment Report: Foreign Direct Investment and the Challenges for Development* (New York: United Nations, 1999), 137–39.

[24] O. Gassmann and M. von Zedtwitz, "New Concepts and Trends in International R&D Organization," *Research Policy* 28 (1999): 231–50.

[25] H. Odagiri and H. Yasuda, "The Determinants of Overseas R&D by Japanese Firms: An Empirical Study at the Industry and Company Levels," *Research Policy* 25 (1996): 1059–79; and Kumar, "Determinants of Location," 164.

[26] Kumar, "Determinants of Location," 165–66

[27] Serapio and Dalton, "Foreign R&D Facilities," 35–36.

[28] Li and Yue, "Managing Global Research and Development in China," 322.

[29] Kumar, "Determinants of Location," 165.

R&D Globalization and Silicon Triangle Dynamics

Kung Wang and Yi-Ling Wei

When we observe the path of semiconductor industry migration from the United States to Asia, we see that the relations among Silicon Valley, Hsinchu Science-based Industrial Park (HSIP) in Taiwan, and Shanghai—the so-called Silicon Triangle—have had significant influence on the global electronics industry. In the 1980s entrepreneurs and venture capitalists who had been working in Silicon Valley brought their experiences back to their home countries, laying the groundwork particularly for the high-tech industries in Taiwan and in Shanghai, China's leading integrated circuit (IC) cluster. Indeed, many of the founders or leaders of the semiconductor companies housed in Taiwan's HSIP gained their professional experience working in Silicon Valley. Further, since China adopted its open-market policies in the 1980s, semiconductor-related companies in Silicon Valley and Taiwan invested aggressively in China, and interactions of people, technology, and capital exploded as never before.

Since the 1980s multinational corporations (MNCs) in the information technology (IT) industry have been choosing locations in a variety of regions to maximize their global advantage, including in production and research and development (R&D). Mainland China has a massive labor force, abundant land and resources, and comparatively lower costs. In the 1990s it acquired a reputation as the "world's factory," and in 1992 Taiwan's government opened the door to Taiwanese enterprises investing there. China has become Taiwan's most favored site for external investment. With standardized logistics, economies of scale, and learning effects, "Get orders in Taiwan and produce in China" has become the mantra governing the division of labor. According to an estimate of the Institute for Information Industry, nearly 80 percent of Taiwan's information hardware products are made in China. On the other hand, nearly 60 percent of the production value of China's information hardware products is attributed to Taiwanese manufacturers. This reveals the interdependence of the cross-strait information industry.

This relationship is not simply bilateral. It is affected by the strategies of MNCs from Silicon Valley and elsewhere. Companies everywhere need to adjust their strategies in an ever-changing environment.

This chapter addresses several questions. Will Taiwan be marginalized by MNCs that are increasing their activities in Mainland China, especially in R&D? What are the motivations and strategies of MNCs' investments in China and elsewhere in Asia? How do these motivations and strategies affect those of Taiwanese enterprises or governmental policies?

Figure 7.1 The Research Framework

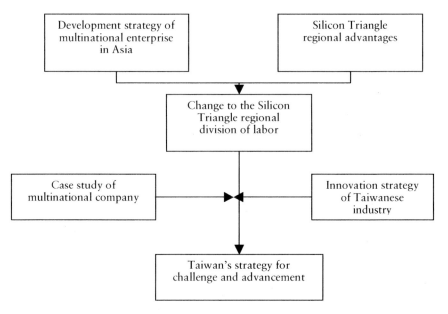

R&D Centers of MNCs in Taiwan

In 1965 Taiwan's government gave priority to attracting color TV manufacturers, including GE, TI, RCA, Philips, Sanyo, and Panasonic. Taiwan produced 1.8 million televisions in 1971, and 7 million in 1980. Foreign companies set up production plants, transferred technology, and developed skilled workers. This laid the foundation of Taiwan's electronic information industry.

In 1981 the Panasonic Taiwan Lab (PTL) was established. The PTL was the first technology research institution in Taiwan that was founded by an MNC. It was also the first overseas R&D organization for the Panasonic Corporation. Because most Japanese enterprises had a centralized R&D strategy, the achievements of the PTL benefited the parent company rather than meeting host market needs. However, in an R&D–resource-scarce era, the PTL acted as an educational institution, developing experts in technology.[1]

At the end of the 1970s, personal computers emerged, and in Taiwan, Acer and several other firms became assemblers. With the emergence of the Wintel structure, the PC market developed rapidly, and its vertically integrated structure, exemplified by IBM, changed to a more horizontal one. Taiwan benefited from

MNCs setting up production bases within its borders, further strengthening the nation's high-tech industry.

During the 1990s Taiwan's labor costs soared, and MNCs moved labor-intensive production to lower-wage regions such as Mainland China and Southeast Asia. Even so, the connections between Taiwan and MNCs were sustained. This is because Taiwanese firms developed high-quality, low-cost production and became very knowledgeable about products. Also, the MNCs have had to put most of their resources into product innovation and marketing. Taiwan has become an important supply location of electronic information products globally.

Taiwan's electronic information manufacturers are located mainly in the "technology corridor," from Taipei to Hsinchu, a counterpart to the IT companies in the Silicon Valley region of northern California. The bilateral relationship between firms in these regions involves not only orders and products but also technical knowledge, personnel, investment funds, and management information.

According to Gao, Lee, and Shih[2] the advantages of Taiwan's electronic information industry include the capacity to carry out global logistics efficiently, using a robust supply-chain management to respond quickly to orders from global customers. Hsu and colleagues[3] have found that Taiwan's information industry has a three- to five-year lead over China in product design, manufacturing skills, and management. However, China has the advantages of a huge market, extensive land, and large numbers of personnel. Therefore, we suggest that the future pattern will be threefold. First, Silicon Valley will focus on technology integration, innovation, new industrial standards, marketing, and service orientation, as well as on an openness to industries. Second, through the tight relationship of overseas Taiwanese with American MNCs, Taiwan will occupy part of Silicon Valley's position. More specifically, Taiwan will develop itself as a product-design R&D center and as a global-supply center for IT products. Third, because China has low labor costs and a huge market and land mass, it will gradually become a capital- and labor-intensive manufacturing center. Meanwhile, China will also develop its own market-oriented R&D and product designs and brands.

Despite Taiwan's remarkable manufacturing development, its technology innovation level does not approach that of advanced countries. Industry still pays huge amounts for technology transfer from foreign companies—about $1.5 billion annually in the IT industry. The same is true in some other industries as well.

In 2001 the Ministry of Economic Affairs (MOEA) initiated the Program to Encourage Multinational Corporations to Set Up Innovative R&D Centers in Taiwan. In all, twenty-seven multinational enterprises have established thirty R&D centers, with over $24 billion invested and over 3,900 employed by the end of 2006.

The R&D centers set up by MNCs in Taiwan help Taiwan's global strategic position. MNCs acknowledge that government encouragement has played a large role in the setup of these centers. Although the MNCs have long had sales offices and partnerships, new concepts are being developed. For instance, IBM xSeries R&D centers and DuPont R&D centers have changed over from sales functions, with reportedly good results. Networks of the R&D centers of multinational and Taiwan enterprises have close interactions with domestic firms. They hire local experts, and social networks among experts are formed. But there are issues, such as fewer Taiwanese students studying abroad and too few specialists available to do advanced research. However, the MNCs are adding R&D resources and are engaged in more—and deeper—skill exchanges with the R&D departments of Taiwan's factories.

The presence of MNCs gives domestic firms opportunities to acquire MNCs' knowledge of product concepts and development processes. For example, the Motorola R&D center has introduced to domestic factories the standard procedures for mobile-phone development, known as M-Gate. Thus, domestic firms that previously did only original design can now engage in the entire development process, from product concept through production.

Most MNCs hire locally, and their centers help develop local R&D internationalization experts. The centers send experts abroad for long postings at one R&D site or for shorter postings at several different sites. These same MNCs send their own experts to Taiwan. For instance, the IBM xSeries R&D center dispatches eight experts—each with over twenty years of experience—to Taiwan for long stays, during which time they exchange knowledge with domestic employees. MNCs' R&D centers also help develop Taiwan's next-generation technology.

Taiwan's IT industry has third-generation (3G) mobile communications, a thin-film transistor liquid-crystal display (TFT-LCD) industry, and similar new strategic industries, which the government actively promotes. Some R&D centers commercialize key technologies that have a potential market in Taiwan and that help the development of next-generation technology. For example, Aixtron, a leading semiconductor equipment provider, cooperates with the domestic factories that produce an organic light-emitting diode (OLED), a technology used for displays. Together, they developed a technology known as metal organic vapor-phase disposition (MOVD).

The AKT Asia Research and Development Center

AKT, a subsidiary of Applied Materials, Inc., specializes in equipment for making TFT-LCDs. It is also the leading supplier of process equipment for plasma-enhanced chemical vapor deposition (PECVD) in the flat-panel display (FPD) industry. AKT has over 70 percent of the world market in chemical vapor displays.

In addition to its business operations in Taiwan, which it has held since the 1990s, in March 2004 AKT announced a plan to cooperate with the Industrial Technology Research Institute (ITRI) to establish the AKT Asia Research and Development Center (ARDC) in Taiwan to target 7G TFT-LCD equipment. A trend in LCD technology today is the replacement of cathode ray tubes in televisions within the next five years. One of several effects of this rapidly growing market is frequent equipment purchases. Keen competition with Korean and Japanese producers is relentlessly driving down costs, which in turn is generating a rapid growth in demand. At the same time, technology is advancing. The result is rapid cycles of equipment purchases and the need for rapid development. Taiwan has become the leading TFT-LCD panel manufacturing site in the world, and this achievement is motivating the industry to get involved in leading-edge technology development.

A second effect is the rapid increase in the size and weight of machines, which presents transportation difficulties. TFT-LCD equipment today is mainly imported—and by air. But while its size is outstripping the capacity of cargo planes, its transportation width poses a problem for roads. So there is an incentive to assemble this equipment locally.

A third effect of LCD technology in televisions is the need for cost control. The price of panels is falling very quickly, implying a need for tight controls on equipment investments and business operations. The production of components in Taiwan should reduce these costs.

A fourth effect is that quick component supply has become urgent, because production-line breakdown could cost several million U.S. dollars per hour. And the final effect is the urgent need to keep advancing the technology. Whereas most TFT-LCD technologies used to come from Japan, now Taiwan needs to advance them, including by acquiring patents. A priority to enable this is cooperation between the TFT-LCD manufacturers and AKT. This implies that AKT needs a better operational mode in order to keep its leading position. Cooperation among the ARDC, ITRI, and the manufacturers will be mutually advantageous.

Alignment between Domestic and International Companies

Most machinery companies in Taiwan are small and widely scattered. Each has its own technology and emphasizes traditional processing and manufacturing, such as found in bicycle- and shoe-making machines.

With the relocation of traditional industry from Taiwan, however, a transformation is needed in ways of doing business. Because most machinery companies have no experience in cooperating with foreign high-tech companies and are unfamiliar with the requirements of the technology specifications of high-tech products, organizational changes are necessary. Moreover, establishing a component supply system for TFT-LCD process equipment will attract more equipment manufacturers in Japan to locate in Taiwan, and this will bring in valuable technology.

Figure 7.2 AKT's Operation Strategy in Taiwan

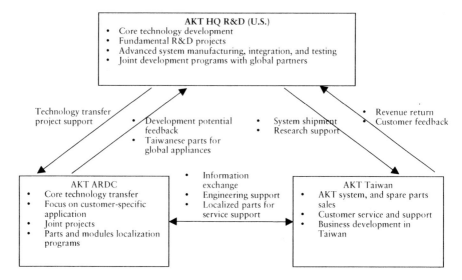

Ericsson's Innovation Center

The leader in the 3G mobile communications technology (GPRS), Ericsson, also has a large market share of the most popular multimedia messaging services (MMS). These achievements come from its R&D and localized applications. The Ericsson Innovation Center in Taiwan not only brings 3G mobile communications to Taiwan but also cultivates local developers and suppliers of local mobile applications and helps them in international markets through the Ericsson network. Ericsson has invested more than $100 million to establish the only wireless communications software simulation and testing environment in Taiwan, which helps local enterprises reduce the critical development time for multimedia applications.

The ratio of mobile phones to people in Taiwan is the highest in Asia. Thus, it has the potential of becoming the pioneer market in Asia, or in the world, through innovations; each domestic telecommunications company seeks successful foreign enterprises to reduce its risk or expand its business opportunities.

Ericsson Mobility World in Taiwan (formerly known as the Wireless Application Center, or WAC) is one of three regional centers in Asia. Ericsson envisions this center becoming a major department for developing mobile video applications and mobile games in Taiwan. The training activity in the Ericsson Innovation Center, which introduces the most advanced wireless technologies, is designed to cultivate domestic talents. Teaching materials will be introduced directly from Ericsson or from the latest standards in the industry.

The research scope of Ericsson Mobility World includes software and hardware components, system equipment, and terminal equipment. If the research is successful, the results could be passed to Taiwanese manufacturers for production and promoted through Ericsson's global marketing channels. The center will also be used for testing and verification; it will be the first and most comprehensive test environment in Taiwan to verify and test mobile video.

Despite Taiwan's good IT environment, until Ericsson Mobility World was established, there was no complete wireless communications software simulation and testing environment to assist manufacturers in developing high-quality wireless communications software. That such an environment now exists will help companies and universities by providing technology consultation, conducting product testing and information sharing, and promoting Taiwan as the wireless technology island.

Taiwanese Efforts to Innovate

The perception that "Made in Taiwan" implies high quality has contributed greatly to the economy—evident in sales of notebook PCs, network equipment, and auto parts. Given the fast-changing global environment, including price competition, industry is moving from its role in original equipment manufacturing (OEM) and original design manufacturing (ODM) toward becoming an innovative brand owner. This requires competencies not only in innovation and integration but also in international branding. Several Taiwanese companies are trying to do this, including Acer, ASUS, and BenQ in the consumer electronics industry, and Trend Micro in software. Acer is the most significant example of an OEM source that has been transformed into a branding business in Taiwan.

Stan Shih started Acer (then Multitech) in his garage in 1976, using innovative techniques to build cloned PCs at a fraction of what PCs were costing in the United States. The PC-clone business model caught on, driving the growth of Taiwan's IT industry. From the beginning, Shih positioned his products further toward the higher end of the market than any other Taiwanese producers. This pricing strategy kept Acer-branded products from being classified as commodities. Acer has always invested heavily in R&D, because Shih wanted cutting-edge products to lead the industry. In the late 1990s the company redefined its main goal to focus on its own brand development.

In 2000 Acer separated its OEM and its branded business and established the Design and Manufacturing Services (DMS) and Acer Brand Operations (ABO). Since spinning off its manufacturing operation by establishing Wistron Corporation in 2001, Acer has focused on marketing its brand-name products and e-business solutions. It aims to shift from being "technology-centric" to adding value by enhancing consumer perceptions of the value of its products based on packaging, design, accessibility, comfort, user-friendliness, and niche solutions.

The appointment of a foreign top manager signaled Acer as a global company. Gianfranco Lanci, an Italian, was appointed president in 2005. This appointment was based on his outstanding performance in the European market and his understanding of cross-regional and cross-cultural differences. As part of its international exposure strategy, Acer is partnering with overseas companies. An example is the new Acer Ferrari 4000 notebook, the third and latest in a partnership series between Acer and Scuderia Ferrari. This partnership could help make inroads into a traditionally tough market for the brand in North America.

In 2005 Acer was ranked as the world's number-four branded PC vendor. Acer's chairman, J. T. Wang, believes that by building strong relationships with its dealers, offering lower prices, and making product innovations, the company will be in the number-three spot worldwide by the end of 2008.

From Advanced Wafer Foundry to System-on-Chip Design Services

Taiwan's semiconductor industry now has a complete supply chain, has spawned other firms with a significant clustering effect, and has strong contract wafer manufacturing abilities (wafer foundry). It is the world's second-largest supplier of fabless IC designs, after Silicon Valley. Also, facing competition from Chartered in Singapore and SMIC in China, TSMC and UMC are developing design services especially for system-on-chip (SoC) customers. The foundries provide not only advanced manufacturing service but also total solutions to fabless IC design customers. For example, in April 2002, TSMC introduced a development architecture, called Nexsys, with an easier design and manufacturing interface and flow integration.

TSMC has had partnerships with seven domestic fabless IC design service companies, and in 2003 it formed the Design Center Alliance (DCA). Including the DCA members, TSMC has twenty-nine partners globally for design service and has gained majority control of Global Unichip Corporation (GUC), a prominent SoC design service firm. The other foundry giant, UMC, had invested in Faraday Technology Corporation for SoC design service business. Collaboration between Taiwan's foundries and SoC design services companies is a solution to the growing complexity of the technology, and it shortens the time-to-market of SoC products. R&D centers of the MNCs in Taiwan focus, for the most part, on short-term innovation and prototype (within one year), mid-term application, and mid-term research (one to two years).[4] There is also engineering support and improving present technology. However, basic research is not done. This pattern matches the ecology of the domestic industries. They, too, focus on developing new products quickly, and they have years of experience in doing this. Given the resources available, these overseas R&D centers in Taiwan have trouble doing basic R&D; additionally, most of them, according to surveys, are sales or regional headquarters instead of R&D headquarters, and their technologies are oriented toward current market demand. The centers in

Taiwan need to explore fields that have been neglected, such as software services and process equipment and test verification, in addition to the enhanced value of information and communications hardware. A closer connection between domestic and foreign enterprises would raise obstacles for competitors.

The R&D centers of companies from America and Japan are focused on semiconductor and information and communications hardware, while Ericsson and Alcatel, from Europe, work on communications industry standards and platform technologies, including testing technology and software development. The latter could help Taiwan's manufacturers make service innovations. Because the boundary between the manufacturing industry and service industry is blurring, an emerging issue for cooperation between Taiwan and these R&D centers is innovations in such services as mobile e-commerce, test verification, and the development of new processes.

Conclusion

There is a complementary relationship between Taiwan and Silicon Valley that is based on technology, talent, and capital. Innovative products and global marketing offered by Silicon Valley are important for Taiwanese manufacturers; furthermore, the process and planning and management skills that are developed make for mutual dependence between them.

Very few firms from Silicon Valley have invested in Taiwan, and since the end of the 1980s Taiwanese manufacturers have moved their production to China. The epoch of Taiwan as the mass production base has passed; thus, R&D and innovations have become urgent. We have seen that AKT established the ARDC in Taiwan to gain access to large TFT-LCD panel manufacturing equipment. It is also filling a value-chain gap in Taiwan. Ericsson sees Taiwan as the regional R&D center of mobile service innovation.

Over seven hundred R&D centers have been established in China by MNCs, but only thirty in Taiwan; in addition to this difference in numbers, there are also differences in missions. Those in China are based on extensive domestic markets and abundant, low-cost personnel; thus, they undertake fundamental R&D in cooperation with local academic institutions and contribute technology to the global R&D network of the MNCs. Because MNCs in Taiwan do R&D on products that could be commercialized in two years, it is rare for them to do advanced technology R&D there. In any case, there are few resources for basic research in Taiwan.

The literature shows that even though the many R&D centers established by MNCs in China open the country to advanced research and cultivate talents, they have not provided much assistance to its ability to innovate. Nor do such R&D centers in Taiwan contribute much to Taiwan's technology. However, close cooperation with Taiwan's enterprises is important for their leading position globally, while Taiwanese firms have to maintain these partnerships to foster their own ambitions internationally.

A developing country should not rely on R&D investment by foreign companies, but needs to promote cooperation and investment among local governments, enterprises, and academic and research institutions to foster innovation abilities.

The pattern of technology division in the Silicon Triangle is unlikely to change significantly in the next five years. In the past two decades, Taiwan has been cooperating with MNCs based on its manufacturing skills; however, there is the threat of its being replaced by China and India due to their vast markets and labor advantage. Taiwan needs to maintain existing cooperative relations with MNCs and strengthen them with companies in Europe and Japan. Furthermore, the growing importance of services requires that Taiwan address the issue of how to use global R&D resources to develop this sector.

Notes

[1] Interview (February 2006) with Dr. Chuie-Ming Shyu, senior advisor of the Industrial Technology Research Institute (ITRI), formerly a researcher in the Panasonic Taiwan Lab (PTL) in Taiwan.

[2] Chang Gao, Ji-Ren Lee, and Hui-tzu Shih, "The Strategy of Cross-Strait Industrial Division of Labor from the Viewpoint of Global Logistics and Integration Advantage" (research report, Ministry of Economic Affairs [MOEA], Chung-Hua Institution for Economic Research, Taipei, Taiwan, 2000).

[3] Ji-Sheng Hsu, Chintay Shih, Chi-Yang Hong, and Chia-Sheng Chen, "The Division of Labor among Silicon Valley, Taiwan and China: Case Study," in the proceedings of The First ITRI Conference on S&T Cluster Development, (ITRI, Hsinchu, Taiwan, 2003), 9-1–9-27.

[4] Chintay Shih, Yuan-chieh Chang, and Shih-Yin Wu, "Evaluation of the Program to Encourage Multinational Corporations to Set Up Innovative R&D Centers in Taiwan" (research sponsored by the Department of Industrial Technology, MOEA, 2005).

COOPERATION IN CHINESE INNOVATION SYSTEMS

Ingo Liefner and Stefan Hennemann*

In the 1990s, China started to give a high priority to foreign and domestic investment in technology-intensive sectors. China also introduced measures to narrow the technological gap with industrialized countries. One major source of new knowledge is foreign companies that supply technology to their Chinese subsidiaries or joint-venture partners, but such knowledge transfer is still restricted to production or assembly-related knowledge.[1] A second source of new knowledge is public research organizations and higher education institutions (PROHEIs), but they also cannot fill this void.[2] The vast majority of Chinese PROHEIs are technologically behind foreign companies, and while PROHEIs can offer scientific knowledge, they usually cannot provide other knowledge that may be in higher demand, such as marketing know-how or means of quality assurance.

Changes are occurring, however. First, more foreign-owned companies are setting up research and development (R&D) departments in their Chinese subsidiaries, and thus are improving the potential for knowledge spillovers. Second, reforms in the science sector will add to the supply of knowledge. And, third, government-funded programs and incentives are increasing Chinese firms' awareness of the positive, long-term effects of R&D and innovation. Indeed, research has shown that high-tech companies in Beijing cooperate both with foreign companies and with PROHEIs.[3]

The aim of this chapter is to present the determining factors and impacts of innovation-related cooperation, with a focus on relations between foreign and Chinese companies. The chapter addresses two questions: What factors determine the cooperative activities of high-tech firms in China? What is the impact of such cooperation on their innovations?

---

* The authors thank the German Research Association (DFG) for providing the research grant that funded the surveys in Beijing and Shanghai. We also thank our local partners, Gang Zeng and Jie Fan, as well as their researchers—Tao Wang, Xin Lu, and many others—for their help carrying out the survey and interpreting the results.

Cooperation, Spillovers, and Systems of Innovation and Learning

The role of linkages among organizations (universities, suppliers, customers, service providers, and so on) is embedded in the concepts of national and regional innovation systems (NISs and RISs).[4] Close interaction allows companies to acquire and use knowledge. The result is a product or process that is new to the world market.

This concept of the NIS was derived from observations in developed countries; it is of doubtful utility for understanding innovation in developing countries because they produce little that is new to the world market. Instead, they incorporate knowledge developed in industrialized countries in their companies' knowledge base and then imitate, copy, or generate products that are technologically similar to existing products.[5] The process does not necessarily require close cooperation with organizations such as universities or banks. The only requirement is a knowledge inflow from technologically advanced parts of the world, either through direct cooperation or through the use of advanced products or production equipment. In theory, new knowledge could be provided by PROHEIs, but the PROHEIs of developing countries also lag and are unable to offer new and commercially relevant knowledge.[6] Recognizing these differences, Viotti calls the innovation systems of developing countries "national learning systems."[7]

In practice, the differences might be smaller than in concept.[8] Asheim and Vang[9] point out that metropolitan regions in developing countries may offer conditions for innovation and learning similar to those in industrialized countries. Their universities and research institutes are usually among their nations' top institutions. Because human capital concentrates in such regions, the technologically most advanced domestic companies—and foreign companies that seek cheap but skilled personnel—choose them. These conditions not only allow for learning but might also allow for innovation.

Asheim and Vang's argument that firms in the most advanced regions of newly industrialized countries (NIEs) can cooperate both with foreign companies and with PROHEIs, has been supported empirically with findings from Beijing,[10] although much remains to be explained. However, this literature does not explain why firms cooperate. Cooperation among firms can contribute to profits by aiding the acquisition of new knowledge and by helping firms gain access to capital—for example, by cooperating with larger companies, banks, or financial service providers. These are two primary functions of cooperation; examples of secondary functions are shortening the time span of innovations and getting state subsidies. The importance of these goals varies with circumstances.[11]

An important precondition for successful knowledge acquisition is an understanding of the kind of knowledge that is being sought. This ability—a firm's absorptive capacity—is closely connected to the resource-based view of cooperation.[12] Cohen and Levinthal define a firm's absorptive capacity as the ability to acquire new information, assess its relevance for the company, and process it.[13]

The literature focuses primarily on R&D and human capital. (Factors that are beyond a single company's control are institutions, government policy, openness to trade, and so on.)[14] Scott[15] states that R&D cooperation helps a firm's absorptive capacity and shows that absorptive capacity is a precondition for successful cooperation and is itself affected by external contacts.[16] There is a strong reciprocal relationship between a firm's innovative capacity and the supporting environment that, in turn, enables firms to increase their absorptive capacity.

For the purpose of empirical studies, indicators are necessary. Examples include R&D expenditure, R&D manpower, and highly qualified personnel. These are measures of inputs into the innovation process. Indicators of throughputs and outputs are patents and new products. Factors that may affect a company's absorptive capacity are size, age, and ownership.[17] The latter set of indicators measure absorptive capacity indirectly. For example, a larger company can more easily take a risk by investing in new technology, and privately run companies may have a higher propensity to seek technological opportunities than government-owned or subsidized opportunities. Cooperative behavior also varies among industries.

A point that is not explicit in the concept of learning systems is the impact of cooperation on the firms involved. The recent literature has focused less on intended results, such as the outcome of a joint development project or the transfer of a certain amount of money, and more on spillover, or unintentional, effects. Blomström and Kokko[18] state that spillover effects either induce technological upgrading, called "technology spillovers," in the recipient firm, or lead to increasing productivity, or "productivity spillovers." Technology spillovers are generated by a transfer of knowledge that may come from the introduction of new products or processes, or are generated by organizational improvements. Productivity spillovers are independent of technology spillovers. They occur when firms enter a certain market, country, or region and challenge the existing companies with a superior product or a lower price. Empirical studies have proved the existence of both types of spillovers.[19]

Survey Methods and Data

Our analyses are based on a survey of high-tech companies in Beijing and Shanghai. Within Beijing, the Zhongguancun Science Park (ZGC), and within Shanghai, the Pudong region, including the Zhangjiang High-Tech Park (ZJP), were selected as survey sites. These are the most knowledge-intensive regions in China.[20] In both regions, the population of our survey was comprised only of high-tech companies. At the time, 7,100 companies were registered with the ZGC board. The many enterprises, the human capital, 39 universities, and more than 200 research institutes make it the most knowledge-intensive region in China.[21] Pudong's ZJP is a top technology zone.[22] One of its main goals is to integrate all activities along the value-added chain within the park. As in the ZGC, public and private R&D facilities are located close to high-tech companies.[23] Local universities, institutes of the Chinese Academy of Sciences (CAS), and offshoots of first-class universities from Beijing were set up in Pudong to support both training and research. Today, more than 300 companies from China and abroad are registered.[24]

The surveys were done in 2003 as part of a project on technological change and regional economic development, sponsored by the German Research Association (DFG) and carried out jointly by the Department of Economic Geography, University of Hanover, in Germany; the East China Normal University in Shanghai; and the CAS in Beijing. In both locations a standardized survey was combined with in-depth interviews of CEOs. Support from local authorities ensured a rather high response rate of around 46 percent for Beijing ($n = 234$) and about 66 percent for Shanghai ($n = 253$). The results were discussed in detail by the European and Chinese researchers to avoid misinterpretations that might arise from cultural differences and background experience.

Factors Affecting Cooperation

Table 8.1 summarizes the results of logit regressions that reveal factors affecting cooperation. The first model explains cooperation with partners abroad (CPA). CPA, the dependent variable, is "0" for companies with no links outside Mainland China (including Hong Kong, Macao, and Taiwan). CPA is "1" for firms with at least one link with a partner outside China. The second model addresses cooperation with PROHEIs (CPROHEI). CPROHEI is "0" for companies that do not work with PROHEIs and "1" for firms with at least one cooperative agreement with a PROHEI.

Model 1: CPA[25]

Six independent variables (and the dummy-variable of Shanghai) are identified as affecting CPA and are statistically significant at the 95-percent level. These factors are "ownership by foreign investors," "turnover," "percentage share of R&D employees," "percentage share of R&D expenses," "cooperation in the central stages of the innovation process," and "manufacturing." Five of them affect cooperation positively; only one has a negative sign. The most important factors are the influence of a foreign investor and the size of the annual turnover: wholly foreign-owned subsidiaries and joint ventures are likely to cooperate with partners abroad, an unsurprising result. The size of a company increases its absorptive capacity and makes it an attractive partner for foreign companies.[26] Firms with a high turnover include manufacturing companies that target foreign markets with cheap products. They are likely to cooperate at least with key customers abroad.

Table 8.1 Logit Models

	Independent variables	Logit (b_k)	Std-error (s_k)	t-value (b_k/s_k)	Sig.	Exp (B)	Std-Exp (B)
CPA	Foreign ownership	1,643	0,369	4,459	0,000	5,171	2,208
	Turnover 2002 (ln)	0,543	0,100	5,448	0,000	1,722	3,042
	R&D personnel > 10%	-0,889	0,361	-2,463	0,014	0,411	0,653
	Share R&D / Sales > 5%	0,930	0,375	2,482	0,013	2,536	1,514
	Throughput[a]	0,273	0,102	2,680	0,007	1,314	1,577
	Manufacturing	1,062	0,529	2,007	0,045	2,892	1,456
	Location Shanghai	-0,644	0,349	-1,844	0,065	0,525	0,748
	Constant	-4,172	0,833	-5,010	0,000	0,015	–
CPROHEI	Foreign ownership	-2,142	0,384	-5,578	0,000	0,117	0,354
	Own R&D activity	0,853	0,336	2,539	0,011	2,346	1,520
	University graduates	0,071	0,017	4,176	0,000	1,074	2,704
	Patents	0,147	0,051	2,882	0,004	1,158	3,246
	Throughput[a]	0,952	0,166	5,735	0,000	2,590	6,297
	Location Beijing	-1,940	0,456	-4,254	0,000	0,144	0,430
	Biotech	2,157	0,667	3,234	0,001	8,649	2,051
	Manufacturing	-1,321	0,536	-2,465	0,014	0,267	0,625
	ICT/Software	-0,884	0,427	-2,070	0,038	0,413	0,642
	Constant	-0,367	0,459	-0,800	0,423	0,693	–

Note: [a] Throughput: Frequency and quality (at least two separate external partners) of contacts in the stages of prototyping and the development of pilot applications.

Three other factors are less important. One is manufacturing. A company's activity in large-scale manufacturing is strongly linked to turnover and thus is statistically linked to CPA. Such companies depend on manufacturing technologies supplied by foreign partners. Another variable, throughput, is defined by the frequency of contacts and the heterogeneity of collaborators in prototype development and pilot applications. Firms that seek assistance in these stages tend to look abroad. A third, less important, factor is the share of R&D–related expenses that exceeds 5 percent of turnover, signaling a company's interest in developing R&D capabilities. In contrast, firms with a share of R&D personnel above 10 percent of the total workforce are less likely to cooperate with partners abroad. This last result can be explained only by particular characteristics of the companies surveyed: R&D–related expenses over 5 percent of turnover characterize firms that are trying to build technological capabilities. Most of them focus on process innovations, as confirmed in our interviews. While expenditure on R&D is mainly invested in fixed assets, such as production facilities—which in turn leads to relatively high shares of investment in relation to the turnover—such production sites need fewer R&D workers, and some of them work with firms abroad. Other companies in this group do more R&D, employing more than 10 percent of their workers. These tend to be young, small, and government-oriented. Examples are firms that modify existing geographic information system (GIS) software and carry out GIS-based analyses for the government. They very seldom cooperate with partners abroad.

The findings of Model 1 (CPA) can be summarized as follows. Cooperation with partners abroad by high-tech firms in Shanghai and Beijing is driven predominantly by foreign ownership, company size, and some efforts to develop new products. These factors go along with large-scale manufacturing, either to serve foreign markets or to produce for the domestic market using foreign technology.

Model 2: CPROHEI

We identified nine variables that affect CPROHEI, all of them significant at least on the 95-percent level, with most of them at the 99-percent level. Foreign ownership is a negative—unsurprisingly, since foreign companies are better linked to other foreign actors and can access premium research partners at home or transfer technologies internally to their subsidiaries. In-house R&D is positively related to CPROHEI. This is for firms that need complementary resources, especially smaller and newer firms.

The "throughput" variable is even more effective in explaining CPROHEI. Intensive collaboration in the central stages of the innovation process is the single most important predictor of such cooperation. Regarding the t-values, the variables "firms with foreign ownership," "throughput," "location Beijing," and "share of university graduates" have high influences, with the other variables being minor. The most relevant factor for increasing the probability of cooperation is still throughput—that is, intensive cooperation during the central stages of innovation. Patent application and the share of university graduates are slightly less important. The latter factor can be viewed as a proxy for complementary knowledge, enabling companies to incorporate external PROHEI technologies. The share of R&D personnel significantly increases the chance to cooperate with a PROHEI.[27] CPROHEI varies among industries and, therefore, in knowledge spillovers as well.[28]

The findings of Model 2 (CPROHEI) can be summarized as follows. Cooperation of high-tech firms in Shanghai and Beijing with PROHEIs is driven predominantly by work performed with others in the central stages of the innovation process, patenting activities, large shares of university graduates, and network contacts between former students and faculty. Company size and age are of minor importance. Economies of scale in manufacturing and production are not relevant to this type of innovation-related cooperation.

Effects of Cooperation

The idea is that cooperation in the innovation process helps a company's capacity to innovate. However, innovation cannot be measured directly, because of complexities. Table 8.2 shows data on the inputs, throughputs, and outputs of the innovation processes. Inputs are measured by R&D personnel and R&D expenditure. Cooperation during the central stages of the innovation process, as well as patent applications, are used as throughput indicators. Outputs are measured by new products (innovating) and the new products' shares of sales (innovative). Table 8.2 gives the percentage shares of companies that exceed certain thresholds or to which the indicators apply. We present numbers for all companies (a), for companies that cooperate *both* with partners abroad and with PROHEI (b), for all companies that cooperate with PROHEI (c), for all companies that cooperate with partners abroad (d), for companies that cooperate *either* with PROHEI *or* with partners abroad (e), and for companies that cooperate *neither* with PROHEI *nor* with partners abroad (f).

From Table 7.2 the following results stand out:

- The variation in R&D inputs varies moderately among the groups of companies. Firms that cooperate with PROHEIs or foreign companies in either form have a higher propensity to invest in R&D, a finding that is particularly relevant for firms that cooperate with both PROHEI and partners abroad. A positive correlation between cooperation and inputs underpins the notion of absorptive capacity, especially for firms cooperating with PROHEI. As shown by the *phi* statistics, having their own R&D personnel is of minor relevance for firms that cooperate with foreign companies.
- Companies that cooperate with both PROHEIs and foreign companies show outstanding results with respect to the throughput indicators—patent applications in particular—and also to cooperation activity and quality. Obviously,

Table 8.2 Cooperation and Innovation Indicators

Cooperation		n	Input				Throughput					Output	
			Personnel[a]	Phi[b]	Turnover[c]	Phi	Coop[d]	Phi	Patents[e]	Phi	Innovating[f]	Phi	Innovative[g]	Phi
(a)	Total	479	51,1		62,0		34,9		45,9		72,4		48,4	
(b)	PROHEI and FC	139	55,4		76,3	0,128	55,4	0,275	64,7	0,232	82,0	0,137	56,1	
(c)	PROHEI	259	56,8		71,8	0,161	46,3	0,261	57,1	0,234	74,9		52,1	
(d)	FC	252	49,6	-0,140	67,9		42,9	0,177	49,2		81,7	0,219	55,6	
(e)	Either	372	52,4		67,5		40,6		48,9		76,9		53,0	
(f)	Neither	107	46,7		43,0		15,0	-0,224	35,5		57,0	-0,185	32,7	

Notes:

[a] Percentage of companies with R&D personnel share above 10%.
[b] Phi statistics are significant on the 0,01 level except "neither" category (0,05 level).
[c] Percentage of companies with R&D expenditure higher than 5% of turnover.
[d] Percentage of companies with more than one partner in the central stage of the innovation process.
[e] Percentage of companies with patent application.
[f] Percentage of companies that introduced new or substantially improved products within the last three years.
[g] Percentage of companies generating more than 25% of turnover with new products.

internal R&D in combination with cooperation generates new knowledge. Patenting is most relevant for companies linked to PROHEIs.
- The impact of R&D activities and cooperation on outputs—generating and selling new products—is, however, not very pronounced. With the exception of companies that do not cooperate at all, companies are alike in this regard. This result may be attributed to two reasons supported by in-depth interviews: first, many companies in the survey were new, and thus, neither their R&D nor their cooperation may have yet produced innovations. Second, most of these companies see innovation as introducing products and processes that are new to their own company but not new to Western firms. Not much R&D is needed for copying, imitating, or transferring from foreign partners. This produces the significant correlation between innovation and cooperation with foreign firms.
- Firms that do not cooperate at all—or that share their knowledge only with other local firms—do not benefit from learning, compared to those who activate PROHEIs or foreign companies. We find moderate significant negative correlations for the former group.

Discussion

Foreign companies and PROHEIs are both important sources of new knowledge for Chinese high-tech companies. Companies seek cooperation with PROHEIs and foreign companies for various reasons. Foreign-invested companies have little difficulty in accessing partners abroad, but foreign ownership affects CPROHEI negatively. Companies in fields closely linked to advances in science and higher education—for example, biotechnology—are greatly in need of knowledge from PROHEIs. In contrast, firms in advanced manufacturing have less need for academic knowledge but need technological and marketing expertise from foreign companies. Moreover, it is apparent that networks evolving around PROHEIs, and around foreign companies are not strongly interlinked; network participation is based not only on ownership but also on the type of industry.

The independent variable "throughput"—signaling the need for complementary technological knowledge—is important in the model; in this respect, Chinese high-tech companies are no different from high-tech companies in industrialized countries. The Chinese companies strive for innovation and learning through internal R&D and cooperation. Internal R&D helps to increase absorptive capacity. The "throughput" indicator shows that high-tech companies in China's advanced metropolitan regions have reached a relatively strong technological position,[29] an investment in ideas and human capital that will pay off in the future. Until now the transformation of knowledge into new products has been weak, as has been cooperation with customers.

This lack of transformation capability points to limitations in comparing high-tech companies from industrialized countries with those in China. A striking result in Table 8.2 is that cooperation affects the outcome of the R&D

process (patenting) no less than it affects the outcome of the innovation process (innovating). One-third of the surveyed companies that cooperate neither with foreign companies nor with PROHEIs report that they are innovating. There are several possible reasons. First, China's factor endowments and strong position in low-technology products create an incentive for firms to be in this market. Thus, many companies with high-tech capacities also make low-tech products that do not require outside knowledge. Second, some of the claimed innovations of the surveyed companies may be imitations of existing products. Third, the benefit of being a registered high-tech firm creates an incentive to complete the registration process without actually doing high-tech work. And fourth, some firms sell other companies' products but count them as their own innovations.

Notes

[1] F. Lemoine and D. Ünal-Kesenci, "Assembly Trade and Technology Transfer: The Case of China," *World Development* 32 (2004): 829–50.

[2] P. G. Altbach, *Comparative Higher Education: Knowledge, the University, and Development*, Contemporary Studies in Social and Policy Issues in Education (London: Ablex Publishing, 1998): 20.

[3] I. Liefner, S. Hennemann, and X. Lu, "Cooperation in the Innovation Process in Developing Countries: Empirical Evidence from Zhongguancun, Beijing," *Environment and Planning A* 38, no. 1 (2006): 111–30.

[4] S. J. Kline and N. Rosenberg, "An Overview of Innovation," in *The Positive Sum Strategy: Harnessing Technology for Economic Growth*, ed. R. Landau and N. Rosenberg (Washington, D.C.: National Academy Press, 1986), 275–305.

[5] E. Viotti, "National Learning Systems: A New Approach on Technical Change in Late Industrializing Economies and Evidences from the Cases of Brazil and South Korea," *Technological Forecasting & Social Change* 69 (2002): 653–80; and J. Mathews, "Competitive Dynamics and Economic Learning: An Extended Resource-based View," *Industrial and Corporate Change* 12 (2003): 115–45.

[6] Altbach, *Comparative Higher Education*. Also, compare L. Leydesdorff and P. Zhou, "Are the Contributions of China and Korea Upsetting the World System of Science?" *Scientometrics* 63 (2005): 617–30; and Zhou and Leydesdorff in chapter 15 of this book.

[7] Viotti, "National Learning Systems."

[8] Organisation for Economic Co-operation and Development (OECD), *OECD Science, Technology and Industry Scoreboard 2003: Towards a Knowledge-Based Economy* (Paris: OECD, 2003): 6.

[9] B. Asheim and J. Vang, "What Can a Learning Region Approach Offer Developing Regions?" (paper presented at the 2nd Globelics International Conference, Beijing, October 2004): 34*ff*.

[10] Liefner et al., "Cooperation in the Innovation Process," 126.

[11] K. Schumann, *Kooperationen zwischen technologieorientierten Gründungsunternehmungen und Forschungseinrichtungen. Erfolgskonzept,*

empirische Untersuchung und Gestaltungshinweise, Schriften zum Management 23 (München, Germany: Hampp, 2005), 43–47, 83; J. Hagedoorn, A. N. Link, and N. S. Vonortas, "Research Partnerships," *Research Policy* 29 (2000): 567–86; and D. J. Teece, G. Pisano, and A. Shuen, "Dynamic Capabilities and Strategic Management," *Strategic Management Journal* 18 (1997): 509–33.

[12] C.-M. Lau, Y. Lu, S. Makino, X. Chen, and R. Yeh, "Knowledge Management of High-Tech Firms" in *The Management of Enterprises in the People's Republic of China*, ed. A. S. Tsui and C. M. Lau (Boston: Kluwer, 2002), 185*ff*.

[13] W. Cohen and D. Levinthal, "Innovation and Learning: The Two Faces of R&D," *Economic Journal* 99 (1989): 569–70.

[14] R. Griffith, S. Redding, and J. van Reenen, "R&D and Absorptive Capacity: Theory and Empirical Evidence," *Scandinavian Journal of Economics* 105 (2003): 99–118.

[15] J. T. Scott, "Absorptive Capacity and the Efficiency of Research Partnerships," *Technology Analysis & Strategic Management* 15 (2003): 247–53.

[16] Compare D. B. Audretsch, B. Bozeman, K. L. Combs, M. Feldman, A. L. Link, D. S. Siegel, P. Stephan, G. Tassey, and C. Wessner, "The Economics of Science and Technology," *Journal of Technology Transfer* 27 (2002): 181.

[17] For example, see Cohen and Levinthal, "Innovation and Learning"; E. W. K. Tsang, "Strategies for Transferring Technology to China," *Long Range Planning* 27 (1994): 98–107; and H. Katrak, "Developing Countries' Imports of Technology, In-house Technological Capabilities and Efforts: An Analysis of the Indian Experience," *Journal of Development Economics* 53 (1997): 67–83.

[18] M. Blomström and A. Kokko, "Foreign Investment as a Vehicle for International Technology Transfer," in *Creation and Transfer of Knowledge: Institutions and Incentives*, ed. B. Navaretti (Berlin: Springer, 1998); and M. Blomström and A. Kokko, "Foreign Direct Investment and Spillovers of Technology," *International Journal of Technology Management* 22 (2001): 435–53.

[19] Blomström and Kokko, "Foreign Direct Investment," 441.

[20] C. J. Dahlman and J.-E. Aubert, *China and the Knowledge Economy: Seizing the 21st Century* (Washington, D.C.: WBI Development Studies, 2001), 43*ff*.

[21] J. Wang, "In Search of Innovativeness: The Case of Zhong'guancun," in *Making Connections*, ed. E. Malecki and X. Oinas (Aldershot, U.K.: Ashgate, 1999), 207.

[22] H.-C. Lai and J. Z. Shyu, "A Comparison of Innovation Capacity at Science Parks across the Taiwan Strait: The Case of Zhangjiang High-Tech Park and Hsinchu Science-based Industrial Park," *Technovation* 25 (2005): 811.

[23] Lai and Shyu, "A Comparison of Innovation Capacity," 806.

[24] Lai and Shyu, "A Comparison of Innovation Capacity," 807–11.

[25] The models and analysis employed in this chapter are described in an appendix, available at <http://sprie.stanford.edu/publications/greater_chinas_quest_for_innovation/>.

[26] Compare Katrak, "Developing Countries' Imports."

[27] Compare M. Fritsch, "Co-operation in Regional Innovation Systems," *Regional Studies* 35 (2001): 297–307.

[28] Compare L. Anselin, A. Varga, and Z. Acs, "Geographical Spillovers and University Research: A Spatial Econometric Perspective," *Growth and Change* 31 (2000): 512.

[29] Compare R. Rothwell, "The Changing Nature of the Innovation Process: Implications for SMEs," in *New Technology-Based Firms in the 1990s*, ed. R. Oakey (London: Paul Chapman Publishing, 1994), 13.

CROSS-BORDER R&D NETWORKS AND INTERNATIONAL R&D: A STUDY OF TAIWANESE FIRMS

Meng-chun Liu and Shih-horng Chen

The internalization of research and development (R&D) is a topic of growing importance and interest. This chapter examines the effects of foreign subsidiaries' cross-border technology and R&D networks on their R&D investments abroad, using Taiwan-based multinational corporations (MNCs) as a case study. It seeks to enrich the literature in several ways. First, we look at Taiwan, whereas most studies have focused on MNCs based in advanced countries. Second, we underline the roles played by both local R&D networks and the technology linkages within MNCs' R&D investments in the host countries. We measure Taiwan-based MNCs' internal and external innovation networks using the perspectives of Zander and Kuemmerle.[1] Third, abundant human capital and indigenous technologies have been emphasized as advantages of hosts in motivating MNC investments, but little attention has been paid to MNC strategies for R&D. Finally, we test Dunning's eclectic framework of the behavior of MNCs' R&D investment abroad, and in particular, his OLI (ownership advantages, location-specific advantages, and internalization advantages) framework.[2]

Drawing on two government data banks for 2004, the Survey on Foreign Direct Investments by Taiwan-based MNCs and the Survey on Overseas Investment to China by Taiwan-based MNCs, both conducted by the Investment Commission of the Ministry of Economic Affairs (MOEA), we take a quantitative approach toward Taiwan-based MNCs' R&D overseas.

MNCs' Technology Linkages and Innovation Networks

Several studies have focused on the location of offshore R&D,[3] while other studies have explored their motivations,[4] and yet others have addressed the management and organization of offshore R&D.[5]

Some previous work has sought to figure out the internationalization strategies of firms' innovation networks[6] in terms of their technology application overseas and technology sourcing[7] by using patent databases[8] or the locations of their offshore R&D centers.[9] Few have directly mapped patterns of overseas R&D networks in terms of their R&D partners in the host countries.

To figure out the patterns of cross-border R&D networks, we take various R&D partners of foreign subsidiaries into account. This is because some partners can share detailed information about the competition and others may facilitate cross-border technology transfers. We use the international innovation network framework of Patel and Vega[10] to explore the technology strategy of the MNCs' overseas subsidiaries at the industry level.

Studies such as those by Kuemmerle, Zander, Patel and Vega, and C. L. Bas and Sierra[11] have shed light on the overseas R&D strategies of MNCs, but they mainly address the experiences of developed countries. Few studies, apart from the 2004 study by Chen,[12] have explored the experiences of recently developed countries. Following Patel and Pavitt, Patel and Vega, and Zander,[13] we have examined the international strategies of multinationals in terms of their foreign subsidiaries' R&D networks. As argued in previous studies, multinationals can be a nexus of knowledge flows across the home and host countries.[14] Generally speaking, the MNCs' international R&D investments are motivated by helping their local manufacturing activities, connecting to regional production networks, or modifying products to serve the local market. Multinationals also organize R&D networks with firms, research institutes, or universities in the host countries.

In the last two decades, foreign direct investment (FDI) has been extended from overseas production and marketing to promote cross-border innovations. Since the 1990s, MNCs' overseas R&D has also been diffused to the developing world for at least three reasons. First, developing countries aggressively seek these investments. Second, the shortage of skilled labor in the home countries is a supply-side driving force. Third, both delinkages in value-added chains and great improvements in global telecommunications enable MNCs to extend their innovation networks beyond national borders.

An important question concerns the cross-border technology linkages of MNCs. Overseas R&D sites perform two main tasks: using knowledge generated by competitors and research institutions in the host country, and supporting offshore manufacturing or marketing activities. These are home-based-augmenting and home-based-exploiting, respectively.

One mission of overseas laboratories is to transfer host countries' knowledge to headquarters.[15] In contrast, as home-based technology exploiters, MNCs usually put offshore R&D units close to large markets or manufacturing sites. This mission entails transferring technologies from headquarters, or modifying the manufacturing process or adjusting products in order to meet host-country conditions.

The home-based-technology-augmenting and home-based-technology-exploiting tasks can be mixed in proportions. Zander[16] outlines technology linkages between a parent and its foreign affiliates in two dimensions: international duplication and international diversification. Duplication refers to the similarity between technologies adopted by the foreign R&D subsidiaries and those used by the parent. International diversification refers to affiliates skilled in technology fields that are different from the parent.

Previous work argues that knowledge specialization depends on three host-country attributes: knowledge endowment, the attraction of local talent, and support policies. Host countries with abundant talent support the mandate of local R&D units to participate in global innovation networks. Such an upgrading of foreign subsidiaries highlights the importance of learning from foreign clients, local suppliers, competitors, and foreign research centers, and attracting talent globally.

Cross-border R&D Networks of Taiwan-Based MNCs

As a socioeconomic phenomenon, innovation takes place when organizations interact and cooperate. In performing R&D investment in a host country, foreign firms clearly intend to cooperate with local enterprises and other organizations. In recent years, Taiwanese firms have started to conduct R&D overseas. In this section, we present the historical development of Taiwanese firms' FDI and overseas R&D, and further outline their R&D networks.

Taiwan's Outward FDI and Overseas R&D

Taiwanese firms have been making FDIs since the late 1970s. Increases in labor costs and currency exchange rates gradually eroded their export competitiveness and motivated them to venture overseas in search of cheap labor. Southeast Asia was the main destination, and small manufacturing firms, especially in low-tech industries, were in the lead. Later the government abolished martial law and in 1987 lifted the ban on kinship visits to China, prompting the beginning of Taiwanese FDI in China. Currently, FDI in China accounts for 70 percent of Taiwan's global FDI.

The history has three periods: the late 1980s, the 1990s, and the period since 2000. In the early period, labor-intensive small- and medium-sized enterprises (SMEs) played the main role, with the Pearl River Delta being the primary region. These FDI firms were mainly in light industries, such as apparel, umbrellas, and footwear. Taiwanese information technology (IT) firms began the second stage by aligning their PC assembly with their branded customers' global strategies. Among the customers were Dell, HP/Compaq, and IBM. From the perspective of global production networks, as championed by Ernst,[17] flagship firms generally conduct such FDI. Following the relocation of PC production, Taiwanese IT original design manufacturing (ODM) and original equipment manufacturing (OEM) firms began assembling notebook PCs in China, in spite of a prohibition by the government. Although that prohibition continued until November 2001, by 2005 all of Taiwan's notebook PCs were assembled in China.

In the third period, Taiwanese FDI in China changed. The Yangtze River Delta replaced the Pearl River Delta as the primary FDI destination. And Taiwanese FDI became larger in scale, capital- and technology-intensive, and oriented toward the local market. For example, the foundry service business began in 2003. These Taiwanese investments and foundry activities gave China several advantages in assembling final goods (computers) and intermediate inputs (integrated circuit, or IC, chips). First of all, the world's top foundry service provider, TSMC, set up an 8-inch fab in Shanghai. This may be because the Shanghai government perceives that foundry services stimulate the growth of IC design houses. Taiwan foundry service providers cannot ignore this emerging market. Second, the growth—and huge potential scale—of the domestic market could be an advantage in building up Taiwan's own foundries, such as SMIC. Since 2004 SMIC has become the world's third-largest provider, overtaking Chartered (Singapore) and behind only TSMC and UMC, although it has not performed well financially. If TSMC fails to compete with SMIC in China's domestic market, it may be at risk.

Taiwan-based Firms' R&D in 2004

According to registrations with the Investment Commission of the MOEA, in 2004 there were 2,662 reported Taiwan-based FDI cases, which involved over NT$10 billion total. Of these, 3,875 were headed for China, which amounted to NT$76.99 billion. The industries were mainly manufacturing, with the top five in 2002–2004 being electronics, basic metals, chemicals, precision machinery, and nonmetals. In contrast, Taiwan-based FDI to other parts of the world were in service, or 50 percent of project values.

Why Do Taiwanese Companies Move Their R&D Offshore?

Table 9.1 shows recent motivations for Taiwan-based companies moving their R&D offshore. China, unsurprisingly, looms large in cutting costs and also, somewhat surprisingly to some people, in product development.

Table 9.1 Motivations of Taiwan-Based MNCs' Offshore R&D, 2004 (%)

Motivation	Outside China	China
Product development	62.50	59.02
Access to new markets	48.39	59.33
Business diversification	14.11	19.27
Production cost reductions	26.21	57.19
Catching up to rivals' technological capabilities	18.55	29.66
Other motivations	10.89	4.28

Source: Adapted from the Investment Commission of the MOEA (Taipei, Taiwan, 2005).

Table 9.1 shows that the market and production orientation outweighs the technology orientation. This finding in the case of a newly industrialized country (NIE) such as Taiwan, coincides with a previous survey by Kuemmerle[18] and other studies, in which home-based-augmenting FDI is an important part of overall FDI in R&D. Most of the overseas R&D by Taiwanese MNCs targets overseas market demand. Such R&D investments can be regarded as demand-side-driven innovations. In addition, a considerable proportion of the local R&D cooperation partners of Taiwanese overseas subsidiaries are upstream material and component suppliers. It can thus be expected that such cooperation works to promote production efficiency and reduce production costs.

Measures of Technology Linkages and Offshore R&D Networks

Earlier, headquarters mainly directed overseas subsidiaries' R&D, but the expansion of production activities abroad and the growth of overseas markets eventually led to more local autonomy.

Differing from previous studies, this study measures the cross-border technology linkages of overseas subsidiaries in relation to their main technology sources. We have used the approach suggested by Bartholomew[19] to do a factor analysis to extract the factor scores.

Technology Linkages in Host Countries

Using the data bank of the Investment Commission of the MOEA, we show the main technological sources of Taiwan-based overseas subsidiaries. Eight technology sources are identified in this survey, including (1) technology transfers from the parent companies, (2) in-house R&D conducted by an affiliate in Taiwan, (3) buying technology abroad, (4) general partnerships, (5) licensing by research institutes in their home country, (6) licensing by research institutes in Taiwan, (7) OEM and ODM, and (8) other sources. Table 9.2 summarizes various technology sources based on 1,146 available observations.

Table 9.2 shows also that technology transfer from their parent companies accounts for 57 percent of FDI firms outside China and for 84 percent of FDI in China. In-house R&D accounts for 40 percent and 25 percent, respectively.

R&D Partners in Host Countries

Rapid changes in global markets and technologies hinder the ability of many firms to convert their inventions into profitable innovations. The innovation process is not a linear system in which each unit of an organization individually and internally does R&D, manufacturing, and marketing. By contrast, a networked innovation strategy engages customers and suppliers as innovation partners to source the best outside talent and ideas.

Table 9.2 Technology Sources of Taiwan-Based Overseas Subsidiaries, 2004 (%)

Host countries	Outside China	China
Technology transfers from parent companies	57.30	84.28
In-house R&D by overseas subsidiaries in host countries	39.55	24.83
Overseas technology purchasing	4.94	1.38
Joint venture firms	4.04	5.24
Technology licensing by research institutes in host countries	3.60	5.24
Technology licensing by research institutes in Taiwan	6.52	1.10
OEM, ODM	2.02	3.72
Other sources	9.89	6.90

Source: Adapted from the Investment Commission of the MOEA (Taipei, Taiwan, 2005).

Table 9.3 shows that Taiwanese innovation networks differ across various countries. The top three R&D network partners of Taiwan-based MNCs in China are their clients (45.8 percent), their suppliers (24.36 percent), and production networking partners (7.64 percent). Similarly, most Taiwan-based MNCs in regions other than China prefer to engage their clients as R&D partners (32.02 percent), followed by their upstream suppliers (17.73 percent) and technology institutes (13.79 percent).

Table 9.3 suggests that the main motivation for offshore R&D investments is to expedite new product development for local markets. Some partner with upstream suppliers, perhaps to reduce their production costs by developing new production methods and adopting local materials.

We can highlight the overseas R&D strategies of Taiwan-based companies by contrasting them with the strategies of advanced and developing countries. In advanced countries, local R&D institutions and technology consultants have much higher shares of the offshore R&D partnerships of Taiwan's companies, by 23 percent and 12 percent, respectively. Taiwan-based companies' R&D in advanced countries has access to abundant local innovation resources for augmenting their own capabilities. As shown in Table 9.3, there are six main offshore R&D partners of Taiwan-based MNCs, including their clients, upstream material suppliers, production network partners, local technology consultants, R&D institutes, and local universities.

An exploration of cross-border technology linkages leads to several observations. First, with the advent of R&D internationalization, innovation activities not only are kept within the firms themselves, but also spill over into their host countries. In the host countries, foreign subsidiaries also organize local R&D networks. Their R&D partners may include material suppliers, clients, and local research institutes and universities.

Table 9.3 Offshore R&D Partners of Taiwan-Based Firms, 2004 (%)

Type	Outside China	China
Clients	32.02	45.82
Suppliers	17.73	24.36
Production networking partners	6.40	7.64
Technology consultants	3.94	1.09
Technology institutes	13.79	6.55
Higher education units	4.43	2.55
Other partners	14.29	10.55

Source: Adapted from the Investment Commission of the MOEA (Taipei, Taiwan, 2005).

Figure 9.1 Establishment of Cross-border Technology Linkages

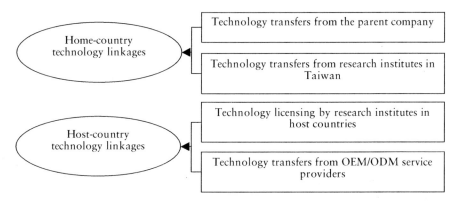

Second, we find that Taiwan-based MNCs with more host-country technology linkages tend to be more R&D–intensive. In contrast, MNCs that depend on the host country's technology have only a weak incentive to make R&D investments there. This may be because they find it difficult to build ties with local technology communities. Moreover, from the perspective of industrial cluster theories, well-developed industrial clusters have advantages in terms of the localization of skills, specialized materials and inputs, and technological know-how. These advantages increase the MNCs' incentives to do R&D in those communities.

Third, the evidence shows that the ties MNCs have with their respective local scientific and technical communities—as illustrated in Figure 9.1—determine offshore R&D more than ties with upstream suppliers or clients. This empirical result is very reasonable. Given the technological capabilities of MNCs, firms have the incentives or the ability to shift their innovation activities when they link up with local universities and R&D institutes. In terms of policy implications, developing countries that are hosting may to some extent regard their universities and R&D institutes as affecting MNCs' R&D investments.

Conclusion

International economic development has changed the value chains of MNCs. R&D internationalization is no longer confined to MNCs based in developed countries. Few studies have examined whether the internalization advantage of MNCs makes it easier for them to do R&D in host countries. Because Taiwan, the home country here, is not characterized by technological leadership, its offshore R&D might seem to give it an advantage in acquiring technology. This chapter goes a step further than previous research by jointly examining the ownership, location, and integration advantages at a corporate level, which appears to provide more insights.

In generating the measures that capture the internalization advantage of MNCs, this study used two data sets maintained by Taiwan's Investment Commission and employs a common factor analysis approach with tetrachoric correlations to measure four latent variables, which serve as proxies for MNCs' technology linkages and R&D networks. They enable us to determine the magnitudes of the MNCs' ties with R&D partners in both host and home countries. This study further empirically examined the effects of MNCs' cross-border technology linkages and R&D networks on their R&D investments in the host countries.

Our empirical results show that Taiwan-based MNCs with higher R&D intensity are more export-oriented and have stronger technological capabilities as well as a higher marketing mandate. The host countries of such MNCs are predominantly in the advanced world. This evidence is in line with the firm-specific assets and location-specific advantages of R&D internationalization.

The internalization advantages of MNCs are confirmed in terms of driving their R&D investments in the host countries. MNCs with higher host-country technology linkages and local R&D networks do more R&D investment than their counterparts with higher home-country technology linkages. This implies that firms that are able to access local innovation systems in host countries can more successfully exploit or augment their home-based technologies. By contrast, those with stronger home-country technology linkages use knowledge from the parent company and have a lower mandate in R&D internationalization.

In the last decade, attracting R&D investments by MNCs has become part of industrial technology policies in many developing countries. These policies are aimed mainly at MNCs with high-technology profiles and take the form of financial support and tax credits. But just as important as financial policies are policies that focus on MNCs' ties with local innovation networks in the host countries. Promoting scientific/technological cooperation and the formation of industrial agglomerations can motivate MNCs to become a part of the host countries' innovation networks by doing more R&D in those countries.

Notes

[1] I. Zander, "How Do You Mean 'Global'? An Empirical Investigation of Innovation Networks in the Multinational Corporation," *Research Policy* 28, nos. 2/3 (1999): 195–213; and W. Kuemmerle, "Foreign Direct Investment in Industrial Research in the Pharmaceutical and Electronics Industries: Results from a Survey of Multinational Firms," *Research Policy* 28, no. 1 (1999): 179–93.

[2] J. H. Dunning, "The R&D Activities of Foreign Firms in the United States," *International Studies of Management and Organization* 25, nos. 1/2 (1995): 39–73.

[3] G. Fors and M. Zejan, "Overseas R&D by Multinationals in Foreign Centers of Excellence" (working paper no. 111, The Economic Research Institute, Estocolmo, Stockholm School of Economics, 1996); T. Gao, "Multinational Activity and Country Characteristics in OECD Countries," manuscript, 2000; and M. Cornet and M. Rensman, "The Location of R&D in the Netherlands: Trends, Determinants and Policy" (CPB Netherlands Bureau for Economic Policy Analysis, 2001).

[4] E. D. Westney, "Organizational Change and the Internationalization of R&D," in *Transforming Organization*, ed. T. A. Kochan and M. Useem (New York: Oxford University Press, 1992), 245–60; M. Paoli and S. Guercini, "R&D Internationalisation in the Strategic Behaviour of the Firm" (Steep Discussion Paper no. 39, Science Policy Research Unit, University of Sussex, 1997); and R. Narula, "Explaining 'Inertia' in R&D Internationalization: Norwegian Firms and the Role of Home-Country Effects" (working paper, Centre for Technology, Innovation and Culture, University of Oslo, 2000).

[5] E. D. Westney, "Internal and External Linkages in the MNC: The Case of R&D Subsidiaries in Japan," in *Managing the Global Firm*, ed. C. Bartlett, Y. Doz, and G. Hedlund (London and New York: Routledge, 1990); and K. Asakawa, "Organizational Tension in International R&D Management: The Case of Japanese Firms," *Research Policy* 30 (2001): 735–57.

[6] Zander, "How Do You Mean 'Global'?"; and Kuemmerle, "Foreign Direct Investment in Industrial Research."

[7] R. Voelker and R. Stead, "New Technologies and International Locational Choice for Research and Development Units: Evidence from Europe," *Technology Analysis and Strategic Management* 11, no. 2 (1999): 199–209; and N. Kumar, "Determinants of Location of Overseas R&D Activity of Multinational Enterprises: The Case of U.S. and Japanese Corporations," *Research Policy* 30, no. 1 (2001): 159–74.

[8] Zander, "How Do You Mean 'Global'?"; and C. L. Bas and C. Sierra, "Location versus Home Country Advantages in R&D Activities: Some Further Results on Multinationals' Locational Strategies," *Research Policy* 31, no. 4 (2002): 589–609.

[9] Kuemmerle, "Foreign Direct Investment in Industrial Research."

[10] P. Patel and M. Vega, "Patterns of Internationalization of Corporate Technology: Location vs. Home Country Advantages," *Research Policy* 28, nos. 2/3 (1999): 145–55.

[11] W. Kuemmerle, "Building Effective R&D Capabilities Abroad," *Harvard Business Review* 75, no. 2 (1997): 61–70; Kuemmerle, "Foreign Direct Investment in Industrial Research;" Zander, "How Do You Mean 'Global'?"; Patel and Vega, "Patterns of Internationalization;" and Bas and Sierra, "Location versus Home Country Advantages."

[12] S.-H. Chen, "Taiwanese IT Firms' Offshore R&D in China and the Connection with the Global Innovation Network," *Research Policy* 33, no. 2 (2004): 337–49.

[13] P. Patel and K. Pavitt, "National Systems of Innovation under Strain: The Internationalisation of Corporate R&D" (SPRU Electronic Working Papers Series, 22, Science Policy Research Unit, University of Sussex, 1998); Patel and Vega, "Patterns of Internationalization;" and Zander, "How Do You Mean 'Global'?"

[14] A. K. Gupta and V. Govindarajan, "Knowledge Flow Patterns, Subsidiary Strategic Roles, and Strategic Control within MNCs" (Academy of Management Proceedings, 1991), 21–25.

[15] P. Almeida and B. Kogut, "Localization of Knowledge and the Mobility of Engineers in Regional Networks," *Management* 45, no. 7 (1999): 905–17; A. B. Jaffe, M. Trajtenberg, and R. Henderson, "Geographic Localization of Knowledge Spillovers as Evidenced by Patent Citations," *Quarterly Journal of Economics* 108, no. 3 (August 1993): 577–98; A. Saxenian, *Regional Advantage: Culture and Competition in Silicon Valley and Route 128* (Cambridge, MA: Harvard University Press, 1994); and S. Feinberg and K. Gupta, "Knowledge Spillovers and the Assignment of R&D Responsibilities to Foreign Subsidiaries," *Strategic Management Journal* 25, nos. 8/9 (2004): 823–45.

[16] Zander, "How Do You Mean 'Global'?"

[17] D. Ernst, "Global Production Networks and the Changing Geography of Innovation System: Implications for Development Countries" (East-West Center Economics Series Working Paper no. 9, East-West Center, Honolulu, HI, 2002).

[18] Kuemmerle, "Foreign Direct Investment in Industrial Research."

[19] See D. Bartholomew, *Latent Variable Models and Factor Analysis* (London: Charles Griffin & Company Limited, 1987).

Talent and Innovative Capacity

China's Emerging Science and Technology Talent Pool: A Quantitative and Qualitative Assessment

Denis Fred Simon and Cong Cao[*]

China's increased prominence in science and technology (S&T) is generating a growing need for a deeper understanding of its talent pool. This chapter analyzes its talent in science and engineering.

There are conflicting stories about China's strategic development and uses of high-level talent. On the one hand, spending on research and development (R&D) and on education has been growing rapidly. The number of scientists, engineers, and other trained professionals has been steadily increasing, and the education pipeline in recent years is filled with literally millions of students entering colleges and universities. On the other hand, everyone—from the Chinese government and local employers to Western and Asian multinational corporations (MNCs)—complains that qualified talent is difficult to find and thus is getting more expensive.

In fact, there is growing evidence that China faces a serious talent challenge. Active professionals are relatively young, even among those holding senior positions, and are often fresh out of school. According to McKinsey, only 10 percent of professionals with up to seven years of experience are capable of working for MNCs.[1] The shortage is most serious among forty-five- to fifty-five-year-olds, especially at the high end, as a consequence of the lingering effects of the Cultural Revolution. This situation has been worsened by the so-called brain drain that kicked in after China opened to the outside world in the early 1980s. Many Chinese people, most likely the best and the brightest, who were sent abroad for advanced education have not returned, although increasing numbers have begun to straddle the Pacific Ocean, keeping a foot

---

[*] The research on which this paper is based was partially supported by a grant from IBM. We also thank Bojan Angelou, Bilguun Ginjbaata, and Howard Harrington, who helped analyze the data. We appreciate as well our many colleagues in China who have provided guidance, access to sources of data, and their insights. The research is a part of the Levin Institute's project, Global Talent Index™. For an in-depth analysis of China's talent in science and technology, please refer to the authors' book, *China's Emerging Technological Edge: Assessing the Role of High-End Talent* (forthcoming).

planted in China's S&T community while pursuing careers in the West or in Japan. In fields with a surplus of talent, there tends to be a significant mismatch between students' preparation and the jobs they are asked to perform. This situation is worsened by the maldistribution both at senior and middle levels and regionally: while the central government tries to foster development in western China and the northeast, the talent continues to move east to cities such as Beijing, Shanghai, and Shenzhen.

With perceived urgency, the Chinese Communist Party (CCP) and the government have become much more determined to address talent problems. The leadership views educated people, along with science, as the key to building a harmonious and well-off society and to solving the nation's problems in such areas as the environment, the need for energy, urban-rural and regional development disparity, social inequality, the aging population, and national security. The leaders have stepped up efforts through the Organization Department of the CCP Central Committee, as well as through governments at various levels, to train, attract, retain, and better utilize skilled people. Most important, they seek to create an environment conducive to achieving the goal of "strengthening the nation with talent" (*rencai qiangguo*) as the key to transforming China into an innovative society by 2020. Given its checkered history of treatment of intellectuals since 1949, this is a major advance, one that could reduce the hesitancy that some scientists and engineers have about returning home. In May 2002, the General Office of the CCP Central Committee and the General Office of the State Council formulated an outline for building China's skill pool. In December 2003, the CCP held its first-ever conference on talent, calling for the creation of a more skilled professional labor force. While these measures followed the tradition that the party rules the cadres (professionals being part of the cadre strata), most significantly, Hu Jintao, president of the People's Republic of China (PRC) and general secretary of the CCP Central Committee, indicated in the conference that the talent issue is on a strategic level in the party and is intimately tied to China's long-term national development.[2] This meeting reinforced a set of changes on the role and status of intellectuals that had begun with the reform program and open policy by Deng Xiaoping in the early 1980s. The main difference was the new emphasis placed on the centrality of China's high-end talent pool within the core goals of the government and the party.

To that end, the CCP Central Committee and the State Council soon issued a circular on further strengthening the talent, while the Party Central Committee's Department of Organization and the government's Ministry of Personnel started to survey the talent situation. In the meantime, as the Medium- and Long-Term Plan for the Science & Technology Development (2006–2020) was being formulated, the skills issue moved to front and center. In proposing to leapfrog and become a so-called innovative nation, the final plan suggested that China should focus on raising the talent's innovative capabilities and achieving better use of existing skills while maintaining a suitable quantitative growth rate.

Science & Technology Talent in China

The precise meaning of the word *talent* is as difficult to define in China as elsewhere.[3] Operationally, four categories have been identified as the core components.[4] The first is S&T human resources, the total number of those with at least two years of college education in an S&T discipline, plus those in the workforce with equivalent professional experience. (This is the same definition used by the Organisation for Economic Co-operation and Development [OECD] in its *Canberra Manual*.[5])

The second category is basically defined as "professionals." It covers those working in seventeen professions, including engineering, agriculture, scientific research, public health, and education. This category as formulated by the Party Central Committee's Department of Organization and the government's Ministry of Personnel for the purpose of managing professionals, and includes only those employed in state-owned institutions and enterprises.

People who spend at least 10 percent of their time in S&T activities fall into the third category, S&T personnel. This counts scientists at independent R&D institutes and universities, engineers working inside enterprise R&D labs, scientists and engineers, and inside institutions dealing with S&T information, graduate students at their thesis or final project stages, S&T administrators, and those who provide services to S&T organizations. (This definition differs from the definition in the *Canberra Manual*.)

The fourth category is R&D personnel who are engaged in conducting, administering, and supporting actual R&D activities. They are counted by their full-time equivalent (FTE)—that is, person-years. (This definition is similar to the definition in the *Canberra Manual*.)

Within the third and fourth categories are scientists and engineers who either hold professional job titles at the middle level or above, or are graduates of four-year colleges and above even if they do not yet have professional standing. Thus, there are scientists and engineers engaged in S&T activities and scientists and engineers engaged in R&D activities, respectively, which are both subsets of total S&T personnel and R&D personnel. This chapter addresses only two subsets: (1) scientists and engineers involved in S&T activities and (2) scientists and engineers involved in R&D activities.

A Stock-Taking of the Science and Engineering Talent Base

Since 1990, China has witnessed a steady increase in its S&T human resources (Table 10.1). In 2006, the latest year for which such statistics are available, China had 4.13 million persons engaged in S&T activities, of whom 2.80 million were scientists and engineers. In that year, China devoted 1.50 million person-years to R&D, with scientists and engineers accounting for 1.22 million person-years, which made China second only to the United States.

Table 10.1 China's Human Resources in Science and Technology
(1,000 persons; 1,000 person-years)

	1999	2000	2001	2002	2003	2004	2005	2006
Human resources in S&T	N.A.	31,500	34,000	38,000	42,000	48,000	54,000	N.A.
Professionals	29,043	28,874	28,477	28,344	27,746	27,504	27,567	27,739
Scientific personnel	2,905	3,224	3,140	3,222	3,284	3,481	3,815	4,132
Scientists and engineers in scientific activities	1,595	2,046	2,072	2,172	2,255	2,252	2,561	2,798
R&D personnel	822	922	957	1,035	1,091	1,153	1,365	1,502
Scientists and engineers in R&D activities	531	695	743	811	862	926	1,119	1,224

Sources: National Bureau of Statistics and Ministry of Science and Technology, *China Statistical Yearbook on Science and Technology* (Beijing: China Statistics Press, various years); Ministry of Science and Technology, *China Science and Technology Indicators 2006* (Beijing: China Scientific and Technical documents Publishing House, 2007).

Characteristics

Chinese scientists and engineers, as do their counterparts elsewhere, work at universities (15.3 percent in 2006), research institutes (11.8 percent), and enterprises (72.9 percent). The distribution of R&D personnel in 2006 was 16.2 percent at universities, 15.5 percent at research institutes, and 65.8 percent at enterprises (2.7 percent in other organizations). The relatively high distribution of Chinese scientists and engineers and R&D personnel in enterprises might suggest that most research happens inside the companies. However, this fact also reflects the impact of reforms, especially the conversion of many government R&D institutes into commercial enterprises.

Geographically, scientists and engineers are heavily concentrated in eastern China, including Beijing, Tianjin, Hebei, Liaoning, Shanghai, Jiangsu, Zhejiang, Fujian, Shandong, Guangdong, Guangxi, and Hainan. These accounted for 61.8 percent of scientists and engineers and 60.8 percent of scientists and engineers doing R&D in 2006. They are further concentrated in seven provinces or municipalities: Beijing, Guangdong, Jiangsu, Zhejiang, Shandong, Shanghai, and Liaoning. Sichuan in the west, and Hebei and Hainan in central China, are among the top ten regions with the most S&T human resources.

Although no data are available on the age structure of S&T talent, clearly this cohort is young. This can be deduced simply by considering science- and engineering-type graduates in recent years. Of those with at least a bachelor's degree in science, engineering, agriculture, and medicine who were working at civilian R&D institutions in 2004, 71.6 percent were under forty-five years old, and 55 percent of the total scientists and engineers with senior titles were under forty-five years old. Those between the ages of thirty-five and forty-nine constituted 51.4 percent of the overall total in 2004, compared with 44.5 percent in 1999.[6] This suggests that many persons in senior positions have less experience than their counterparts in the West.

Talent in the Pipeline

Since 1999 China has witnessed an exponential explosion of enrollment in higher education (Table 10.2).[7] Undergraduate enrollment reached 25 million in 2006, including 18 million in the regular institutions of higher education and 7 million in programs of adult education, TV, and broadcast education, and so on. About one-half of the enrolled undergraduate students were at the baccalaureate level.

In particular, undergraduate engineering majors increased by about three times between 1999 and 2006, keeping pace with the overall growth in total enrollments. There are also considerably more engineering than science graduates at all levels, from undergraduate to doctorate. In 2006 China enrolled 1.1 million students in master's and doctoral programs, with over 60 percent in science, engineering, medicine, and agriculture.[8]

In 2006 China graduated 1.7 million students at the bachelor's level and produced 256,000 masters and PhDs, of which 36,200 received doctorates; 70 percent were in science, engineering, agriculture, and medicine. These students are largely found not only in universities but also in research institutions such as the Chinese Academy of Sciences (CAS).

Table 10.2 Total Higher Education Enrollment in China (1,000 persons)

Year	Undergraduate*	Graduate
1991	2,044	88
1992	2,184	94
1993	2,536	107
1994	2,799	128
1995	2,906	145
1996	3,021	163
1997	3,174	176
1998	3,409	199
1999	4,086	234
2000	5,561	301
2001	7,191	393
2002	9,034	501
2003	11,086	651
2004	13,335	820
2005	15,618	979
2006	17,388	1,105

Source: Department of Development and Planning, Ministry of Education, *Educational Statistical Yearbook of China* (Beijing: People's Education Press, various years).
Note. * Enrollment in regular institutions of higher education only.

The Demand and Supply of Scientists and Engineers

Demand and Its Drivers

According to a regression underlying this research, the demand for scientists and engineers in China is driven by four main factors: (1) increasing integration into the world economy, which among other things makes foreign investment more accessible; (2) the growth of indigenous S&T and R&D activities; (3) increasing technological sophistication of the economy and society; and (4) the rising level of Chinese participation in the global value chain as reflected in the expansion of high-technology exports by Chinese firms and MNCs. As the economy grows, the demand for scientists and engineers will steadily increase. China is on an explosive trajectory toward becoming a knowledge-intensive information society; already it has the world's largest number of mobile subscribers, 547 million as of the end of 2007, and in mid-2008 replaced the United States as the country with the most Internet users (the number was 253 million). Increasing household PC ownership and telecommunications infrastructure will give the information and communication technology (ICT) industry further room to grow. The coming launch of the third-generation (3G) mobile service will be another stimulant for foreign direct investment (FDI) and China's ICT industry. Informatization is embedded in the changing composition of FDI and the growth of China's high-tech industries.

There is likely to be a concomitant rise in the demand for more S&T talent, a demand also likely to be heightened by the increased emphasis on indigenous technological development. China's leaders say they want the country to be a so-called innovative nation by the year 2020. Since 1999 it has increased spending on S&T, R&D, and education. At 1.49 percent of gross domestic product (GDP) in 2007, China's gross expenditure on R&D (GERD) was below that of most developed economies (2.7 to 3.0 percent) and short of the goal set for 2005 (1.5 percent). Nonetheless, it has the highest S&T spending level among economies that are at a similar level of development, and China is on its way to achieving its goal of spending 2.0 percent of its GDP on R&D by 2010. Most impressive is the fact that much of the recent R&D growth is in companies. The contribution from them—more than two-thirds in 2007—has reached the average Organisation for Economic Co-operation and Development (OECD) level. Similarly, China has increased its education spending over the last decade and will continue to do so. It has much room to grow in this regard. Its education spending as a percentage of public expenditure, at 2.82 percent in 2005, is among the lowest in the world (compared to 4.1 percent for India and 4.3 percent for Brazil in 2002–2003).

An important development is the upgrading of universities, which are becoming more central in the R&D system instead of serving primarily as teaching institutions. New facilities have been built, many with advanced equipment and research space. The principal shortage is qualified faculty.

The growing technological sophistication of the Chinese economy and society has also been reflected in the increasing share of high-tech exports in trade. In 2006, they were $281.5 billion, an almost 100-fold increase over the level of high-tech exports in 1991. This sector is also moving up the value chain. Finally, on average, workers in the ICT sector are among the best paid in China.

Because of the increasing mobility of talent globally, and in response to market forces, wages in advanced sectors have been steadily converging on international levels. Headhunters in Beijing and Shanghai report that firms such as Huawei, Datang, and Lenovo find themselves paying closer to U.S. wages to attract qualified talent to their R&D labs and engineering-design units. However, with the continued expansion of enrollments, wages may not be the most accurate barometer for understanding the supply and demand for S&T human resources in China, because the leaders do not seem prepared to retreat from their "higher education push."[9]

In summary, China's increased spending on education, research, and development will drive a growing demand for scientists and engineers, while structural changes in FDI will "supercharge" the demand. The growing informatization of society and the growth of the domestic high-end market will serve as key intervening factors—fostered and shaped by the changing face of FDI and indigenous innovation efforts. The domestic market will become a more important force in demand, especially when the supply-demand gap becomes narrower.

Supply

The future supply of scientists and engineers looks impressive, but some problems have become obvious. The growth in enrollment occurred with no guidelines in terms of specialization or niches, no consideration of demand, nor any preparation for dealing with the quality of these graduates. The economy has not yet created enough jobs for them, while this surge coincides with reforms of state-owned enterprises (SOEs)—which involve employee layoffs and the migration of rural youths to urban areas. Inevitably, a surplus of graduates has emerged. In recent years, only 79 to 89 percent of the graduates found employment, and there were poor matches in specific fields.

Demographics

The supply of scientists and engineers is tied integrally to a nation's demographics. The eighteen to twenty-two-year-old, college-bound cohort will begin to shrink after 2010, reducing the numbers of scientists and engineers entering the workforce (Figure 10.1). China probably has a window of opportunity from now until 2020 to achieve a better match between the demands of the economy and the types of people coming from the country's institutions of higher education.

Forecast

In this section we project supply-and-demand trends for China's high-end talent pool to 2010, using a linear-regression-based methodology. Along with assessing the dynamics of supply and demand for high-end talent, we look specifically at scientists and engineers in R&D activities, to gauge the technological potential and innovative capacity from a people perspective.

The Demand Forecast

In addition to the historical trends, our projections for the next five to fifteen years take into account the following information:
- Forecasts of China's economic growth by Goldman Sachs (known as the BRIC report)[10]
- Forecasts of China's population growth by the United Nations Population Division
- The Chinese government's policy initiatives in the areas of S&T, as well as in higher education, talent growth and development, high-tech development, and several other areas, such as ICT

Figure 10.1 China's Population Change by Age Group

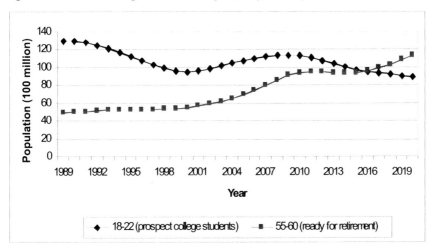

Source: Based on China's population growth projection by the United Nations Population Division (2005).

Using our historical and discontinuity models, we have come up with three scenarios—historical trends where no regressions analysis was performed, a downward trend, and an upward trend—for each of the demand models between 2005 and 2010.

- **Historical:** Values of the independent variables, forecasted using their historical trends
- **Model 1, upward:** FDI 8 percent increase and S&T expenditure (or GERD) 15 percent increase annually
- **Model 1, downward:** FDI 2 percent decrease and S&T expenditure (or GERD) 5 percent increase annually
- **Model 2, upward:** High-tech exports historical trend and R&D relative wage 10 percent increase annually
- **Model 2, downward:** High-tech exports only 10 percent increase and R&D relative wage flat after 2004

From these extrapolations, we assess a broad range of possibilities, from the lowest and highest outcomes on the demand side, and a narrower possibility, which we believe to be the most likely (Table 10.3).

Our demand forecast is this: China needed some 3.06 million scientists and engineers in 2007 and will require 3.77 million in 2010. China needed 1.7 million FTE scientists and engineers engaged in R&D activities in 2007, and this will increase to 1.90 million in 2010. This forecast is significantly higher than the forecast in the Medium- and Long-Term Plan for Science & Technology Development (2006–2020), based on data up to 2001. The model showed a need of between 848,000 to 1.75 million FTE scientists and engineers in R&D activities in 2010.[11]

Forecast of Supply of Scientists and Engineers

To forecast the supply of scientists and engineers, we have taken the following factors into consideration:

- Some of the existing workforce are retiring.
- On average, 88.7 percent of men between sixteen and sixty years of age, and 78.3 percent of women between sixteen and fifty-five years of age, are in the labor force.
- Seventy-five percent of those with at least a masters' degree have become working scientists and engineers.
- For undergraduates who are trained in science, engineering, agriculture, and medicine, the 2007 rate of placement in these fields was the average of the rates between 2001 and 2006.[12] Thereafter, the placement rate will increase 5 percent annually until 2020.

Based on this data, the model shows a supply of 2.4 million scientists and engineers in 2007 and 3.83 million in 2010. We could not forecast the supply of scientists and engineers involved in R&D activities, because they are counted only as FTE.

Science & Technology Talent Shortage

The preceding forecast implies that China's demand for scientists and engineers between 2007 and 2009 will not be met by the current supply pool (see Table 10.3). However, this gap will decrease over time, especially as current graduates gather more experience or improve the learning curve. The placement rates of undergraduates should improve over time. Supply and demand could reach equilibrium by then.

Given the present demographic shift, which is centered on 2015, China will have more people in the fifty-five-to-sixty age group than in the eighteen-to-twenty-two age group. This implies a flattening or even a decrease in S&T graduates. Together with a smaller percentage of graduates majoring in science and engineering (still high by U.S. standards), there could be a real impact on the supply side. This would be ironic in view of the huge absolute numbers entering S&T fields, past and present.

Table 10.3 Forecast of the Demand and Supply of Scientists and Engineers in China (1,000 persons)

Year	Supply	Demand				Shortage or oversupply versus most likely
		Historical trend	Most likely	High	Low	
2007	3,011	2,900	3,062	3,300	2,781	-51
2008	3,253	3,001	3,298	3,780	2,874	-46
2009	3,523	3,103	3,534	4,286	2,976	-11
2010	3,825	3,205	3,771	4,820	3,089	55

Source: Authors' projection.

An Important Unknown—The Chinese Diaspora

This diaspora is a potential source of "brain gain"; a return home of overseas Chinese and various programs targeting overseas Chinese have had some success. For example, the number of university presidents and professors in Chinese universities are overwhelmingly returnees (though mostly scholars visiting from abroad for the short term, rather than degree holders).[13] An increase in returnees would counteract to some extent the tendency for Chinese "high fliers" to go and remain overseas. It still remains to be seen whether more high-level talent will make a permanent move back home.

Quality Assessment

Science & Technology Workforce and Productivity

Of the current 2.80 million scientists and engineers, some 20 percent are estimated to have at least master's degrees. This reflects a gradual improvement in output of the S&T workforce. For example, China ranked second in publications indexed by the major international citation systems in 2006, a large increase over the past fifteen years.

Lack of High-Quality Talent

Field interviews conducted in several cities suggest that China still lacks an adequate pool of qualified talent capable of leapfrogging in S&T and education. The lack of major breakthroughs in scientific research and technological innovation is still frustrating Chinese leaders, who keep looking for signs that the technological gap with the West and Japan is decreasing, especially because of the huge emphasis placed on worldwide control of intellectual property rights (IPRs) and technical standards. This is why the government increasingly emphasizes programs that recruit overseas Chinese, as well as measures that improve the quality of higher education, especially in the elite universities.

With the damage of the Cultural Revolution fading, the impact of the brain drain has become relatively more serious. (Our forecasts do not formally account for its role.[14]) Potentially, it has large benefits, especially in the transfer of knowledge and in attracting foreign capital for high-tech endeavors and local entrepreneurship. Many Chinese abroad are connected to China's S&T system and are often a source of scientific and technical information. Chinese scientists in the United States frequently run labs in China as a way to access high-quality graduate students or to do research with grants from the Chinese government. While this has very positive aspects, it also produces conflicts and petty jealousies from those who feel that these "outsiders" receive special treatment while not having to deal with the challenges of working inside China's R&D system.

The most serious problem facing China's innovation system today is the experience gap. People fresh out of universities have trouble handling tasks that require knowledge and skills beyond formal education. Based on discussions with many MNC executives as well as Chinese R&D managers, there is an apparent mismatch between students' training and the jobs they are asked to perform. In addition, many students simply were not trained for the jobs being created. Last but not least, there is a concern about the lack of creativity and entrepreneurial behavior among recent graduates. A lack of creativity is revealed in discussions about recent graduates' risk aversion and the low level of failure tolerance inside the innovation system.

Higher Education

Key universities, with increasing funding, can now offer higher salaries and research support to highly qualified faculty members, mainly of Chinese origin, from abroad. At Tsinghua University, for example, the School of Economy and Management has recruited Chinese economics professors from such institutions as the University of California (UC), Berkeley; the Massachusetts Institute of Technology (MIT); Columbia University; and the London School of Economics, as endowed professors. Some universities have started to recruit nonethnic-Chinese scholars. As early as 2001, Tsinghua's Department of Industrial Engineering appointed Professor Gavriel Salvendy from Purdue University as its chair. Most recently, the Ministry of Education has appointed some non-ethnic-Chinese professors to its Cheung Kong Scholar Program. Now, new hires at key universities need a doctoral degree—and the competition has become fierce. On average, 10 percent of Chinese professors today have doctorates, with 20 percent at leading institutions of higher education and more than half of those at Beijing and Tsinghua universities. This low level of doctorates is one reason for the unevenness in the quality of higher education. It also helps to explain why China, despite many new graduates, still is not able to leverage this seemingly large pool of talent to achieve a higher level of innovation.

Some schools now have improved international standing. The enrollment of many Chinese in the world's leading graduate programs in the United States and Europe reflects the quality, certainly of their native abilities and perhaps of their education, especially in math and science. Between 1999 and 2003, Beijing University and the University of Science and Technology of China in Hefei were the two largest baccalaureate-origin institutions of U.S. doctorates in physical science (558 and 461 doctorate recipients, respectively), surpassing both MIT and UC Berkeley by well over 100. In engineering, for the same period, Tsinghua University was the largest baccalaureate-origin institution, with more than twice as many graduates earning U.S. doctorates than the largest U.S.-origin institution, MIT (863 versus 344).[15]

Nonetheless, China's key universities still lag behind many of their counterparts in developed countries, as well as in some Asian countries and regions, including Hong Kong and Taiwan. According to the academic ranking of world-class universities by Shanghai Jiaotong University in 2008 (based on four criteria—quality of education, quality of faculty, research output, and per capita performance), both Tsinghua and Beijing universities, China's most prestigious universities, ranked somewhere between 201 and 302.[16]

Furthermore, the huge increase in enrollments in higher education has hurt quality. As graduate schools have an incentive to accept ever-increasing numbers of students to work on research grants that bring money into universities and attract new professors, it is quite common for one professor to have ten doctoral students. Concern about the quality of graduate education is also heightened by students from leading universities going abroad for advanced study, leaving graduates from lower-status universities to fill the advanced-degree slots at the leading universities. That is, the higher the level on the educational ladder, the lower the quality of students. This could become a serious problem.

Conclusion

Our forecast of China's demand for S&T talent has captured projected shifts in economic growth—away from FDI-dependent growth and toward higher levels of informatization, and ultimately toward more indigenous efforts to drive scientific and technological advances. The demand for scientists and engineers was most likely around 3.06 million in 2007 and will increase to 3.77 million in 2010. However, in all likelihood, China supplied only 3.01 million scientists and engineers in 2007 and will be able to supply only 3.83 million in 2010. China will therefore likely face a shortage of such skills and experience, even if there is improvement in quantitative terms.

This shortage, which is likely to be larger than the numbers suggest, will largely derive from a poor fit between skills trained and those wanted in the market, along with the difficulty of getting skilled people to move to regions where they are in the greatest demand. Skilled people continue to move East, while demand in the West is not being met. While tertiary education numbers have significantly increased, quality problems could hurt indigenous innovation and technological advances.

Our forecast models expect that the demand for scientists and engineers will most likely come from new types of higher-value-added FDI, growing manufacturing-driven high-tech exports, and indigenous innovation efforts. (Higher wages will affect this.) However, a perception that the country is facing a talent crisis could signal MNCs to slow investments and thus reduce their potential contributions. Demographics show a comparatively young, although seemingly quite well educated, cohort coming. Among the 2.25 million scientists and engineers deployed in 2004, half are so-called new entrants, having graduated during the past ten years, with about 350,000 having graduate credentials. For those working in civilian R&D institutions in 2004, 71 percent were under forty-five years old. This "youth" trend is going to continue over the next fifteen years.

The young group will be outnumbered by those in the fifty-five-to-sixty-year-old category (approaching retirement) in about 2015. This shift could prove to be greatly disadvantageous. China may have started on the path to becoming a richer society, but it is more likely to become aged before it truly becomes wealthy.

Exacerbating all this is the absence of what the Chinese call a "culture of creativity." Frustration about the innovation system is growing in the face of policy measures to foster talent. This suggests that even more attention will be paid to the talent issue at the highest levels. Perhaps the best indicator of sensitivity on this topic is the announcement by the National Bureau of Statistics (NBS) that it is creating indicators to track progress toward becoming an innovative nation.

The environment in which scientists and engineers work will have a large effect. While there is greater transparency and a higher tolerance of risk and failure, many barriers remain. The greatest threat to the ambitious S&T plan may have less to do with an external brain drain and more to do with an internal brain drain. In the latter, Chinese companies and research units may find skilled people moving to new, high-value-added foreign manufacturing operations or foreign R&D centers rather than to domestic organizations because of differences in operating culture, business environment, and career potential. However, the history of Taiwan suggests that this worry may be misplaced. People, in time, move out of the foreign sector and become technological entrepreneurs and innovators. In the meantime, there will continue to be many inhibitors—cultural, political, and social—that will moderate China's capacity to capture the full potential of its talent pool.

Notes

[1] Diana Farrell and Andrew J. Grant, "China's Looming Talent Shortage," *The McKinsey Quarterly*, no. 4 (2005).

² Nevertheless, it is easier to identify the problem than to solve it. For one thing, even though the talent issue has been on top of the agenda of the Chinese leadership, there is no interministry coordinating agency to tackle the urgent talent challenge.

³ Melissa Pollak, *Counting the S&E Workforce—It's Not That Easy* (Issue Brief, NSF 99-344, Arlington, VA: Division of Science Resources Studies, National Science Foundation, May 3, 1999).

⁴ See Qian Du and Weiguo Song, "The Definition of S&T Talent and the Related Statistical Issues" (in Chinese), *China's Forum on Science and Technology*, no. 5 (2004). Scientists and engineers in enterprises where they are primarily doing production, technical sales and customer services, testing and quality control, and so on, are excluded in this chapter. There are statistical reasons for this; Chinese data is not organized to fit into our analysis without significant consistency issues.

⁵ Organisation for Economic Co-operation and Development (OECD), *The Measurement of Scientific and Technological Activities: Manual on the Measurement of Human Resources Devoted to S&T ("Canberra Manual")* (Paris: OECD, 1995).

⁶ An Overview of S&T Personnel at Independent Civilian R&D Institutes of the Natural Sciences and Technologies in 2004 (in Chinese), 2005, <http://www.sts.org.cn/tjbg/kjjg/documents/2005/051012.htm>.

⁷ Increasing enrollment at universities was one of the measures that China adopted after the 1997 Asian financial crisis to stimulate domestic consumption. It has raised the educational level of the Chinese people as a whole.

⁸ The increasing enrollment of graduate students may be due to some extent to the difficulty that college seniors have in finding jobs.

⁹ However, an increase in the enrollment of China's higher education since 1999 has to some extent hedged the coming drain of college-bound youth. See <http://edu.china.com/zh_cn/1055/20051118/12862674.html>.

¹⁰ BRIC refers to the countries Brazil, Russia, India, and China.

¹¹ Xiaoxuan Li and Jie Yu, "A Forecast on the Number of R&D Scientists and Engineers in China" (in Chinese), *Science Research Management* 25, no. 3 (2004): 124–30.

¹² Our analysis indicates that China has had a low and inconsistent record in placing undergraduates in science, engineering, agriculture, and medicine, but we expect this record to gradually improve.

¹³ Cheng Li, "Coming Home to Teach: Status and Mobility of Returnees in China's Higher Education," in *Bridging Minds Across the Pacific: U.S.-China Educational Exchange, 1978–2003*, ed. Cheng Li (Lanham, MD: Lexington Books, 2005), 69–109.

¹⁴ According to the criteria used in China's S&T statistics, those who spend 10 percent of the time in S&T–related activities and have the right credentials are supposed to be counted. But we do not know whether scientists and engineers from overseas who work part-time in China are counted.

[15] Thomas B. Hoffer, Lance Selfa, Vincent Welch, Jr., Kimberly Williams, Mary Hess, Jamie Friedman, Sergio C. Reyes, Kristy Webber, and Isabel Guzman-Barron, *Doctorate Recipients from United States Universities: Summary Report 2003* (Chicago, IL: National Opinion Research Center, 2004), 78–79.

[16] "2008 Academic Ranking of World Universities," <http://www.arwu.org/rank2008/ARWU2008_c(EN).htm>.

CAN CHINESE IT FIRMS DEVELOP INNOVATIVE CAPABILITIES WITHIN GLOBAL KNOWLEDGE NETWORKS?

Dieter Ernst*

China's opportunities to build innovative capabilities in the information technology (IT) industry differ from those faced earlier by Japan and East Asian newly industrialized economies (NIEs). China has a unique combination of advantages in a booming market for electronics products and services, the world's largest pool of low-cost and easily trainable knowledge workers,[1] the emergence of sophisticated lead users and test-bed markets, and policy efforts to strengthen China's innovation system. As a late-latecomer, China can learn from the achievements and mistakes of earlier latecomers.

The international environment is also dramatically different than it was even a few years ago. Most important is the expansion of global knowledge networks, which have extended beyond markets for goods and finance into markets for technology and knowledge workers.[2]

Taking into account these important differences, this chapter addresses two questions: Does integration into global knowledge networks facilitate the efforts of Chinese IT firms to develop innovative capabilities? If yes, precisely what type of capabilities are they developing?

The findings of this chapter can be summarized as follows:

- Integration into global knowledge networks exposes Chinese IT firms to leading-edge technology, "best-practice" management, and sources of knowledge.
- Knowledge about their own markets and production sites helps Chinese IT firms to exploit these opportunities.
- Successful Chinese firms have not attempted to compete head-on with global leaders through radical innovations. Instead, they have focused

---

* I gratefully acknowledge ideas, comments, and suggestions from Henry S. Rowen, William Miller, Naushad Forbes, Greg Shea, Duncan Clark, Richard P. Suttmeier, Barry Naughton, Ron Wilson, Xielin Liu, Xudong Gao, Lan Xue, Chintay Shih, Kane Wang, T. C. Tu, Shin-horng Chen, Eng Fong Pang, Poh Kam Wong, Eric Baark, Ted F. Tschang, Xiaobai Shen, and Adam Segal. At the East-West Center, I am grateful to Charles Morrison and Nancy Lewis for supporting this research. The Volkswagen Foundation provided generous funding. Kitty Chiu, Peter Pawlicki, and Rena Tomlinson provided excellent research assistance.

on incremental and architectural innovations that support technology-diversification strategies.
- Integration into global networks needs to be supported by a strong domestic innovation system.

These findings contradict a pessimistic literature that appraises China's innovative capabilities as weak.[3] They also contradict fears, sometimes played up for political purposes, that Chinese firms could make radical innovations that would challenge U.S. technology leadership.[4] A central proposition here is that Chinese IT firms make the most progress in areas that escape the attention of both pessimists and proponents of an emerging technology threat.

China's Integration into Global Knowledge Networks

China is far more integrated into global knowledge networks than were Japan and Korea at a similar stage of their development. *Formal corporate networks* link Chinese firms to global customers, investors, technology suppliers, and strategic partners through foreign direct investment (FDI), as well as through venture capital, private equity investment, and contract-based alliances. *Informal global social networks* link China to more developed overseas innovation systems, primarily in the United States, through the international circulation of students and knowledge workers.

The Role of Foreign Direct Investment

Since 2003 China has become the world's largest recipient of FDI, overtaking the United States. FDI is nowhere more important than in high-technology exporting. In 2005, foreign-invested enterprises (FIEs) produced 58 percent of China's total exports but 88 percent of high-technology exports. And Taiwan-owned FIEs produced 60 percent of China's exports of computers and handsets.

In addition, practically all global IT industry leaders have begun to do research and development (R&D) in China.[5] By 2004 China had become the third most important offshore R&D location after the United States and the United Kingdom, followed by India (sixth) and Singapore (ninth).[6] Much of the R&D offshoring to Asia is concentrated in the electronics industry, with China dominating R&D of hardware.

As for nonequity forms of R&D internationalization (offshore outsourcing), according to one survey, China had become the third most attractive location for future foreign R&D, behind the United States and the United Kingdom and ahead of Germany and France.

Venture Capital and Private Equity Investment

More recently, venture capital and private equity investment have added an important dimension to China's integration into corporate networks.[7] Venture capitalists in Silicon Valley now often require start-ups to present an offshore outsourcing plan as a condition for funding. This model keeps strategic management functions such as customer relations and marketing, finance, and business development in Silicon Valley, while moving product development and research work to offshore locations.[8]

An example is a start-up company in Beijing that specializes in mixed-signal chip design.[9] Chinese engineers with PhD degrees from leading U.S. universities, who have worked as senior project managers in U.S. semiconductor companies, founded the company, which has venture capital funding for developing chip designs in both China and Silicon Valley. A fully integrated design team in Beijing develops decoder chips customized for the new Chinese audio-video signal (AVS) standard. Of the more than sixty engineers in Beijing, 90 percent hold at least master's degrees. Five senior managers based in Santa Clara handle customer relations and provide design building blocks (the so-called silicon intellectual properties, or SIPs) and tool vendors for design automation, testing, and verification.

Private equity funding also plays an important role. Much of it takes place behind the scenes and is difficult to document.[10] A recent survey estimates that $1.3 trillion has been invested in global private equity.[11] While the global credit crisis, since mid-2007, has slowed down the pace of leveraged buy-out deals, private equity firms remain as active as ever in Asia. One of these firms, the Texas Pacific Group (TPG), established a strong presence in Asia through its Hong Kong–based Newbridge affiliate, well ahead of other leading players. As discussed later in this chapter, the TPG has played a key role in Lenovo's acquisition of IBM's PC division.

Informal Social Networks

Equally important is China's integration into informal networks through the international circulation of students and knowledge workers. As a result, China has become intricately linked to overseas innovation systems through a massive brain drain of its students and, more recently, through a reverse brain drain that brings graduates and overseas knowledge workers home.

In 2005 China had more than 61,000 students in American universities, more than any other country except India.[12] Associations of U.S.-based Chinese engineers and managers, such as Monte Jade, the Chinese American Semiconductor Professional Association (CASPA), and the North America Chinese Semiconductor Association (NACSA), channel information back and forth. China is also involved in U.S.-centered professional peer-group networks, including the Institute of Electrical and Electronics Engineers (IEEE) and its many specialized working groups. China's large diaspora of skilled migrants helps to diffuse complex and often tacit knowledge about technology and management. Equally important are IT "mercenaries" from Taiwan, Hong Kong, Singapore, Malaysia, and the Philippines and more recently from Japan, the United States, and Europe. Informal social networks provide much-needed experience and links with markets and financial institutions and can become an important source of reverse brain drain.

Defining Innovative Capabilities

An important challenge is to define *innovative capabilities* in a way that reflects globalizing markets for technology and knowledge workers and the resultant changes in the international innovation system.[13] Unfortunately, much of the literature on Asian innovation systems is based on empirical evidence that predates these transformations.[14] However, it does provide important insights. It demonstrates, for instance, that distinctive economic structures and institutions offer quite different possibilities for learning and innovation, and hence should affect the design of innovation strategies. Specifically, the economic structure (of a country, an industry, and a region) affects a firm's specialization (that is, its product mix) and its learning needs, as well as the breadth and depth of its capabilities, while institutions shape how things are done and how learning takes place. An important concern is the congruence of various subsystems, which is necessary for creating a *virtuous* rather than a *vicious* circle.[15]

International Knowledge Sourcing

Through integration into global production networks (GPNs), Asian firms have tapped into the world's leading markets, especially in the United States, and compensated for the initially small size of their domestic markets.[16] This integration also has provided access to leading-edge technology and best-practice management approaches. This access, in turn, has created new opportunities, pressures, and incentives for Asian network suppliers to upgrade their technical and management capabilities and worker skills.[17]

Asian economies, even the most successful ones, are constrained by weak domestic research capabilities and a narrow portfolio of homegrown intellectual property.[18] Hence, they need ways to attract R&D by global firms. In a case study of Malaysia's electronics industry, I demonstrated that attracting foreign R&D not only compensates for initial knowledge weaknesses, but also helps adaptation to abrupt changes in technology and markets. The study concludes that, under certain conditions, attracting R&D by global firms may catalyze the development and the diffusion of innovative capabilities ahead of what the market would provide.[19]

DIETER ERNST

Foreign R&D Labs

In considering the roles of foreign R&D labs, it is important to distinguish between those that are "home-base-exploiting" and those that are "home base-augmenting."[20] Home-base-exploiting R&D overseas has been around for a long time. It adapts technology from the company's home base for commercialization overseas. In contrast, home-base-augmenting overseas R&D in Asia has become considerably more important since the 1990s. It taps into new knowledge from abroad, transfers it back to the home base,[21] and combines diverse technologies in new products and processes.[22]

The following taxonomy[23] helps to clarify this topic. "Satellite" R&D labs, the least developed type, combine elements of home-base-exploiting and home-base-augmenting R&D. They are of relatively low strategic importance. "Contract" R&D labs are the pure-play version of "innovation offshore outsourcing." For them, China's role is to supply lower-cost skills and infrastructure. Dense information flows link these labs with R&D teams at headquarters and other affiliates, and knowledge exchange is tightly controlled and highly unequal. The highest level, "(more) equal partnership," is reserved for R&D labs of global firms charged with a regional or global product mandate. For them, barriers to knowledge exchange are supposed to be much lower and may eventually give way to full-fledged mutual knowledge exchange.

Although satellite and contract R&D labs continue to dominate,[24] there are examples of (more) equal partnerships, especially for the development of China's alternative standards in mobile telecommunications, open-source software, and digital consumer electronics.[25] Governments are playing key roles as promoters and funders of innovative capabilities.[26]

A Broad Definition of Innovative Capabilities

I suggest defining *innovative capabilities* to include not only technology knowledge but also complementary "soft" entrepreneurial and management innovations. Especially useful here is work on measures of firm-level innovations. Ernst, Ganiatsos, and Mytelka developed the first comprehensive taxonomy of firm-level capabilities required for production, investment, minor changes, strategic marketing, establishing interfirm linkages, and major changes.[27] Their emphasis on strategic marketing is supported by recent case studies of Lenovo and China's handset industry,[28] which highlight the roles of distribution channels and close interaction with end users as preconditions for developing innovative capabilities. Ariffin[29] documents the role of minor change capabilities and interfirm linkages, while the importance of major change capabilities is emphasized by Amsden and Tschang,[30] who classify R&D by technological complexity.

Here, I define *innovative capabilities* broadly to include the skills, knowledge, and management techniques needed to create, change, improve, and successfully commercialize products, services, equipment, processes, and business models.[31] R&D is important, but so are complementary soft capabilities. Research on successful innovations demonstrates that "the technology is the easy part to change. The difficult aspects are social, organizational, and cultural."[32] In short, in addition to R&D, I emphasize the following complementary soft innovative capabilities:

- Sense and respond to market trends before others take note (entrepreneurship).
- Recruit and retain educated and experienced knowledge workers who are the carriers of new ideas.
- Provide global knowledge sourcing for core components, reference designs, tools, inventions, and discoveries.
- Raise money required for quickly bringing an idea to the market (the litmus test of innovation).
- Deliver unique and user-friendly industrial designs (especially for fashion-intensive consumer devices, such as mobile handsets).
- Develop and adjust process management (methodologies, organization, and routines) in order to improve efficiency and time-to-market.
- Manage knowledge exchange within multidisciplinary and cross-cultural innovation projects.
- Participate in and shape global standard-setting.
- Combine protection and development of intellectual property rights (IPRs).
- Develop credible and sustainable branding strategies.

Lenovo: Leveraging Linkages with Global Industry Leaders

Lenovo shows how integration into global knowledge networks can influence the evolution of a company's business model and its development of innovative capabilities. Network integration exposes Lenovo to best-practice management approaches, intellectual tools, and leading-edge technologies. Its recent expansion into global markets has culminated in its purchase of IBM's PC division, opening the door for private equity investors. Implementing this new business model poses major challenges but also provides new opportunities for learning. As long as its focus was on the China market, informal social knowledge networks remained China-centered, but "going global" has forced Lenovo to globalize these networks.

Ownership

Lenovo is a state-owned, privately run company whose state oversight virtually disappeared since the company was listed on the Hong Kong Stock Exchange in 1994.[33] In 2003 the name Lenovo was chosen as one that could be used without restrictions in global markets.

Lenovo Group Ltd is part of the Hong Kong–listed conglomerate Legend Holdings Ltd, together with Digital China Holdings Ltd, which focuses on distribution and services, and Legend Capital, one of China's leading corporate venture capital investors.

Liu Chuanzhi, the founder of Lenovo, was influenced by the experience of his father, who after working at the Bank of China had become a lawyer and had moved to Hong Kong to become a specialist in intellectual property law. In 1984 Liu Chuanzhi, stuck in a low-paid computer scientist position, was frustrated that the institute's research had not turned into something practical. His main motivation in founding the company was to commercialize information and communications technology (ICT) research.

When Legend was founded in 1984, all the ingredients for creating a successful start-up company were missing, and there were no established role models. This meant starting small and trading in whatever could be sold. Rather than proceeding from manufacturing to sales and then to R&D, the company had to start as a trading company, a sequence typical of Hong Kong traders. It focused attention on peculiar features of the China market. After the mid-1980s, the China market expanded rapidly and by 2005 it was still growing seven times faster than the U.S. market. Despite the entry of Dell and Hewlett-Packard, Lenovo has remained the market leader for eight consecutive years, with about one-third of China's market in 2005.

Lenovo owes its success to having user-friendly products and services that address peculiar needs of China's market and are less overengineered and less expensive than products and services provided by global market leaders. The Lenovo business model combines the following:

- Familiarity with specific market characteristics and user requirements
- A superior domestic distribution network and information management
- Advanced industrial designs
- Strong brand recognition
- Reliance on a low-cost structure
- Access to well-educated and trainable knowledge workers

Learning from Global Market Leaders

Much of Lenovo's success came through linkages with global market leaders. It entered the China market as a distributor of foreign products rather than as a manufacturer.[34] In 1985 the company began to distribute computers by HP, IBM, and AKT. Especially from HP, Lenovo learned the basics of modern business management; as Feng and Elfring note: "Legend learned to be more sensitive to the market and to market trends, and it learned the value of working with established procedures."[35]

Lenovo's current chairman, Yang Yuanqing, learned much from his close association with HP, including formal decision-making procedures, the transition from a functional organization to business units, and the use of performance evaluation and incentives. Generous share options were used to recruit top university graduates and to poach aggressive young Chinese executives from leading global corporations. But the most important lesson was in the establishment of a superior distribution network and the early introduction of an efficient IT control system for inventory and accounts receivables.

Another important source of knowledge was through contact with Taiwan's IT industry, including brainstorming discussions at Acer and other leading Taiwanese PC companies that convinced Lenovo to focus on the China market. In contrast to Taiwan, where a small market has forced companies to embark on global subcontracting, the China market was big enough to sustain a brand. It was hoped that the experience gained in China could be used to develop a brand internationally.

In 2000 Legend began talks with a group of private equity investors at General Atlantic. Initially, the purpose of these talks was to get advice on how to spin off its distribution and software activities. This resulted in the 2001 listing of Digital China Holdings Ltd. on the Hong Kong Stock Exchange.[36] The chairman of General Atlantic, Steven A. Denning, also serves on the advisory board of the School of Economics and Management at Tsinghua University, and this apparently helped to establish contact with Liu Chuanzhi, who as chairman of Legend Holdings Ltd. retains control behind the scenes.

For Lenovo's founding generation, education and work experience was overwhelmingly China-centered. Among the eight most senior Chinese executives of today's Lenovo group, only one, Mary Ma (senior vice president and CFO), had studied abroad and had other international experiences. Another strength of Lenovo comes from its origin as a Chinese Academy of Sciences (CAS) spin-off; its senior managers have retained strong ties to the academic research community. The new generation of managers has been much more exposed to international networks through overseas graduate studies, travel, and work.[37]

Over time, Legend developed close relations with a set of technology suppliers. In addition to IBM and HP, this included Intel, Microsoft, Hitachi, Siemens, and Texas Instruments (TI). Legend decided to use only the best of Intel's microprocessors, and Intel selected it as its first strategic partner in China. As a result, most of the company's executives have participated in Intel's organizational training programs.

Another strength is in industrial design and materials. Legend's success in the China market owed much to its decision in 1998 to introduce a laptop (the Tianxi model) for consumers and small businesses, melding a stylish design with performance features that suited this market.[38] It was introduced after two years of work that involved some of the leading global players in materials, such as GE Plastics and Nike, and well-known design firms such as Palo Alto–based IDEO and Portland-based Ziba Design.

Going Global

In about 2004, Lenovo realized that it had to expand beyond its China base if it wanted to avoid a slowdown. Its domestic market leadership was under attack from Dell and HP, which benefited from support by Taiwanese contract manufacturing partners. This exposed two fundamental weaknesses. The first was a heavy reliance on China's price-sensitive and low-margin markets. Together with insufficient size and limited economies of scale and scope, this reliance could squeeze the company's profits. Second, this reliance on China's markets limited the funds that were available for developing new products and services and for developing global brand recognition.

With revenues of HK$22 billion (roughly US$2.9 billion) in 2005, the company was much smaller than global industry leaders. A telling indicator of its lack of exposure to global markets was that, as of the end of March 2005, only 57 of its 9,682 employees were employed outside Mainland China (mostly in Hong Kong).

Lenovo's response was to develop a distinct global brand and to expand its global market share.[39] The strategy focused on what the company saw as the weak spot of global market leaders and their Taiwanese partners; Lenovo offered less expensive products that were not overengineered, with performance features that reflected specific characteristics of the target markets.[40] This required gaining insights into how to penetrate the extremely demanding lead markets, especially in the United States.

Acquiring IBM's PC Division

In December 2004, Lenovo bought IBM's PC division—a move that shook the industry. This takeover of an American icon, albeit a highly unprofitable part of Big Blue, provided "the best laboratory in which to study a blended Chinese and Western management model."[41] Private equity investors were instrumental in bringing about this deal by providing information and helping Lenovo navigate stormy U.S. political waters. According to one source, "Liu [Chuanzhi] called General Atlantic, and asked for help determining if such an acquisition would make sense. If Lenovo were to bid and win, Lui told GA, the private equity firm would be welcomed in as an equity participant on the deal. GA agreed, conducted due diligence, and told Liu that the acquisition would, indeed, make sense."[42]

Lenovo paid $1.25 billion for IBM's PC division and assumed debt, which brought the total cost to $1.75 billion. Initially, IBM was reported to gain an 18.9 percent stake in Lenovo, but this soon changed. As part of its acquisition of IBM, Lenovo received a $350 million private equity commitment from Texas Pacific Group ($200 million); General Atlantic ($100 million); and Newbridge, TPG's Asian affiliate ($50 million). The entry of these new investors will shrink IBM's eventual share in Lenovo to 13.4 percent, down from the 18.9 percent first reported.[43] Upon full conversion of preferred shares and after share issuance to IBM, the private equity investors are expected to hold around 12.4 percent of Lenovo's capital.

As a result, private equity investors are now involved "in much of the decision-making . . . [and] . . . are treated as partners instead of like minority investors."[44] It is too early to judge how this will affect the long-term development of Lenovo's innovative capabilities.

Implementation Challenges

The challenges of implementing this acquisition are huge: to compete in a business with slim margins and fierce competition while patching together complex supply chains, sales networks, and information systems. The new Lenovo needs to retain IBM ThinkPad customers, while developing its own brand.

An equally important challenge is keeping cost leadership. The new company has four times the revenue of the old Lenovo, but six times the staff cost of the China group alone and only twice the profit. With one stroke, Lenovo's workforce more than doubled, from around 9,000 to almost 21,000. Of those 10,000 originally from IBM, 40 percent were then in China and 25 percent in the United States. It did not take long for management to respond. The company soon announced staff layoffs, mostly ex-IBM employees outside China.

Another challenge is bringing together two vastly different workforces, both in terms of demographics and behavior. The average age of Lenovo employees is 27.5 years. Many senior Lenovo managers are promoted within a few years—some even received three promotions within a year—while IBM executives are older and more experienced. In addition, IBM employees are accustomed to a management style that is highly structured (some would say "bureaucratic") and may not be easily reconciled to Lenovo's peculiar blend of authoritarian top-down decision-making and Silicon Valley–type remuneration packages.

A clash seemed unavoidable. The first CEO of the new Lenovo, an IBM veteran, left after only a few months The new one, William J. Amelio, was hired by TPG. He had worked in IBM's PC division from 1979 to 1995, but arguably his greatest attraction for Lenovo was his position with Dell since 2001, where as senior vice president for Asia-Pacific and Japan he had shaped Dell's aggressive market-penetration strategies.

Opportunities

However, this diversity of management cultures could also become a source of learning. For instance, Lenovo was eager to strengthen its position in the markets for laptops and corporate customers. In fact, only 18 percent of Lenovo's pre-acquisition revenues came from laptops, while that share for IBM's PC division was 60 percent. For IBM, corporate customers represented almost 60 percent of revenues, while 83 percent of Lenovo's revenues came from small businesses and consumers. Most important, Lenovo needed to overcome its lack of experience in overseas markets, and expected to benefit from IBM's global presence.

A greater exposure to global networks starts with the top executives. Seven out of the fifteen most senior executives of Lenovo now are non-Chinese, two educated in India, one in Canada, and one (of Greek origin) in Australia. Of particular importance is their work experience. For instance, Ravi Marwaha, senior vice president for geographies, has worked in India and Australia. Bill Matson, who as senior vice president for human resources bears responsibility for blending the merged company's diverse labor forces, has 24 years of experience with IBM in global human resources management.

Lenovo expects to gain access to a worldwide network of skilled computer sales and distribution employees who know tax laws and invoicing practices in sixty-six countries. The same is true for R&D. After the acquisition, Lenovo's R&D workforce went from 1,100 to 1,800. This represents about 9.5 percent of Lenovo's total workforce and about 18 percent of its nonmanufacturing workforce.

Noteworthy is the addition of two IBM R&D labs, in Yamato, Japan, and North Carolina. Before the acquisition, Lenovo's R&D was almost completely China-centered, with main centers in Beijing, Shanghai, and Shenzhen.

Yamato is credited with developing IBM's highly rated ThinkPad family of laptops, and Lenovo will use this experience in notebook development in addition to its work on radio frequency identification (RFID) technology. In North Carolina, Lenovo decided to invest $84 million in a new R&D campus that will house more than 2,200 employees, increasing Lenovo's North Carolina head count by about 400.[45] The new campus will focus on desktop computers, software, product definition, and quality control.

An important part of the takeover agreement is that Lenovo can use the IBM brand for five years, including the successful ThinkPad brand. IBM has promised to support Lenovo with marketing and with its IBM corporate sales force.

In addition, Lenovo can now add about fifteen hundred IBM patents to its own one thousand patents. The fact that most of the former IBM patents are registered at the United States Patent and Trademark Office (USPTO) could help Lenovo to overcome its almost exclusive reliance on patenting in China, and to gradually internationalize its patent portfolio.[46]

Immediate benefits include technical support from IBM R&D labs, access to global market intelligence and distribution know-how, and learning for further development of ThinkPad laptops. The acquisition of fifteen hundred IBM patents might also generate additional income for Lenovo. It is an open question, however, to what degree this will enable Lenovo to strengthen and upgrade its own innovative capabilities.

Assessment of Innovative Capabilities

What specific innovative capabilities have Lenovo and other Chinese IT companies developed? To address this question, I distinguish among incremental, modular, architectural, and radical innovations.[47] I use this taxonomy to examine why the more successful Chinese IT firms tend to focus on a combination of incremental and architectural innovations, why modular innovations are less frequent, and why radical innovations are limited to state-supported megaprojects.

Incremental Innovations

Incremental innovations take both the dominant components and architecture for granted but improve on cost, time-to-market, and performance. They do not require science inputs but do require skill and ingenuity, especially in what I earlier called complementary soft entrepreneurial and management capabilities.

Some examples are improvements in the organization of manufacturing, distribution, and support services, such as Dell's "direct sales" model and its integration of factory automation and supply-chain management. Other examples are new ways of subcontracting, pioneered by Taiwanese IT firms, such as original design manufacturing (ODM), foundry services (for fabrication of integrated circuits, ICs), and design implementation services.[48] They can also involve continuous improvements in industrial design.

Chinese IT firms are well placed to pursue incremental innovations across all stages of the value chain. They operate in extremely price-sensitive markets, especially in China, but also as suppliers to global industry leaders. Hence, they are under tremendous pressure to improve on cost, time-to-market, and performance. These normally can be achieved through relatively minor changes to the existing product or production process.

Barriers to such improvements are relatively low, as tools and methodologies are familiar, and investments tend to be low and predictable. Most important, they build on existing operational and engineering skills, as well as on the management of supply chains, customer relations, and information systems.

An example is the set of easy-to-use Lenovo Care support tools in Lenovo's 3000 series desktops and laptops that provide automatic updates and offer one-button system recovery. Drawing on IBM's ThinkVantage technology, Lenovo has improved these tools, with the result that the J series has been described by reviewers as a smart, versatile, and affordable choice for small businesses—one of the few market segments in the PC industry that display reasonable growth.

Modular Innovations

Modular innovations introduce new component technology and plug it into a fundamentally unchanged system architecture. This type has been a defining characteristic of the PC industry; within each generation of the Wintel architecture (combining the Microsoft Windows operating system with Intel's microprocessors), specialized suppliers have introduced new component technology, for purposes such as memory, storage, and display devices.[49] These have been made possible by a division of labor in product development: "[M] odularity is a particular design structure, in which parameters and tasks are interdependent within units (modules) and independent across them."[50] One consequence has been the disintegration of the innovation value chain as well as its dispersion across firm boundaries and geographic borders, giving rise to innovation offshoring through global innovation networks.[51]

Although modularity has created opportunities for industrial latecomers, the barriers to it are substantial. High technological complexity requires top scientists and experienced engineers in various fields. In addition, investment requirements can be very substantial (around $3 billion for a state-of-the-art semiconductor fabrication plant), as are risks of failure. This probably explains why it is difficult to find modular innovations by Chinese IT firms. Huawei provides a few examples of projects for developing new component technology that can be plugged into unchanged system architecture. It has substantially strengthened its capabilities in software development, with its R&D lab in Bangalore playing an important role.[52] Huawei has also invested heavily in the development of application-specific integrated circuit (ASIC) chips, embedded software, and shared platforms for communication and networking equipment.[53] Until recently, its internal semiconductor design unit supplied no more than 10 percent of the chips the company needs, a share that is now expected to increase substantially. After the company had spun off its independent chip design company Hi-Silicon, Huawei reported the completion of design projects for nearly one hundred types of ASIC chips, including so-called Internet Protocol cores for 3G mobile systems.

For Lenovo, an innovation that relies on the introduction of new materials is the "roll-cage" technology developed by the company for its ThinkPad Z series laptops. This technology provides extra physical protection, by fusing a magnesium alloy skeleton into the chassis of the laptop. Coupled with the existing titanium exterior, the laptops are supposed to better survive falls, bumps, and other shocks.

Architectural Innovations

Architectural innovations are "innovations that change the architecture of a product without changing its components."[54] They use existing component technologies but change the way those technologies work together.

What enables industrial latecomers to pursue architectural innovations? By definition, latecomers such as Chinese IT firms continue to lag behind industry leaders in the breadth and depth of their innovative capabilities. Their strength, however, is their familiarity with China's markets and institutions and their exposure to user requirements that global industry leaders have neglected. Chinese firms might be able to use this knowledge to penetrate the nation's large mass markets. Doing this does not require new components, but it does call for a change in the architecture of a product or service; the new components can be purchased from specialized suppliers.

An early example is the development of China's electronic switching system, HJD04, a system architecture that optimizes the specific features of the national telecommunications network to match specific needs of service providers.[55] Other examples are the development of Chinese-language electronics publishing systems by the Founder Group Company,[56] the development of the unique Chinese video compact disk (VCD) technology, and the successful transition to Chinese DVD system technology.[57]

Two recent examples of new features being added to existing architectures are Lenovo's Tianxi laptop and Huawei's development of its new integrated Internet Protocol service platform, ME60. Lenovo's Tianxi laptop computer is tailored for private consumers and small businesses in China. Global firms were uninterested, and for apparently good reason; in the mid-1990s, an ordinary PC cost RMB 13,000 ($1,900), the equivalent of one or two years' salary for an ordinary family.[58]

Legend started by developing a home computer for the Chinese family. The cheapest model, priced at RMB 3,000 ($440), did not even have a hard disk. Through trial and error the company found what Chinese consumers wanted. Using Intel's new Pentium processors, while global companies still shipped PCs with the older 486 processors to China, Lenovo provided user-friendly features through in-house design of ASICs and add-on cards.

The Tianxi project aimed to provide Internet access at an affordable price within an attractively designed box. A defining characteristic was its "one-touch-to-the-net" button that enabled national roaming. Through an arrangement with China Telecom, the Internet registration process was dramatically simplified, which meant that users no longer had to go to the push-to-talk for an account—a process that was infamously complex. Buyers of the Tianxi received a year of free Internet access, and a special pen allowed Tianxi users to write Chinese-character email messages.

Current Internet Protocol networks do not offer the security and quality of service that operators want, while traditional networks are incapable of supporting bandwidth-hungry multimedia services such as IPTV. Such products need to be aggregated and run over a common Internet Protocol core.[59] The ME60 is the "Swiss army knife" that does this. Technically, this system is quite an achievement. As a 10-gigabyte-multiservice control gateway,[60] the ME is an edge router that sits between the Internet Protocol core and the access network (which may be fixed or mobile). It can deliver tailor-made products in response to customers' specific needs. Such integrated solutions are not widely offered in the network equipment industry, where incumbent industry leaders typically provide standard solutions. In addition, many of their products provide leading-edge technology that far exceeds the needs of most users. These high R&D costs necessitate a business model that seeks to reap economies of scale through "mass manufacturing" of standard and fairly inflexible solutions.

Radical Innovations

Finally, radical innovations involve both new component technology and changes in architectural design. They require breakthroughs in both architectural and component knowledge. Examples include the discovery of new drugs or the invention of the Internet. In China, an interesting example is the development of the "pebble-bed" reactor, which offers the hope of cheap and safe nuclear power stations.[61]

Radical innovations require dense interaction with leading-edge science and require a broad base of capabilities in a society; top scientists and engineers are needed who work at the frontier of basic and applied research in a broad range of disciplines.

In short, such innovations are costly and risky, and failure can destroy even large, well-endowed companies. They are beyond the reach of most IT companies in China, with the possible exception of state-supported megaprojects such as military and space projects.

An example is Lenovo's supercomputers. The first project was the DeepComp 1800 supercomputer, introduced in 2001, which, based on 526 Intel Xeon processors, was ranked fifty-first by 2002. This was followed in 2003 by the DeepComp 6800 model, which was ranked fourteenth worldwide. The most recent one, the 1000 TFLOPS supercomputer, was started in 2005 and is scheduled for completion before 2010. It is expected to be nearly ten times more powerful than the world's fastest supercomputer. The underlying rationale was clearly more political than commercial: "China will need more supercomputing power in the years ahead to maintain its economic growth and development.... China cannot rely on other countries to develop a supercomputer that meets its needs."[62]

In short, for Chinese IT firms, radical innovations pose a difficult challenge: investment requirements are huge and require substantial government support, while markets are likely to be small. However, there may be indirect commercial benefits, because the successful completion of a radical innovation project may help to establish a company as a serious player and foster its brand image.

Technology Diversification

In addition to incremental innovations, a focus on architectural innovations makes more sense for Chinese firms than modular or radical innovations. An ability to profit from architectural innovation may seem counterintuitive, but it follows from their knowledge of Chinese markets and their capacity to complement missing pieces through their integration into global networks. A combination of incremental and architectural innovations should enable Chinese IT companies to pursue a strategy of technology diversification.[63] Defined as the expansion of a company's or a product's technology base into a broader range of technology areas,[64] this strategy focuses on applied research and the development of products that draw on component and process technologies that are not necessarily new to the world or difficult to acquire.

Technology diversification builds on strengths in distribution, supply-chain management, and manufacturing, and can use experience in providing lower-cost integrated solutions to customers with limited resources and who require strong support services. Most important, it could enable them to leverage their participation in global networks.

Conclusion

Integration into global knowledge networks helps industrial latecomer societies such as China by exposing them to best-practice management approaches, intellectual tools, and sources of knowledge on leading-edge technology.

An important finding here is that successful companies have not attempted to jump into technology leadership roles, trying to compete head-on with global technology leaders through radical innovations. Instead, they have focused on incremental and architectural innovations that allow them to pursue technology diversification strategies.

This chapter does not address government policies. It is clear, however, that there is ample scope for policies to help the development of firm-level innovative capabilities.[65] In fact, integration into global knowledge networks increases the need for a strong national innovation system. The needed policies are very different from earlier, top-down "command economy"–type policies that were characteristic of the East Asian development model. Such policies can succeed only if they fulfill two conditions: they need to balance the protection of IPRs with incentives for knowledge diffusion to local firms, and they need to provide a sufficiently large pool of people with the skills necessary to benefit from integration into global knowledge networks. There are no quick-fix solutions. Successful innovations require fundamental adjustments in institutions and behavior. They take time in any country, and even more so in China, because of the uncertainty and instability created by its recent transition to capitalism.

Notes

[1] "Knowledge workers" include science and engineering personnel, as well as managers and specialized professionals (in areas such as marketing, legal services, and industrial design) that provide essential support services to research, development, and engineering.

[2] D. Ernst, "Innovation Offshoring: Asia's Emerging Role in Global Innovation Networks," East-West Center Special Reports, no. 10, jointly published with the U.S.–Asia-Pacific Council, July 2006 <http://www.EastWestCenter.org/pubs/2006>.

[3] See, for instance, P. Nolan, *China and the Global Business Revolution* (London: Palgrave), 2001.

[4] See, for instance, the congressional hearing of the U.S.-China Economic and Security Review Commission, on China's High Technology Development,

held at Stanford in April 2005, published June 6, 2005 <http://www.uscc.gov/hearings/2005hearings/transcripts/05_04_21_22.pdf>.

[5] Ernst, "Innovation Offshoring."

[6] United Nations Conference on Trade and Development (UNCTAD), "UNCTAD Survey on the Internationalization of R&D," *Occasional Note*, December 12, 2005. The UNCTAD sample consists of the first three hundred firms of the R&D scoreboard of the seven hundred top worldwide R&D spenders, published by the U.K. Department of Trade and Industry (DTI).

[7] Private equity investment is medium- to long-term finance provided in return for an equity stake in potentially high-growth companies that are not listed on a major public stock exchange. According to the British Venture Capital Association website, private equity investment encompasses both venture capital (from the seed stage to the expansion stages of investment) and management buy-outs and buy-ins. See "An Introduction to Private Equity," <http://www.bvca.co.uk>.

[8] Ernst, "Innovation Offshoring."

[9] D. Ernst, "Complexity and Internationalization of Innovation: Why Is Chip Design Moving to Asia?" *International Journal of Innovation Management* 9, no. 1 (special issue in honor of Keith Pavitt, 2005): 47–73.

[10] Sources include the U.S. and the U.K. National Venture Capital Associations and consulting firms such as Greenwich Associates; Private Equity Intelligence (PEI), a London-based specialized consulting firm; and the Shanghai-based Zero2IPO for China.

[11] Courtesy of PEI (July 6, 2006).

[12] Data courtesy of the U.S. Council of Graduate Schools, March 2006.

[13] See D. Ernst, "The New Mobility of Knowledge: Digital Information Systems and Global Flagship Networks," in *Digital Formations: IT and New Architectures in the Global Realm,* ed. R. Latham and S. Sassen (Princeton, NJ: Princeton University Press, 2005), 89–114; and D. Ernst, "Innovation Offshoring."

[14] R. R. Nelson and H. Pack, "The Asian Miracle and Modern Growth Theory," *The Economic Journal* 109 (1999): 416–36; and S. Lall, "Technological Change and Industrialization in the Asian Newly Industrializing Economies: Achievements and Challenges," in *Technology, Learning and Innovation: Experiences of Newly Industrializing Economies*, ed. L. Kim and R. R. Nelson (Cambridge: Cambridge University Press, 2000), 13–68.

[15] C. Freeman, "Innovation Systems: City-State, National, Continental and Sub-National" (unpublished, Science Policy Research Unit, University of Sussex, December 1997), 13.

[16] D. Ernst, "From Partial to Systemic Globalization: International Production Networks in the Electronics Industry" (report to Sloan Foundation, published as "The Data Storage Industry Globalization Project Report 97-02," Graduate School of International Relations and Pacific Studies, University of California at San Diego, 1997).

[17] D. Ernst and L. Kim, "Global Production Networks, Knowledge Diffusion and Local Capability Formation," *Research Policy* 31, no. 8/9 (2002): 1417–29.

[18] D. Ernst, "Global Production Networks and the Changing Geography of Innovation Systems. Implications for Developing Countries," *Economics of Innovation and New Technologies* 11, no. 6 (2002): 497–523

[19] D. Ernst, "Global Production Networks in East Asia's Electronics Industry and Upgrading Perspectives in Malaysia," in *Global Production Networking and Technological Change in East Asia*, ed. S. Yusuf, M. A. Altaf, and K. Nabeshima (Washington, D.C.: World Bank and Oxford University Press, 2004), 89–155.

[20] W. Kuemmerle, "Home Base and Foreign Direct Investment in R&D" (PhD dissertation, Harvard Business School, 1996).

[21] W. Kuemmerle, "Building Effective R&D Capabilities Abroad," *Harvard Business Review* (March–April 1997): 66.

[22] For example, see O. Granstrand, P. Patel, and K. Pavitt, "Multi-Technology Corporations: Why They Have 'Distributed' Rather Than 'Distinctive Core' Competencies," *California Management Review* 39, no. 4 (1997): 8–25.

[23] Ernst, "Innovation Offshoring."

[24] See, for example, M. Von Zedwitz, "Foreign R&D Laboratories in China," *R&D Management* 34, no. 4 (2004): 439–52.

[25] D. Ernst. and B. Naughton, "China's Emerging Industrial Economy: Insights from the IT Industry," in *Capitalism in the Dragon's Lair*, ed. Chris McNally (London: Routledge, 2007), chapter 3.

[26] Ernst, "Innovation Offshoring."

[27] D. Ernst, T. Ganiatsos, and L. Mytelka, eds., *Technological Capabilities and Export Success: Lessons from East Asia* (London: Routledge Press, 1998).

[28] W. Xie and S. White, "Sequential Learning in a Chinese Spin-Off: The Case of Lenovo Group Ltd," *R&D Management* 34 (2004): 407–22; and Xie and White, "Windows of Opportunity, Learning Strategies and the Rise of China's Handset Makers" (INSEAD working paper, Singapore, 2005).

[29] N. Arrifin, "Internationalisation of Innovative Capabilities: The Malaysian Electronics Industry" (PhD thesis, Science Policy Research Unit, University of Sussex, 2000).

[30] A. H. Amsden and F. T. Tschang, "A New Approach to Assessing the Technological Complexity of Different Categories of R&D (with Examples from Singapore)," *Research Policy* 32, no. 4 (April 2003): 553–72.

[31] Ernst, "Innovation Offshoring."

[32] D. A. Norman, *The Invisible Computer* (Cambridge, MA: MIT Press, 1998).

[33] On the origins and evolution of the company, see Q. Lu, *China's Leap into the Information Age: Innovation and Organization in the Computer Industry* (Oxford, U.K.: Oxford University Press, 2000); and Z. Ling, *The Lenovo Affair: The Growth of China's Computer Giant and Its Takeover of IBM-PC* (Singapore: John Wiley & Sons [Asia], 2006).

³⁴ This decision was imposed by the company's failure to secure a manufacturing license from the Ministry of Electronics Industry (MEI). Instead, licenses went to companies that were part of the "MEI family"—companies such as Great Wall, which was then the largest domestic producer.
³⁵ S. Feng and J. Elfring, *The Legend Behind Lenovo: The Chinese IT Company That Dares to Succeed* (Hong Kong: Asia, 2000, 2004), 37.
³⁶ D. Primack, "Lenovo Gives TPG, General Atlantic a China Presence," *Buyouts* (March 6, 2006).
³⁷ Interview in Beijing, May 26, 2006.
³⁸ *Plastics News* [Crain Communications], December 9, 2005.
³⁹ According to Lenovo Group chairman Yang Yuanqing, "[w]e will make money but we don't focus on profit, we focus on growth." He added that this phase may last at least two years. (Quoted in *Network World*, December 9, 2005).
⁴⁰ An example is Lenovo's first computers sold under its own brand outside China—the Lenovo 3000 J series desktops and 3000 C notebooks. These models target lower performance at a lower price for overseas markets.
⁴¹ J. McGregor, *One Billion Customers: Lessons from the Front Lines of Doing Business in China* (New York: Free Press, 2005), 289.
⁴² Primack, "Lenovo Gives TPG."
⁴³ *China IT Weekly*, April 1, 2006.
⁴⁴ Primack, "Lenovo Gives TPG."
⁴⁵ Lenovo is expected to benefit from a $750,000 grant from the One North Carolina Fund, as well as up to $8.4 million in tax incentives from a Joint Development Investment Grant if it meets employment and performance targets.
⁴⁶ As of June 2004, Lenovo had registered 787 patents in China, entering for the first time the top ten list of the most competitive intellectual property owners in China. But a search in the USPTO patent database showed that, from 1976 to the present, Lenovo had registered only 34 patents in the United States.
⁴⁷ R. M. Henderson and K. B. Clark, "Architectural Innovation: The Reconfiguration of Existing Systems and the Failure of Established Firms," *Administrative Science Quarterly* (March 1990): 9–30.
⁴⁸ D. Ernst, "Developing Innovative Capabilities in Chip Design: Insights from the U.S. and Greater China" (manuscript, East-West Center, Honolulu, 2006).
⁴⁹ See, for example, R. N. Langlois and P. L. Robertson, "Networks and Innovation in a Modular System: Lessons from the Microcomputer and Stereo Component Industries," *Research Policy* 21 (1992): 297–313; and C. W. Baldwin and K. B. Clark, *Design Rules: The Power of Modularity* (Cambridge, MA: MIT Press, 2000).
⁵⁰ Baldwin and Clark, *Design Rules*, 88.
⁵¹ Ernst, "Innovation Offshoring."
⁵² According to China's Ministry of Information Industry (MII), Huawei is now one of the three largest domestic Chinese software enterprises (together with ZTE and Haier).

[53] D. Ernst, "Building Innovative Capabilities within Global Knowledge Networks: The Case of Huawei" (manuscript, East-West Center, Honolulu, HI, 2006).

[54] Henderson and Clark, "Architectural Innovation," 9.

[55] X. Shen, *The Chinese Road to High Technology: A Study of Telecommunications Switching Technology in the Economic Transition* (New York: St. Martin's Press, 1999).

[56] Lu, *China's Leap*, chapter 4.

[57] L. Fang and L. Mu, "Indigenous Innovation, Capability Development and Competitive Advantage: The Origins and Development of the Competitiveness of Chinese VCD/DVD Industry" (paper presented at the Annual Meeting of the Business History Conference, Univ. of Mass. at Lowell, 2003).

[58] Feng and Elfring, *The Legend Behind Lenovo*, 61.

[59] The IP core, also sometimes called the backbone, is the primary path of IP network traffic. It connects smaller segments of a network and has a high concentration of traffic.

[60] A gateway is the entrance to another network. The gateway allows equipment with different protocols to communicate.

[61] "China in Drive for Nuclear Reactors," *Financial Times* (February 8, 2005): 4.

[62] *People's Daily*, May 3, 2005.

[63] D. Ernst, "Pathways to Innovation in Asia's Leading Electronics-Exporting Countries: A Framework for Exploring Drivers and Policy Implications," *International Journal of Technology Management* 29, nos. 1–2 (special issue, "Competitive Strategies of Asian High-Tech Firms," 2005): 6–20.

[64] O. Granstrand, "Towards a Theory of the Technology-Based Firm," *Research Policy* 27, no. 5 (1998): 465–89.

[65] D. Ernst, "Pathways to Innovation."

TOWARD A BETTER UNDERSTANDING OF CHINA'S SCIENTIFIC ELITE

*Cong Cao**

In 1955 the Chinese Academy of Sciences (CAS) founded Academic Divisions (*xuebu*), which appointed 233 Academic Division members (*xuebu weiyuan*), including 172 natural scientists, to lead the scientific research of the academy and the nation. Two years later, the Academic Divisions appointed 21 additional members (18 from the natural sciences), some of whom would have been appointed in 1955, while others were recent returnees from abroad. Afterward, however, the Academic Divisions and their members became political targets and existed in name only. China's scientific enterprise was almost destroyed during the Cultural Revolution between 1966 and 1976. The CAS Academic Divisions were restored in 1979 and resumed adding members. In November 1980, 283 scientists were elected. There was then another ten-year interruption; not until 1991 did the CAS elect new members (210 this time) and formalize the election process. Since 1993 the CAS has elected no more than 60 new members every two years. In that year, their title

---

* This essay is based on my book *China's Scientific Elite* (London and New York: RoutledgeCurzon, 2004). Interviews were conducted in China between 1995 and 1997, and interviewees include 79 CAS members of different ages, from different regions, different institutions, and disciplines, and elected at different periods. Research was supported by the U.S. National Science Foundation (NSF-9521358 and NSF-9800174), which is gratefully acknowledged. I would like to thank Professors Jonathan R. Cole, Harriet Zuckerman, Richard P. Suttmeier, and others for their advice, and to colleagues in and outside China for their encouragement, help, and suggestions.

"China's scientific elite" refers only to members of the Chinese Academy of Sciences (CAS) working in China and does not include those foreign members elected since 1994 nor members of the Chinese Academy of Engineering, another honorific Chinese institution in engineering and technology.

One of the CAS Academic Divisions was devoted to philosophy and the social sciences when the divisions were set up in 1955. However, members of that division have not participated in the CAS since 1960; in fact, the division was separated from the CAS to form the Chinese Academy of Social Sciences (CASS) in 1977. In 2006 the CASS finally founded its own Academic Divisions, not related to the CAS Academic Division of Philosophy and Social Sciences, and selected 47 Academic Division members (*xuebu weiyuan*) and 95 honor Academic Division members (*rongyu xuebu weiyuan*). This chapter does not cover CASS Academic Division members.

Table 12.1 Distribution of CAS Members by Year of Election and Academic Division

Year	Academic division						Total
	Mathematics and physics	Chemistry	Life science and medicine	Earth science	Information technology science	Technical science	
1955	30	22	60	24	1	35	172
1957	6	2	5	3	0	2	18
1980	51	51	53	64	12	52	283
1991	38	35	34	35	24	44	210
1993	10	10	11	10	5	13	59
1995	10	9	12	10	7	11	59
1997	9	10	12	10	8	9	58
1999	10	8	11	10	8	8	55
2001	10	10	12	9	4	11	56
2003	10	10	11	10	7	10	58
2005	8	9	12	7	6	9	51
2007	6	6	7	4	1	5	29
Total	198	182	240	196	83	209	1,108

Source: Author's records.

was changed from "CAS Academic Division member" to "CAS member" (*yuanshi*), making it a Western-style academic honor. A total of 108 Chinese scientists have become CAS members (Table 12.1); and as of March 2005, 669 of these, including 117 senior members, were alive. Counted two different ways (by geography and by institution), they are concentrated in Beijing (358), Shanghai (83), Jiangsu (40) and the economically developed eastern regions; in the CAS (273) (elite CAS members do not have to work at the CAS) and institutions of higher education (211).[1]

According to its bylaws, "CAS membership is the highest academic title that the State sets up in science and technology and an honor of a lifetime." CAS members have a reputation similar to that enjoyed by their counterparts in other countries, such as members of the National Academy of Sciences in the United States, fellows of the Royal Society of London in England, and so on.[2]

The Formation of China's Scientific Elite

The formation of China's scientific elite has been influenced by many factors. Through an analysis of the careers of the 970 CAS members elected between 1955 and 2001 and interviews with some of the members, the following aspects seem to be salient.

First, such eastern provinces and cities as Jiangsu, Zhejiang, Shanghai, Fujian, Guangdong, and Beijing have each produced more than 50 CAS members, and Hunan, Shandong, Hebei, and Sichuan more than 40. In one extreme case, 42 CAS members were born in one county: Dongyang, in Zhejiang province.[3] It is a combination of factors—a region's development, flourishing education and high-quality students, and access to new ideas—that gave some people a head start. This, in turn, might have concentrated China's future scientific elite in these regions. In addition, the educational attainment of the parents of elite scientists is more important than their socioeconomic background as measured by their fathers' occupations: of the parents whose educational attainments were known, not only 32.7 percent of the fathers, but also 45.1 percent of the mothers, had received formal higher education.

Second, more than 85 percent of the elite scientists have attended "leading" (*zhongdian*) or prestigious universities in China and, if possible, have gone abroad for advanced studies and earned graduate degrees from leading foreign universities. Of the CAS members elected in 1955–1957 and 1980, 77.4 percent and 66.4 percent had gone to foreign graduate schools. Although about half of those elected between 1991 and 2001 had only a Chinese undergraduate education, reflecting China's closed-door policy and the interruption of graduate education, the proportion of those elected after 1999 with a domestic or foreign graduate degree has been rising. As a whole, American universities awarded more doctoral degrees (197) to future CAS members than did all other countries combined (189). More interesting is the fact that political turmoil has caused discontinuities in education but has not changed the educational attainment pattern of elite scientists.

Third, Chinese mentors behave as their counterparts do elsewhere—teaching students the norms of international elite science and imparting to them an appropriate attitude toward learning and scholarship. They also make great efforts to transmit moral values, such as patriotism, to students. This aspect of mentoring differs from that of Western elite scientists. In any case, following the Chinese tradition of respecting teachers and the aged, junior scientists adopt a respectful and cautious attitude toward their mentors, which might have a negative impact on their thinking. That is, students have been less likely to challenge the scholarship of their senior mentors and to do innovative work.

Fourth, although the Chinese Communist Party (CCP) and the government have emphasized applied and military research, the majority of the CAS members elected between 1955 and 2001 are affiliated with the CAS institutes (374, or 38.6 percent) and universities (335, or 34.5 percent), which suggests that they are likely to be doing basic research. Only 106 (10.9 percent) came from national-defense-related institutions. The number reaches 225, or less than one quarter of the total, if those who had done such research sometime in their career are added. This suggests that the scientific elite, similar to its foreign counterpart, is composed mainly of basic scientific researchers.

Fifth, although being "both red and expert" (*you hong you zhuan*) is the highest requirement for Chinese intellectuals, expertise is more important than redness, or political correctness, for the scientific elite. Except for a few "dual elites" who represent the party in leading the nation's scientific enterprise, most have been accorded CAS membership because of their academic achievements rather than their political loyalty. This status has in turn brought them political distinctions such as membership in the National People's Congress (NPC), the Chinese People's Political Consultative Conference (CPPCC), and even the CCP. Party membership has not been a prerequisite for scientific elite membership. Although 461 of the 669 CAS members who were alive in March 2005 were party members (68.9 percent),[4] some had joined a democratic party before being recruited into the CCP. In fact, some believed that being a CCP member could help their scientific careers—they might have joined the party for the sake of advancing their careers and so do not necessarily believe in communism.

Finally, since 1991, the CAS has elected new members every two years, based on a four-step process. The first step is recommendation, in which three or more CAS members may nominate a candidate. Each member can nominate no more than two candidates, and at least two nominators should be from the Academic Division to which they would recommend candidates. Research institutes, universities, and academic societies can also make recommendations, so as to prevent outstanding scientists from being excluded.[5] Nominated candidates then provide credentials: ten representative publications, citation statistics and academic awards if available, and other supporting documentation.

The second step of the election process is preliminary selection. Candidates who are nominated by CAS members go directly to the next step—evaluation of credentials—while others must first be evaluated and selected by their respective higher-level academic organizations, such as the CAS, the Ministry of Education, the China Association for Science and Technology, and so on. Only those who pass the preliminary selection become effective candidates (*youxiao houxuanren*).

The third step involves assessments of candidates by CAS members according to disciplines. There might be several rounds, with each one reducing the number of candidates by anonymous voting until a final list of formal candidates (*zhengshi houxuanren*) is produced.

In the final step, CAS members cast anonymous and differential votes to elect new members to their own Academic Divisions. Differential voting means that the number of candidates is more than the number of members to be elected. The number of votes that a CAS member can cast depends on the number of new members that an Academic Division is supposed to elect, which guarantees that the number of new members elected in any division does not exceed the quota allocated to it. A CAS member can vote only if he or she participates in more than two-thirds of the evaluation sessions. Those who receive at least half the votes from the members present become new members (the 2006 revision of the CAS membership election procedure stipulates that for a candidate to be elected, he or she must receive at least two-thirds of the votes cast). The final results are acknowledged by each Academic Division, approved by the Presidium of the CAS Academic Divisions, and reported to the State Council for the record.

The rigorous and meritocracy-based electoral process (Table 12.2) should prevent the less qualified from being elected, although it cannot always guarantee the very best. Personal relations, or *guanxi*, and especially mentor-student and colleague relations, have played a role in the CAS membership elections, as they do in the daily life of the Chinese and in the evaluation of scientific work in other countries. But their influence seems limited, because the scientists who are involved work to maintain the integrity of the institution.

Table 12.2 Ratios of Candidates to Elected CAS Members

Year of election	Number of effective candidates	Number of elected members	Ratio
1991	1079	210	5.1
1993	733	59	12.4
1995	587	59	9.9
1997	418	58	7.2
1999	356	55	6.5
2001	337	56	6.0
2003	309	58	5.3
2005	295	51	5.8
2007	287	29	9.9

Source: Author's records.

Roles of the Elite in Science

In addition to doing research, training students, and participating in international science and technology (S&T) exchange and collaborations, the scientific elite also renders professional opinions, provide information, and engage in debates on issues related to its knowledge. The CAS Academic Divisions, from the beginning, have also evaluated research results of significance to the national economy and cultural enterprises and made suggestions on the directions of further research and applications. The bylaws of the CAS, passed in recent years, state that "it is the right for CAS members to put forward suggestions on the decision-making of the state in major science and technology issues."[6] China's scientific elite has been trying to do this.

Advising Scientific Research and Science Policy

Soon after the CAS Academic Divisions were established, their members and other Chinese scientists developed and implemented the twelve-year plan (1956–1967) for the development of S&T. Also thanks to their efforts, the nation's science award system was in place in the mid-1950s.

When the Cultural Revolution ended in 1976, surviving CAS members helped China to restore research and education systems. They also initiated many of the important changes in science policy. In 1982, for example, the physicist Xie Xide and 88 other CAS members, inspired by the U.S. National Science Foundation, proposed that new approaches to the funding of research be introduced. That led first to a peer-reviewed quasifoundation within the CAS, which evolved into the National Natural Science Foundation of China (NSFC) in 1986. CAS members have held the NSFC directorship and have institutionalized mechanisms for identifying and supporting the best research projects. As a result, the NSFC represents the most fair and open mechanism for funding research in China.

Initiating and Promoting Major National Science Programs

CAS members can, and do, submit suggestions to the Politburo of the CCP Central Committee, China's highest decision-making organ. In March 1986, in response to the growing importance of high technology, four CAS members who were involved in the nation's strategic weapons program suggested a program—the State High-Technology Research and Development Program (863 Program)—for tracking the world's high-tech trends. It was approved by Deng Xiaoping and the political leadership immediately. In recent years, the State Science and Technology Commission and its successor, the Ministry of Science and Technology (MOST) have sought opinions from the CAS Academic Divisions on the choice of critical R&D projects such as the "Climbing Program" and the State Basic Research and Development Program (973 Program).

Participating in Discussion on Important Issues Related to the Nation's Development

CAS members make their talents and counsel available on important national issues. These have ranged from China's development, national security, sustainable development, industrial development after its accession to the World Trade Organization (WTO), and science education, to the development of new medicines, mechanisms for the national sharing of geological data, and exploration of oil and natural gas resources.

Political Participation

Most members of the Chinese intellectual community, CAS members included, have been involved in decisions on educational, scientific, and cultural affairs where their expertise is useful. But for intellectuals, independent thought in societal matters is more critical than mere involvement in areas of their particular knowledge. Chinese intellectuals as a group unfortunately do not have such a tradition. From the founding of the People's Republic in 1949 to the Anti-Rightist Campaign in 1957, most Western-trained intellectuals were incorporated into the governing body of the new state, where they played a visible yet clearly subordinate role, motivated by a feeling of patriotism.[7] In the "blossoming and contending" period, CAS members and other Chinese intellectuals advocated autonomy and freedom in research and teaching as well as in political affairs. But the year 1957 taught most of them the virtues of conformity and compliance.[8] Thereafter, intellectuals were considered important mostly for their technical knowledge. The Cultural Revolution further reduced the incentive, if there was any, of intellectuals to voice even constructive opinions.

In the reform era, the issue of professional freedom emerged again. Even then, few intellectuals spoke out, even in their own interests and still less on other matters. Only in the mid-1980s was the political participation of intellectuals mobilized. Motivated by near-term economic needs, the leadership made a trade-off, taking a short-term risk (giving scientists a bigger role) for a long-term benefit (ensuring scientific and technical independence and ultimate superiority, and regime consolidation).[9] Seizing the opportunity, scientists then expressed opinions on such issues as the organization of science, the utilization of scientific talent, and freedom and autonomy in research, as they had done before the Anti-Rightist Campaign. They advocated a national science funding system, peer review, and the restoration of the CAS Academic Divisions and their evolution into an honorific institution and "brain bank" of the country. Moreover, they sought a democratic political environment, which they viewed as fundamental to the development of science. It was under these circumstances that a representative, the astrophysicist Fang Lizhi—a CAS member elected in 1980 and fired in 1989—challenged the party-state.

Increasing Roles of Scientists in the State Political Machine

As noted, some CAS members are also members of the NPC or the CPPCC. Those CAS members were the first to articulate the interests of the scientific community. Along with other scientists, CAS members have, unsurprisingly, advocated increasing science funding year after year. State Council allocations to the NSFC have increased steadily since its inception in 1986, reaching over RMB 4 billion in 2007. The Ninth Five-Year (1996–2000) Plan stipulated that the nation's research and development (R&D) spending be increased to 1.5 percent of the gross domestic product (GDP) by 2000, from 0.5 percent in 1995, of which 15 percent would be devoted to basic research, although both goals have yet to be realized. The 973 Program also resulted from the campaign of representatives of the scientific community in the NPC and CPPCC.

CAS members have been active on both sides of the debates over the Three Gorges Project of building a dam across the Yangtze River. While some, especially those in the water resources and electric power fields, including Zhang Guangdou, a hydraulics engineer and a 1955 CAS member, supported the project, its ecological, cultural, and social impacts brought objections from the scientific community. CAS members in the CPPCC requested that the project be reevaluated. In 1983, led by Zhou Peiyuan, a CAS member and then CPPCC vice chairman, a group of scientists and engineers undertook a major feasibility study that concluded that the project was not feasible scientifically and economically. They succeeded in holding up the project and forcing the government to revise it.[10]

Changing State-Society Relations and the Scientific Elite

The reform period has entailed important changes in state-society relations. In the intellectual community, the control and penetration of the party-state has gradually given way to autonomy and professionalism. Will intellectuals in general, and the scientific elite in particular, therefore become an important force to change China more dramatically?

The party-state believes that the middle class—intellectuals included—can pose a political threat. A study by the Organization Department of the CCP Central Committee published in May 2001 noted, "as the economic standing of the affluent stratum has increased, so too has its desire for greater political standing," which would inevitably have a "profound impact on social and political life."[11] The scientific community has been gaining social autonomy from the party-state,[12] and as long as the regime supports it as essential to the nation's future, "some in the scientific community will continue to find inevitable connections between the ideals that motivate them as scientists and those that inspire their political lives," and it is hopeful that one day elements of democracy will reemerge in China. In other words, the impetus for future change may well come from "the ranks of the scientific establishment itself."[13]

But in reality, the assumption of a link between the rise of a new elite—economic or scientific—and democratization is logically flawed and empirically unsupported. China's new business elite, for example, while relatively autonomous and often holding politically liberal beliefs, has not emerged as a strong independent force; rather, the state's co-optation of this group and its ability to navigate the business environment using nondemocratic means have denied the new business elite the political significance many would wish for it.[14] Members of the urban elite, who, as a whole, stand to benefit the most from the political status quo, are not necessarily the progenitors of radical change, not only because many have bought into a highly authoritarian and narrowly nationalistic ideology, but also because they owe their elite status to the closed and corrupt system of the CCP. Therefore, it is not in their best interests to promote democracy.[15] Furthermore, because of their age, higher level of education, and intimate relationship with the state, members of the scientific elite seem to be unlikely to campaign for individual freedoms.

According to Zhang Dongsun, a late professor of philosophy and the father of two CAS members, Zhang Zhongsui and Zhang Zhongye, the educated class is supposed to be the conscience of society, the paragon of reason and morality, and the motor of political change.[16] Yet during the Cultural Revolution, as Yü Yingshi, a scholar of Chinese intellectual history, sees it, the intellectuals were pushed passively to the periphery, while the "lower elements" in the society, whose number greatly swelled as a result of social disintegration, ascended and occupied the central stage with the help of party ideologies and organization.[17] Now, intellectuals see their political standing and economic situation improving significantly, and some have become "establishment intellectuals."[18] At the same time, their public role has given way to the protection of vested interests. Therefore, the leading intellectuals are unlikely to risk undertaking activities that challenge the party-state[19] or raise the "communal critical self-consciousness" as in the early reform period.[20] In this sense, the political marginalization of Chinese intellectuals is still an issue.

Nevertheless, members of the Chinese scientific community are likely to adopt a modest approach, urging the party-state to be more responsible, accountable, and effective. Elite scientists are expected to enhance their role as the nation's "brain bank," providing the state with advice in the areas of their expertise and related to the economy. Their responsibility is to offer scientific and independent judgments. In doing so, scientists will further legitimate their status in society and will advocate and promote professionalism, but they will be hesitant to pursue causes beyond the profession.

Ambivalence toward the Elite

Members of the CAS (and of the Chinese Academy of Engineering) are highly respected. But the scientific community has given rise to grievances against its *yuanshi*, whose reputation seems to be in jeopardy. For one thing, Chinese *yuanshi* are entitled to a stipend from the state and to de facto privileges (although not stipulated by the government) equivalent to a vice governor in housing, medical care, and transportation; they are also entitled to receive other benefits from regional governments or from the institution with which they are affiliated, as well as lifetime employment.[21] In contrast, membership in Western academies is just an honorary title; further, its holders have to pay membership dues to maintain their membership. The Royal Society, when it was established, stipulated that every fellow was liable to pay an admission fee, and it encouraged its fellows to make further contributions toward particular costs. In the U.S. National Academy of Sciences, any member "whose dues fall in arrears for three successive years shall be transferred to the roll of emeritus members."[22] Even Nobel laureates do not enjoy special treatment; at the University of California, for example, a Nobel laureate gets only a permanent on-campus parking place.

In the Chinese scientific community, elite membership also means a high likelihood that its holder will be recruited to serve on expert panels and to chair national research programs, and thus be in a better position to secure individual research funding and support for his or her work unit (*danwei*). As a result, there is something of a mania surrounding academicians. Some provinces and *danwei* offer lucrative start-up packages, relocation help, higher salaries, and housing to lure *yuanshi*. News reports portray privileges for *yuanshi* as a way of respecting elite scientists and their knowledge; elections to the academy are said to be similar to the imperial civil service examination (*zhongju*). Some potential candidates for membership have launched public relations campaigns to promote themselves, turning the election process into one in which candidates play active roles. With the huge benefits that *yuanshi* could bring to a *danwei*, it is not rare that Chinese institutions of learning promote their candidates. This new elitism reflects the strengthening of meritocratic values and increased academic autonomy, and yet some *yuanshi* utilize their status to engage in rent-seeking activities, which compromises these meritocratic values.

Some *yuanshi* become scientific and social activists, speaking on topics beyond their fields of expertise and taking advantage of their status to lobby on behalf of themselves, their *danwei* or students, or becoming "vases" by holding many positions while being paid lucratively. Some disclose to candidates information about evaluation and election sessions for CAS membership, violating the ethical code. Some behave differently after becoming members—becoming more arrogant, overweening, and supercilious. A small number of *yuanshi* have been bribed by scientists wanting to become new members and have even promoted pseudoscience and engaged in scientific misconduct. This has further damaged the reputation of CAS members and explains why some have advocated overhauling, if not abolishing, the *yuanshi* institution.[23]

CAS members have tried to discipline themselves and to improve their somewhat tarnished image. As early as 1982, they were among the first scholars to call for an ethics code for scientific and technical personnel. In 1993, fourteen CAS members made a similar call for the entire scientific community. Recently, several CAS members—former CAS president Zhou Guangzhao included—have criticized the mania surrounding *yuanshi*, calling for abolishing their privileges.

A science ethics committee was established within the CAS Academic Divisions in 1997. In the same year, when a student of CAS member Chen Minheng was found to have committed plagiarism in his doctoral dissertation, the CAS Academic Divisions investigated Chen's responsibility and finally removed his membership.[24] Posting the information of candidates for membership in their *danwei* encourages colleagues to submit opinions about candidates and thereby safeguards the elite institution. Establishing senior membership, or *zishen yuanshi*—in Western terminology, emeriti—was also intended to decrease the negative influence of some of the aged members. Because the elite membership carries so many responsibilities that have gone beyond its honorific status, the academy believed that it should put some constraints on its members. In late 2001 the Presidium of the CAS Academic Divisions passed an ethics code to self-discipline the behavior of its members. And in April 2006, the CAS Academic Divisions convened presidents from several foreign academies to help improve CAS membership elections. The 2006 CAS Member Assembly further revised the bylaws of CAS membership, stipulating that a candidate must receive at least two-thirds of the votes cast to become a member, which was reflected in the result of the 2007 CAS membership election.

China's Scientific Elite in a Comparative Perspective

We are now in a position to examine this group of elite scientists from a comparative perspective. Where do they stand academically compared with their predecessors and international counterparts?

It is certain that current CAS members are not at the same level as members in some of the most advanced countries. According to a 1999 assessment by Zhu Lilan, then minister of science and technology, China was internationally competitive in only about 5 percent of basic science fields and enjoyed a relatively high status in another 20 percent of these fields.[25] The situation has not fundamentally changed. In 2000 and 2001, scientists from Mainland China published 28 papers in *Science* and *Nature*, two leading international science journals, representing about 1 percent of the articles in those publications.[26] With a *Science* article in early 2001, the biochemist Zhang Yonglian secured her position in the elite group in that year.[27] By this measure, many authors of the *Science* and *Nature* articles could have been elected CAS members if they worked in China. In fact, some CAS members could not meet the requirement to have their works published in high-quality international journals. The recent push for "indigenous innovation" (*zizhu chuangxin*) also points to the lack of homegrown innovation.

In terms of publications, CAS members have also underperformed compared to some of the Chinese scientists who have gone abroad for advanced studies and stayed there, holding teaching and research positions. In some extreme cases, CAS members are not even comparable to some Chinese postdoctoral fellows abroad. For example, one such researcher at an American lab run by an overseas Chinese life scientist had two first-author papers in *Cell*, the most prestigious journal in life science, and one second-author paper in *Science*. The significance of his research is shown by the many citations to the papers; by 2001 the two *Cell* papers were cited 1,217 and 624 times, respectively, and the *Science* paper received 1,399 citations. Presumably, his advisor has achieved even more. In general, those scientists who stay abroad are likely to be the best. For example, among about three hundred China-born life scientists who are as outstanding as their peers in terms of their appointments at prestigious institutions, their leadership of laboratories, their reputation at the international research frontiers, and their grants, only five have returned to China, and none of them is among the top 20 percent in international academic circles.

A comparison of the CAS members appointed between 1955 and 1980 with those elected thereafter also suggests a change in quality. The former group did much better in terms of educational attainment; while many in the 1955–1980 group were trained abroad and had doctoral degrees, CAS members elected since 1991 are more likely to be China-trained and have only undergraduate education. This is because of the reality of higher education in China. According to Zou Chenglu, a Cambridge-trained biochemist and CAS member who passed away in late 2006, the generation of scientists who graduated before the reform and open-door policy and published in "home" journals has difficulties in understanding the "rules of the game" of international science.[28]

Many of the early generations of Chinese scientists are internationally renowned. For example, the physicist Wu Youxun, who helped his mentor, the 1927 Nobel laureate Arthur H. Compton, prove the Compton Effect experimentally, was regarded by Compton as one of his two best students (the other student, Luis Alvarez, won a Nobel Prize in 1968 himself); the aeronautics scientist Qian Xuesen held full professorship at two first-tier institutes of technology, the Massachusetts Institute of Technology and the California Institute of Technology, before he was deported from the United States; the mathematician Hua Luogeng was a full professor at the University of Illinois before returning to China; and the neurologist Zhang Xiangtong was associate professor at Johns Hopkins University before moving back to China in the mid-1950s. But few CAS members elected since the beginning of the 1990s have achieved similar status. Also, most of the first-class prizes of China's Natural Science Awards have gone to those earlier returnees.

Conclusion

CAS members play significant roles in the Chinese scientific community and society. They have been elected through a process that has been improved constantly so as to guarantee that elected members are the best among Chinese scientists.

However, CAS members are not at the same academic level as their international counterparts, and those who have been elected recently are generally inferior to their predecessors. In addition, the *yuanshi* system has become a target of criticism in recent years. While reaffirming autonomy and freedom in evaluating and rewarding scientists, this academic honor has also had unintended consequences that may invite corruption and may compromise its elite status. There is much room for improvement in the *yuanshi* system.

Notes

[1] Yongwei Liu, Zhenzhen Li, and Hongjuan Chen, "An Analysis of the Structure and Social Roles of the Members of the Chinese Academy of Sciences and Suggestions" (in Chinese), *Bulletin of the Chinese Academy of Sciences* 20, no. 3 (2005): 179–94.

[2] See Philip M. Boffey, *The Brain Bank of America: An Inquiry into the Politics of Science* (New York: McGraw-Hill, 1975); and Harriet Zuckerman, *Scientific Elite: Nobel Laureates in the United States* (New York: The Free Press, 1977).

[3] *Renmin Ribao*, overseas edition (November 21, 1992), 4.

[4] Liu, Li, and Chen, "An Analysis of the Structure and Social Roles."

[5] In the post-perestroika Russian Academy of Sciences, nominees for membership could also be put forward by Russian scientific, educational, social, and state organs. See Stephen Fortescue, "The Russian Academy of Sciences and the Soviet Academy of Sciences: Continuity or Disjunction?" *Minerva* 30, no. 4 (1992): 459–78.

[6] Chinese Academy of Sciences (CAS) Academic Divisions, "By-laws on Members of the Chinese Academy of Sciences," adopted by the Sixth General Assembly of CAS Academic Division Members in 1992, and amended by the subsequent general assemblies of CAS members in 1994, 1996, 1998, 2000, 2002, 2004, and 2006, <http://www.casad.ac.cn>.

[7] Merle Goldman and Timothy Cheek, "Introduction: Uncertain Change," in *China's Intellectuals and the State: In Search of a New Relationship*, ed. Merle Goldman with Timothy Cheek and Carol Lee Hamrin (Cambridge, MA: Council of East Asian Studies, Harvard University, 1987), 1–20.

[8] See Gregor Benton, and Alan Hunter, eds., *Wild Lily, Prairie Fire: China's Road to Democracy, Yan'an to Tian'anmen, 1942–1989* (Princeton, NJ: Princeton University Press, 1995); and Boffey, *The Brain Bank of America*.

[9] Wendy Frieman, "People's Republic of China: Between Autarky and Interdependence," in *Scientists and the State*, ed. Etel Solingen (Ann Arbor, MI: University of Michigan Press, 1994), 127–44.

[10] See Alana Boland, "The Three Gorges Debate and Scientific Decision-making in China," *China Information* 13 (1998): 25–42; and Dai Qing (edited in English by Patricia Adams and John Thibodeau), *Yangtze! Yangtze!* (Toronto: EarthScan, 1994).

[11] "To Get Rich is Glorious; China's Middle Class," *The Economist* (January 19, 2002): 33–34.

[12] See Edward Gu, "Social Capital, Institutional Changes, and the Development of Non-governmental Intellectual Organizations," in *Chinese Intellectuals between State and Market*, ed. Merle Goldman and Edward Gu (London: RoutledgeCurzon, 2004), 21–42.

[13] See H. Lyman Miller, *Science and Dissent in Post-Mao China: The Politics of Knowledge*, (Seattle, WA: University of Washington Press, 1996): 283; Jonathan Spencer, "The Limits of Authority," *New York Times Book Review* (August 4, 1996): 20.

[14] See Margaret M. Pearson, *China's New Business Elite: The Political Consequences of Economic Reform*, (Berkeley, CA: University of California Press, 1997).

[15] Bruce Gilley, ". . . But Groups Outside the System Should," *Asian Wall Street Journal Weekly Edition* (February 19–25, 2002), 17.

[16] See Edmund S. K. Fund, "Socialism, Capitalism, and Democracy in Republican China: The Political Thought of Zhang Dongsun," *Modern China* 28, no. 4 (2002): 399–431.

[17] See Weili Ye, *Seeking Modernity in China's Name: Chinese Students in the United States, 1900–1927* (Stanford, CA: Stanford University Press, 2001), 44.

[18] See Goldman, Cheek, and Hamrin, *China's Intellectuals and the State*; and Shiping Hua, "One Servant, Two Masters: The Dilemma of Chinese Establishment Intellectuals," *Modern China* 20, no. 1 (1994): 92–121.

[19] John Israel, "Foreword," in *China's Establishment Intellectuals*, ed. Carol Lee Hamrin and Timothy Cheek (Armonk, NY: M.E. Sharpe, 1986), ix–xix.

[20] Wei ming Tu, "Intellectual Effervescence in China," *Dædalus* 121, no. 2 (1992): 251–92.

[21] Chinese *yuanshi* receive a monthly stipend of RMB 200 ($25) from the state and are not supposed to receive any other benefits. Senior members, those who are over eighty years old, receive an annual stipend of RMB10,000 (US$1,250). However, while elite membership is more than a lifetime honor, Chinese *yuanshi* enjoy the benefit of lifetime employment.

[22] National Academy of Sciences (NAS) of the United States, "Bylaws of the National Academy of Sciences," *Membership Listing* (Washington, DC: NAS, 1997), 94–103.

[23] Xiantang Zhang, "On the Reform of China's *Yuanshi* System: Dialogue with Professor Gu Haibing from the People's University of China" (in Chinese), *Newsweek* 35 (2003).

[24] Probably for the sake of "saving face," the CAS Academic Divisions did not make the news known to the public until it issued the self-discipline ethics code in 2001.

[25] Lilan Zhu, "Basic Research in China," *Science* 283 (January 29, 1996): 637.

[26] *Beijing Qingnianbao*, April 6, 2003.

[27] *Renmin Ribao—Huadong Xinwen* (December 21, 2001): 1.

[28] Chen-lu Tsou, "Science and Scientists in China," *Science* 280 (April 24, 1998): 528–29.

CHINA'S RETURN MIGRANTS AND ITS INNOVATIVE CAPACITY

Claudia Müller and Rolf Sternberg*

Much recent research on innovation activities in industrialized countries has focused on regional innovation systems (RISs),[1] a concept derived from that of national innovation systems (NISs), developed by Lundvall[2] and Nelson.[3] The most important factors are knowledge spillovers, which are most effective over short distances,[4] the similarly distance-sensitive transfer of tacit knowledge,[5] and the significant differences in innovation intensity across regions. The concept of systems of innovation rejects the linear model of innovation in favor of aspects of evolutionary economics.[6]

This focus on geographical proximity, however, leaves several questions unanswered that are pertinent to regions competing with each other in an increasingly globalized world. First, more empirical research on innovation processes in institutionally thin RISs (for example, old industrialized regions, peripheral regions, and fragmented metropolitan regions) is needed.[7] Second, most studies concentrate on RISs in Western Europe or North America;[8] there has been little written on East Asian RISs.[9] Third, almost no empirical research considers extraregional linkages and networks (including international networks) of RISs.[10] Finally, the debate on RISs has not had much to say about entrepreneurship.[11] This chapter intends to fill some of these gaps. The next section attempts to illuminate the role of new knowledge-based firms (NKBFs)

---

* The authors want to thank all the interview partners for the information they provided. Not all the respondents' statements have been incorporated directly into this paper, but they all have contributed to an understanding of the situation. Because all the returning entrepreneurs want to remain anonymous, the verbatim statements in this article refer to "R1," "R2," and so on, instead of referring to names of people or firms. The author talked to more than 100 people in Shanghai, including returning entrepreneurs, domestic entrepreneurs, experts, and intermediaries who helped to arrange an interview. Therefore, the expression "interview R111," for example, does not indicate "interview with returning entrepreneur number 111." Rather, it indicates that the respective interview is number 111 of *all* interviews conducted in Shanghai. The following experts in Shanghai are cited in this paper: Wang Rong (Shanghai Hi-Tech Business Incubator Network), "Expert A" (Roche Pharmaceuticals, Shanghai), and Xue Zi (Shanghai Semiconductor Association).

and their entrepreneurs within RISs, and the remaining sections address the following questions:

- To what extent does Shanghai constitute an RIS?
- Do NKBFs founded by returnees benefit from the Shanghai innovation system?
- What kinds of linkages or networks (extra- and intraregional) are important for the innovation activities of firms?
- How do NKBFs that were started by returnees contribute to the Shanghai innovation system? In particular, do they induce knowledge spillovers?

The Role of Entrepreneurship

A great weakness of the RIS concept, as of all other innovation system concepts, is its neglect of entrepreneurship. This is surprising, since much recent research shows that entrepreneurial activities (defined as start-up activities) significantly affect national growth. Recent intercountry comparisons based on data from the Global Entrepreneurship Monitor show that growth-oriented start-ups (so-called opportunity entrepreneurship) significantly impact economic development,[12] especially high-growth-potential entrepreneurship. A special issue of *Regional Studies* published in 2004, and especially the articles by Acs and Armington[13] and by Fritsch and Müller,[14] argue that entrepreneurial activities are largely "regional events"[15] and that local determinants are more relevant than national or supranational determinants. The current debate on regional-sectoral clusters in particular shows correlations between the existence of such clusters and regional growth. Rocha and Sternberg[16] show for German regions that clusters (that is, geographically proximate groups of interconnected firms and associated institutions in related industries) have an impact on entrepreneurship, whereas industrial agglomerations (that is, clusters without networks) do not.

These results suggest that clusters foster entrepreneurship. Firms neither operate in an atomistic fashion nor interact with others based only on business considerations. Any business activity is embedded in a broader socio-institutional context, and therefore the economic dimensions or relationships cannot be separated from the socio-institutional ones. Accordingly, without socio-institutional dimensions, economic activity declines.[17] Connections of entrepreneurs across regions increase the survival rate of start-ups. From the regional perspective, clusters provide a richer industrial dimension than industrial agglomerations as defined earlier, because they include not only colocation but also interfirm linkages. Sternberg and Litzenberger[18] show that having one or several industrial cluster(s) in a region has a positive impact on the number of start-ups and attitudes within it. Thus, an interdependent relationship between the regional environment (or regional growth) and entrepreneurial activities exists, which makes entrepreneurship a highly relevant topic to the efficacy of

RISs. Space obviously matters for entrepreneurship. However, so far most of the empirical research has been restricted to industrialized economies.

In colloquial English, the word *entrepreneurship* is used in several ways. Here, it means the creation of new firms. New innovative firms, NKBFs in particular, positively influence an RIS in various ways.[19] First, NKBFs are the main innovative actors within a region. Second, they enjoy stronger growth on average and have higher survival rates than other new firms, so that they are a stabilizing element in an RIS. Third, there is a close correlation between the learning capability of (other) regional innovation actors and the number of NKBFs within an RIS. This promotes regional learning.[20] As Lawson[21] and Lorenz[22] have shown, as soon as the capabilities to learn and to forget "old" knowledge have been developed, continuous and collective learning fosters an ongoing extension of the regional knowledge stock. Finally, start-ups in general and NKBFs in particular have the strongest intraregional connections of all innovation actors. Because they rarely change their location, their mainly intraregional innovation linkages, and their personal networks, they contribute significantly to the endogenous development potential of a region. The dangers of a regional lock-in (that is, only intraregional linkages that constrain and limit the process of regional change)[23] must not be overlooked, although entrepreneurial returnees are potentially able to reduce its probability.

This issue is related to the question of how knowledge flows across the boundaries of innovation systems. Howells[24] distinguishes three ways: patents,[25] trade in knowledge-intensive products,[26] and—perhaps most important in our context—the mobility of highly qualified individuals. Florida's work on the geography of talents, as well as the research of others[27] on the correlation between high-tech industries and the locations of star scientists,[28] is relevant. These studies show that a particular regional environment greatly influences the decision of creative people to migrate. They may be self-employed or may work for others, but the former is more interesting here, because only the self-employed can make independent innovation-related decisions. These creative people help deepen connections among RISs, affecting issues of embeddedness, institutions, and conventions,[29] and not only the economic aspects of interaction.

The international background of returnee founders can have strong positive effects on the RIS in which they settle. Figure 13.1 shows these relationships and their effects. The figure differentiates between local NKBFs and those founded by returnees.

Here, the phrase "entrepreneurship by returnees" refers to the work of entrepreneurs who have returned home after studying or working abroad for at least five years. The RIS in the country of origin benefits from the returnees by receiving new knowledge from the outside. These entrepreneurs bring home knowledge from high-tech regions in industrialized countries, resulting in an efficient kind of travel of tacit knowledge.[30] Entrepreneurial migrants are important to the health of active and extroverted firms, which are (besides multinationals) responsible for the external relations of an RIS and for a spatial innovation system.[31] They can contribute to reducing the danger of a regional lock-in in the sense described earlier. They and their interpersonal networks connect an RIS with other RISs and NISs.

Figure 13.1 Entrepreneurial Return Migrants within a Regional Innovation System

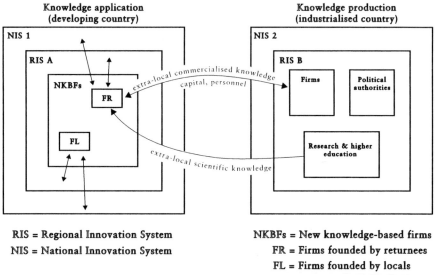

RIS = Regional Innovation System
NIS = National Innovation System
NKBFs = New knowledge-based firms
FR = Firms founded by returnees
FL = Firms founded by locals

Source: Authors' illustration.

Although it is generally accepted that entrepreneurship in the form of knowledge-based start-ups plays an important role in RISs, there has been little empirical research so far that focuses its attention on this area.

NKBFs are not only an advantage to the RIS; they also benefit from being in one. They need the stimulating support of the regional environment more than other firms. More individuals can decide to take the path of self-employment,[32] and existing NKBFs do better. There are various intralocal connections between NKBFs and the surrounding elements of the RIS. NKBFs (and their founders or owner-managers) may also have many connections in other RISs. In contrast to large, established firms or large public research institutions, founder and/or owner-manager connections play a more important role in smaller NKBFs and their linkages within and outside the RIS. These in turn depend on previous experience (for example, connections to previous employers or professors at universities). Here, too, space and place are of central importance.

CLAUDIA MÜLLER AND ROLF STERNBERG

The Shanghai Innovation System

Until the founding of the People's Republic of China in 1949, Shanghai was the nation's economic hub in trade, finance, and light industries.[33] However, after the Communist Party took power, private industry disappeared. Simultaneously, the ratio of light to heavy industry in the city fell from 71:29 in 1957 to 49:51 in 1978. Shanghai lost its economic autonomy and had to remit the majority of its revenues to the center. One of the results of the resource outflow was a chronic underfunding of the local infrastructure.[34]

During the first few years of the reforms, Shanghai was left out because the central government was reluctant to allow experiments that might threaten its revenues. Being excluded seriously damaged its economy. In 1978 the city topped the list of all provinces and regions in contributions to national income; by 1986 Shanghai had fallen to number six, and by 1990 it was number ten.[35]

Shanghai's recovery began with the plan of February 1985.[36] In the early 1990s, the Pudong Policy was devised to revitalize Shanghai with the help of foreign direct investment (FDI) by granting Shanghai a status similar to that of the special economic zones.[37] Another important step was the urban Tenth Five-Year Plan (2001–2005), which proposed that Shanghai become a high-tech hub and business center. To support such a strategic transformation, the traditional six pillar industries of the 1990s (such as automobiles) were replaced by new ones, including high-tech industries.[38]

Of the six high-tech parks in Shanghai, the Zhangjiang High-Tech Park (ZJP) in the Pudong New Area is the locus of new investments in high-tech industries, especially semiconductors, software, and biotechnology. The park administration and the municipal government have aggressively pursued investment—for example, by offering subsidized loans and generous tax exemptions. More recently, Shanghai has paid more attention to the role of entrepreneurship. At the beginning of the reform process, there was doubt about the role of small-scale enterprises. High-tech parks were established mainly to attract FDI and the creation of joint venture firms, but not to promote entrepreneurship.

However, because of the success of nongovernmental enterprises, notably in Beijing, Shanghai began to create innovation centers and incubators for private enterprises within the high-tech development zones and near major universities.[39] By the end of 2004, there were a total of 28 incubators.[40] Additionally, Shanghai established several organizations that finance research and development (R&D) activities of private companies. One of them is Shanghai Venture Capital, Ltd., which was founded in 1999 and supports local projects in information technology (IT), biotechnology, and new materials.

Policies fostering high-tech entrepreneurship focus on two types of entrepreneurs. First, universities encourage faculty members to commercialize their R&D achievements. Second, the city supports entrepreneurship by returnees. Among other measures, the city has established "overseas student parks" within the high-tech development zones reserved exclusively for companies started by them. They not only offer low rent, tax breaks, and so on, as do other science parks in China, but they also address the special needs of returnees, such as by accelerating the bureaucratic process of establishing residency and ensuring access to housing.[41]

Partly as a result of these policies, the number of companies founded by returnees in Shanghai has increased at the rate of one per day since 2002, reaching three thousand and amounting to a total investment of €330 million ($40 million) by the end of 2004.[42] In 2003, in ZJP—the most popular site in Shanghai for entrepreneurs in high-tech sectors—there were approximately five hundred companies founded by returnees.[43]

Shanghai now has the second-highest number of research institutes within the Chinese Academy of Sciences (CAS)—nine CAS institutes, or 9 percent of the national total[44]—and it has two of one hundred "premium" universities. Of the two, Fudan University ranks number six in the number of key state laboratories where "research of strategic importance or high technology" is carried out.[45] Shanghai is among the top five regions in patent applications granted.[46] Table 13.1 shows data about high-tech enterprises in the industrial parks in Beijing and Shanghai in 2003. Beijing outpaces Shanghai in the number of high-tech firms and employees, and in revenue, but Shanghai leads Beijing in exports.

Table 13.1 Indicators of High-Tech Enterprises in Development Areas in Beijing and Shanghai (2003)

	Beijing	Shanghai
Number of firms	12,030	550
Number of employees	488,561	115,009
Revenue (in billion €)	35	20
Export (in billion €)	3.3	6.6

Source: National Bureau of Statistics (NBS), *China Statistical Yearbook* (Beijing: China Statistics Press, 2004), 818.

The Shanghai innovation system has been developing rapidly and is in transition from a site of heavy industry to one with a range of industries (including high-tech) and services. However, it exhibits a weakness typical of old industrial regions: the dominance of large (mostly state-owned) firms, which tend to be concentrated in traditional industries. That state-owned enterprises (SOEs) and large business groups still dominate is demonstrated, for example, by the fact that the six largest SOEs generated 8 percent of total output value in the IT sector in 2000.[47] Table 13.1 also shows that Shanghai's high-tech industries are mostly large enterprises (SOEs and multinationals), whereas Beijing's high-tech output is generated by many smaller enterprises. Thus, although by the mid-1990s there were over seven thousand private companies in Shanghai, technological development there is still not driven by what happens in these enterprises.

Large state-owned firms do not innovate, however. The reason for this is that most SOEs lose money but are kept alive for political reasons through massive loans from the state banking system. These loans are not used to finance R&D activities but to pay employees' salaries. At the same time, private companies run short of capital because the banking system prefers SOEs, thus constraining the innovation activities of private high-tech enterprises.[48] This situation is called a political lock-in and, in industrialized countries, refers to strong relationships between public and private actors who hamper industrial restructuring. In the case of China, *guanxi* (networks of personal relationships) between the government and SOEs constrain the development of the private sector.

Data

Our goal has been to supply some data on the Shanghai innovation system rather than to test hypotheses, so we used systematic instead of random sampling. We requested interviews with founders of high-tech companies. Finding companies founded by returnees required searching for them by establishing networks in the study region, and this meant establishing trust in the course of several visits. These findings are based on in-depth interviews with 31 returning entrepreneurs in the semiconductor and software industries, and with 22 industry and other experts (such as managers of high-tech parks and investment managers), conducted in Shanghai between September and December 2003 and between August and October 2004.[49]

The Semiconductor and Software Industries

In 2001 China had ten domestic integrated circuit (IC) chip plants and approximately 370 IC-design companies.[50] However, most conventional domestic IC companies focus on low-tech chips for consumer goods. In 2000 a new type of company, targeting the middle and high ends of chip design, began to emerge, often started by returnees. Although the technology gap has been narrowing, China is expected to need ten more years to catch up with industrialized nations.[51] The Chinese IC market is characterized by a huge gap between supply and demand. From 2001 through 2004, Chinese IC demand registered a compound annual growth rate of 46 percent. In 2005 China was forecast to have 20 percent of the world's IC demand (with a high proportion met by imports that, in turn, are exported in assembled products), up from 8 percent in 2001.[52] In 2003 indigenous semiconductor production satisfied only 24 percent of the domestic demand.[53] Shanghai stands out as the emerging center of the industry, the only place in China where foundries have been established.

Beijing is by far the most important software region. This is partly due to its prominence as a center for government and leading research institutions.[54]

Shanghai is among the leading regions but ranks fourth in software companies and seventh in the number of employees in the industry. This relatively low position in the hierarchy is surprising, since Shanghai is leading in other high-tech industries, notably in semiconductors.

The Role of Returnees

The activities of returning entrepreneurs are classified here with respect to the level of technology involved, the targeted market, and the motive for starting a company in Shanghai or in China. The goal here is to assess whether their products and services are superior in terms of technology compared to those of domestic companies. Additionally, by examining the difficulties that entrepreneurs encounter when they develop high-tech products, we show how entrepreneurs contribute to the technological upgrading of high-tech industries in Shanghai.

Returnees in the Integrated Circuit Industry

Nine of eleven entrepreneurs surveyed started an IC-design house (see Table 13.2). They are in China to develop a product—the chip and the application software—for either the national or foreign market (activity types 1a and 1b). The remaining two companies are engaged in other types of activity.

The strategy of IC-design house entrepreneurs is to try to offer a product of world-class quality at lower-than-world-class costs. This is unusual because until recently, companies invested in labor-intensive, low-labor-cost, low-tech fields. But salaries for qualified personnel in China are low as well, which makes starting a high-tech company in China very attractive. For example, an engineer's salary in China is only 20 percent of an engineer's salary in the United States—although the average revenue per head in China is also only 20 percent of that in the United States.[55]

The benefits of low labor costs are partly offset by difficulties, however. One difficulty is that IC-design is talent-driven and requires a high degree of creativity and innovation, which makes it challenging to find the right people, because locals are too conservative and inexperienced. It is particularly hard to find experienced engineers, because the Chinese IC industry is so young. Furthermore, the attitude of most workers is rather passive. (To solve these problems, some entrepreneurs try to engage better employees by offering stock options.) Human resource issues can take up to 80 percent of entrepreneurs' time.[56] One major problem for IC design companies is that the training of an engineer takes several years, but fierce competition leads to a high level of employee turnover. To the individual firm this leads to a loss of investment in training activities, although from the point of view of the region, employee turnover can be positive because it results in knowledge spillovers across companies.

There are additional difficulties. Two companies that originally wanted to sell their chips to Chinese customers did not succeed because, as they stated, "the market is not mature," that is, the buyers of chips do not understand the technology. The founders now design chips in China for customers worldwide but hope to sell in China in the future. As a result, one founder's strategy is to offer services for foreign companies in China (activity type 2) and to design chips at the "low end of high tech," because "high tech in China does not sell at this point."[57] Another company is in relatively low-tech equipment manufacturing (activity types 3a and 3b). Both companies are primarily market-driven. The last case is a material manufacturer (activity types 3a and 3b), producing silicon wafers for international customers. The founder imported state-of-the-art equipment worth several million dollars and applied it in a new manner in order to manufacture high-tech material at low costs. Similar to the IC companies, he is pursuing a "cost-driven high-tech strategy."

Nearly all these entrepreneurs returned from Silicon Valley and chose to come to Shanghai because its infrastructure is very good and ZJP is (supposedly) similar to Silicon Valley in style. But according to the founders, ZJP does not have the size—that is, a critical mass of companies—and it lacks the "creative buzz"[58] of Silicon Valley. Therefore, as one entrepreneur noted, "some of the technology has to be done in Silicon Valley, because the environment is still different. In the Valley there are good guys, they get information from other good guys, and they have lunch together every day. Here, it is still different, it looks like S.V., but it is not yet."[59] The challenges seem to be more than offset by the tremendous growth of the Chinese IC market, so returning entrepreneurs have found it relatively easy to raise funds abroad. Also, once established, they receive more support from the government than they would receive in the United States.

Thus, returnees develop products that are superior to those of the domestic industry, and their education and training of employees result in knowledge spillovers.

Returnees in the Software Industry

Seventeen of eighteen entrepreneurs started a company in order to develop a product or set of services for the Chinese market (see Table 13.3). Only two companies consider their products to be an innovation "new to the world." Of those, two are offering systems software (type 2) and the other two are offering sophisticated application software (type 1).

Ten other companies are developing application software that is new in China. For example, one company develops statistics software in Chinese, and another develops enterprise resource-planning software. This requires basic technology from abroad, adapted to the needs of Chinese customers. This is an organizational innovation; the business model was unknown in China at the time the company was started (1999).

Table 13.2 Typology of Integrated Circuit Companies*

Market	Type of activity	Activity and degree of innovation	Motivation for being in China
China			
	Product development (1a)	IC chip design: • Cutting-edge technology (5 cases) • Technology is new in China: better quality and lower costs than domestic companies (1 case) • "At the low end of high tech" (1 case)	Market- and cost-driven
	Technical support (2)	Customer service: High-end services (1 case)	Market-driven
	Production (3a)	Equipment manufacturer: Low-technology content (1 case)	Market-driven
World	Product development (1b)	IC chip design: Cutting-edge technology (2 cases)	Cost-driven
	Production (3b)	Material manufacturer: New application of state-of-the-art technology in order to cut the costs of manufacturing excellent quality (1 case)	Cost-driven

Source: Interviews with returned entrepreneurs in Shanghai.
Note: *The total number of IC companies is 11. One company develops chips and offers services at the same time.

Nine founders are offering software services (Table 13.3, activity types 4 through 6) of varying degrees of technological sophistication; three of them were offering software-outsourcing services. However, developing software for foreign customers is the main business area for only one of the companies.

Table 13.3 Typology of Software Companies

Type of activity	Activity*	Technology level / degree of innovation
Product		
(1)	• Application software (e.g., statistics software, business applications) (12 cases)	• Product is new in China (10 cases) • Product is an innovation new to the world (2 cases)
(2)	• Systems and networking software (including operating systems) (2 cases)	• Product is new in China
Services		
(3)	• Dot-com company (2 cases)	• Medium-level technology, but business model new in China
(4)	• Consulting services (2 cases)	• Low- and high-end services
(5)	• System integration (2 cases)	• Low to medium (e.g., total solutions)
(6)	• Outsourcing for foreign customers (3 cases)	• Low-technology content

Source: Interviews with returned entrepreneurs in Shanghai.
Note: The total number of companies is 18. However, some of them engage in more than one type of activity.

Thus, compared to the world market, the products or services of return migrants are not very innovative, but compared to the domestic market they are very innovative. Quotes from interviewees include:

- "My product is one of the best local products."[60]
- "We are trying to reach Silicon Valley level."[61]
- "Compared to world standard, the technology is medium; in China, I am leading."[62]

As in the IC industry, an immature market and the lack of highly qualified personnel are the main barriers for companies that want to develop innovative products. Many Chinese customers do not appreciate the benefits of modern software and do not buy such products. Returnees need to spend a lot of time educating potential customers and training buyers in using their software.[63] These activities produce knowledge externalities.

Entrepreneurs complained in particular about the difficulties of changing employees' mind-sets, which they perceive as "characterized by a lack of conceptual and designing issues and innovativeness."[64] Software entrepreneurs find it especially difficult to divide the work among various people and to build a team.[65] Competition for employees is less fierce, and training does not take as many years as in IC design. Again, employee turnover is a problem for the individual firm: "After six months' training they can do real work; after two years they leave. It is difficult to keep people because there are so many opportunities."[66]

In contrast to the IC industry, low salaries are not an important motive for starting a company in this sector. Moreover, the entrepreneurs develop products that are new to China but not to the world market. Thus, returnees in the software industry do not pursue a cost-driven, high-tech strategy.

Companies that were started by returnees in these two industries have shown that they predominantly develop products and services that are superior in technology to those developed by domestic companies: IC-design companies develop high-end chips, while most domestic companies sell "low-end" chips. Software companies develop products (systems and application software) that are new in China, while domestic companies mostly offer relatively simple services or application products.

The fact that returnees encounter many difficulties—such as lack of qualified employees and lack of "buzz"—in developing and selling their products is an additional indicator that they are engaging in innovation activities that are so far rare in China.

Thus, returnees contribute in two ways to the technological upgrading of their industries. They contribute *directly* by starting a company in industries that are, so far, underdeveloped. They contribute *indirectly* because many of their activities result in knowledge spillovers.

Returning entrepreneurs need to bring a basic technology (such as a patent) and to integrate in regional and extraregional (national and international) networks. At the same time, the integration of returnees' international networks fosters a continuous flow of knowledge into the region. Regional networks support the daily operations of companies and help them survive in an unpredictable environment. National networks are much less important than either international or regional networks and serve mainly to open up new markets and to outsource development work.

Extraregional Networks of Entrepreneurs in the Integrated Circuit Industry

In the case of international networks, we need to distinguish between formal and informal networks.

Formal Networks

Formal international networks facilitate cross-border R&D collaborations and also help entrepreneurs acquire venture capital from abroad. Two major strategies can be identified: the presence of a Chinese company in the United States and collaboration with investors or partners in the United States.

Some entrepreneurs started an IC company in the United States before returning to China to start another IC company. In these cases, the company in the United States is engaging in top-quality research and is often coordinating the Chinese company's activities as well, while less technology-intensive activities are located in the China-based company. However, the founders stated that the U.S. side was becoming less important over time and that they planned to use more resources in Shanghai in the medium term.

In other cases, chips are partly designed in China and the team is supervised by a returnee. Dividing R&D work between the United States (mostly in Silicon Valley) and China not only allows the entrepreneurs to cut costs but also enables them to develop products more quickly—an important success factor in IC design.[67] As one interviewee commented, "The core development is teamwork between the U.S. and China."[68] Moreover, any company wanting foreign venture capital needs to be registered in China as a subsidiary of a foreign company, because foreign venture capitalists prefer not to invest in Chinese companies due to a lack of exit opportunities.

Collaboration with investors or partners overseas takes various forms. Licensing, consulting, investment (or coinvestment), and joint research—all of these enable the entrepreneurs to get access to knowledge. One IC-design house licensed technologies from abroad and then developed a proprietary technology based on the licensed technology. This requires close collaboration with the technology donor and frequent travel to both the United States and China. According to the founder, it is important to be useful to the licenser "because only this way [can you] promote the company."[69]

Another entrepreneur in the IC industry is also a consultant for a large Japanese company that has subsidiaries in Silicon Valley and wants to enter the Chinese market. Serving in this additional role gives the entrepreneur continued access to current industry information. "First I suffered from the loss of networks (although communication technology is so advanced), but not on a daily basis, face-to-face. That's why I do consulting. I get formal, high-level information, formal networks with the company I do consulting for."[70] Formal collaboration obviously is a means of creating informal, innovation-relevant contacts.

Geographical proximity does not play a role in the informal R&D collaborations of entrepreneurs in the IC industry, a finding supported by the fact that the entrepreneurs are not well integrated into national networks.

To implement the strategies we just described is complex. It is not easy, especially for a small start-up company, to coordinate sites in two countries or to engage in international research collaborations. Increased management complexity also leads to higher expenses, thus countervailing the cost-savings of locating in China.

Informal Networks

Informal networks enable entrepreneurs to access tacit knowledge and to participate in the "buzz" of other RISs. Entrepreneurs also rely on informal networks when recruiting capable people from abroad. A company that does not have a site in the United States is especially likely to assign returnees or foreigners to its senior R&D and management positions. Most entrepreneurs travel several times per year to their former host country, or they participate in international conferences. As one entrepreneur said, "Communication with the U.S. is crucial. I will fly a lot in the future."[71]

Entrepreneurs in the IC industry emphasized the importance of contacts with actors in Silicon Valley. One entrepreneur, for example, said, "I don't want to cut ties with Silicon Valley. It is still the high-tech center. I go back once a month and visit lots of companies there."[72]

Extraregional Networks of Entrepreneurs in the Software Industry

Entrepreneurs in the software industry who sell their products mainly in China need to adapt foreign technologies to that market and do not need to be well informed about developments in foreign markets. One entrepreneur said, "If the knowledge base is not enough, I go back to the States, but it does not happen very often."[73]

Only four entrepreneurs who are developing products new to the world" maintain R&D collaborations with partners overseas (for example, with universities), or they start their company in the United States first in order to acquire venture capital for the company in China and stay informed about developments in foreign markets. "I want to keep in touch with the Silicon Valley community . . . to have a window," said one of these entrepreneurs.[74]

Moreover, entrepreneurs utilize informal networks to recruit software engineers from abroad. In one of the surveyed companies, all seven managers are former employees of Microsoft in the United States.

Although international networks are generally less important for software companies, many entrepreneurs in the software industry rely strongly on international networks for finance; nine of eighteen software companies were financed with foreign capital, seven with foreign venture capital. In contrast to entrepreneurs in the IC industry, for entrepreneurs in the software industry national networks are very important. The main motive for national-level cooperation is to open up the Chinese market or to outsource development work to locations within China where salaries are lower than in Shanghai.

Results

Table 13.4 shows that, in the IC industry, international networks play a much more important role for the entrepreneurs than national networks. This is in contrast with entrepreneurs' experience in the software industry.

Table 13.4 Functions and Importance of Extraregional Networks for Return Migrants in the Integrated Circuit and Software Industries

Type of networks and functions for entrepreneurs	IC	Software
Formal international networks		
• R&D cooperations	+++	O
• Accessing capital (including venture capital)	+++	+++
Informal international networks	+++	
• Recruiting qualified personnel	+++	O
• Accessing tacit knowledge and industry information		O
Formal national networks		
• Opening up the market	+	+
• R&D collaborations (including outsourcing of development work)	+	+
• Accessing capital from Ministry of Science and Technology	+	+

Source: Interviews with returned entrepreneurs in Shanghai.
Notes: +++: Very important; +: Not important; O: Important for companies developing cutting-edge technology

Shanghai's RIS benefits in two ways from the international networks of return migrants. First, it stays open for new knowledge outside the region, which reduces the danger of a technological lock-in. Second, the entrepreneurs' networks foster Shanghai's integration into the global economy, by means of the following:

- Capital imports
- Technology imports
- Exports of products to overseas markets
- Frequent travels of return migrants that connect Shanghai to other parts of the world
- Recruitment of personnel

Types of Regional Cooperation

There are basically two types of regional cooperation: (1) cooperation with universities and R&D institutions, and (2) attentiveness to the party, local government, or bureaucracy. Entrepreneurs in both the IC and software industries collaborate little with local universities and R&D institutions. According to founders, universities do not provide relevant knowledge ("University resources are far from what I need," said one founder),[75] and students do not possess practical knowledge. If entrepreneurs work with universities, it is mainly to recruit personnel or to inform the universities of new products.

However, entrepreneurs of all industries are involved in developing teaching materials for local universities, setting up new courses of study, or building new universities to counteract the lack of qualified personnel. Thus, returning entrepreneurs pursue their own interests and foster the development of the RIS at the same time.

All entrepreneurs try to establish relationships with the party, the local government, or the bureaucracy, because good relations with the government and its bureaucracy are the key to success in China. They are necessary to lower taxes, to get a license faster, and to make business operations easier. As one founder said, "Government officers are humans. They need some fun.[76] I think that's reasonable because they support you and give you some funds. But if your project is not good and you make them approve an unreasonable project, that's dangerous for both sides."[77] Opportunities to develop *guanxi* arise when entrepreneurs are invited to advise the government on high-tech issues, or when highly qualified persons are asked to participate in party training sessions.

Collaborations with Other Actors

Generally, entrepreneurs have few relationships with domestic companies. Those who do have them outsource only simple operations, because of the weak enforcement of intellectual property rights (IPRs) and because of fears that domestic companies might not perform.

Software entrepreneurs work with potential customers mainly to market their products. Returnees in all industries collaborate with regulatory authorities and ministries. For instance, a developer of Chinese statistics software is training employees of the National Bureau of Statistics (NBS) in using his software. Another entrepreneur is involved in developing a new standard for the security of wide area networks (WANs).

Results

Figures 13.2 and 13.3 show the cooperation partners of entrepreneurs and their motivations in the two industries. Fields that are shaded in gray indicate where knowledge spillovers occur. A comparison shows, first, the importance of relationships with the government or bureaucracy. Second, returnees almost always produce knowledge spillovers (with the exception of relationships between returnees and domestic companies, which are mostly pure outsourcing or market relationships).

Figure 13.2 Regional Integration of Returning Entrepreneurs in the Integrated Circuit Industry in Shanghai

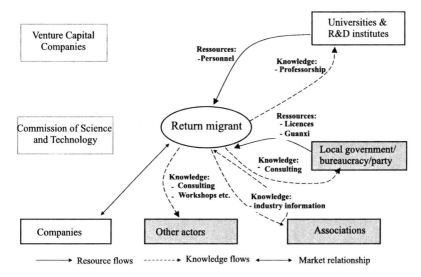

Source: Authors' illustration.

Figure 13.3 Regional Integration of Returning Entrepreneurs in the Software Industry in Shanghai

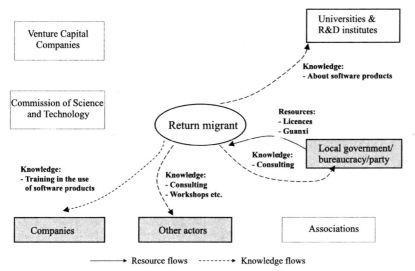

Source: Authors' illustration.

Conclusion

Shanghai is an RIS, albeit one that is still in transition from a manufacturing site and that is hampered by a dominance of SOEs. The roles of entrepreneurial returnees are as follows:

- Most entrepreneurs in the software and IC industries develop products and services that are better in terms of technology than those of domestic companies. Thus, returnees contribute directly to the technological upgrading of high-tech industries in Shanghai.
- Entrepreneurs encounter numerous difficulties. Their activities for the purpose of tackling these difficulties (for example, training employees and educating customers) result in knowledge spillovers and contribute indirectly to the development of high-tech industries.
- Poor framework conditions, such as a lack of IPRs, require that returnees bring core knowledge from abroad, and they need a continuous flow of new knowledge through international networks. This fosters the global integration of the Shanghai innovation system.
- Returning entrepreneurs integrate into regional networks for two main reasons. First, entrepreneurs still face many constraints and need government support. As a side effect, knowledge spillovers occur. Second, regional cooperation is provided mainly by universities and research institutions. However, whereas in industrialized countries the flow of knowledge runs largely from universities to companies, in Shanghai it is the reverse.

Notes

[1] See P. Cooke, M. Heidenreich, and H. J. Braczyk, eds., *Regional Innovation Systems: The Role of Governance in a Globalized World* (London, New York: Routledge, 2004).

[2] B. A. Lundvall, ed., *National Systems of Innovation* (London: Pinter, 1992).

[3] R. R. Nelson, ed., *National Innovation Systems: A Comparative Analysis* (New York, Oxford: Oxford University Press, 1993).

[4] See L. Bottazzi and G. Peri, "Innovation and Spillovers in Regions: Evidence from European Patent Data," *European Economic Review* 45 (2003): 687–710.

[5] See J.R.L. Howells, "Tacit Knowledge, Innovation and Economic Geography," *Urban Studies* 39, nos. 5/6 (2002): 871–84.

[6] See R. Rutten and F. Boekema, "A Knowledge-based View on Innovation in Regional Networks: The Case of the KIC Project," in *Entrepreneurship and Regional Development: A Spatial Perspective*, ed. H.L.F. de Groot, P. Nijkamp, and R. R. Stough (Cheltenham, U.K.: Edward Elgar, 2004): 157–97.

[7] See F. Tödtling and M. Trippl, "Like Phoenix from the Ashes? The Renewal of Clusters in Old Industrial Areas," *Urban Studies* 41, nos. 5/6 (2004): 1175–95.

[8] Exceptions include M. Fromhold-Eisebith, "Effectively Linking International, National, and Regional Systems of Innovation. Insights from India and Indonesia," in *Asian Innovation Systems and Clusters*, ed. P. Intarakumnerd, B. Lundvall, and J. Vang (Basingstoke, U.K.: Palgrave, forthcoming); S. Radosevic, "Regional Innovation Systems in Central and Eastern Europe: Determinants, Organizers and Alignments," *Journal of Technology Management* 27, no. 1 (2002): 87–96; and P. K. Wong, "The Re-making of Singapore's High Tech Enterprise Ecosystem," in *Making IT: The Rise of Asia in High Tech*, ed. H. Rowen, W. Miller, and M. Gong Hancock (Stanford, CA: Stanford University Press, 2006), 123–74.

[9] For an exception, see P. L. Chang and H. Y. Shih, "The Innovation System of Taiwan and China: A Comparative Analysis," *Technovation* 24, no. 7 (2004): 529–39.

[10] See T. G. Bunnell and N. M. Coe, "Spaces and Scales of Innovation," *Progress in Human Geography* 25, no. 4 (2001): 569–89; and P. Oinas and E. J. Malecki, "The Evolution of Technologies in Time and Space: From National and Regional to Spatial Innovation Systems," *International Regional Science Review* 25, no. 1 (2002): 102–31.

[11] Excepting the special issue of *Regional Studies*, no. 8, 2004.

[12] P. K. Wong, Y. P. Ho, and E. Autio, "Entrepreneurship, Innovation and Economic Growth: Evidence from GEM Data," *Small Business Economics* 24, no. 3 (2005): 335–50; and A. van Stel, M. Carree, and R. Thurik, "The Effect of Entrepreneurial Activity on National Economic Growth," *Small Business Economics* 24, no. 3 (2005): 311–21.

[13] Z. Acs and C. Armington, "Employment Growth and Entrepreneurial Activity in Cities," *Regional Studies* 38 (2004): 911–28.

[14] M. Fritsch and P. Müller, "Effects of New Business Formation on Regional Development over Time," *Regional Studies* 38 (2004): 961–77.

[15] M. P. Feldman, "The Entrepreneurial Event Revisited: Firm Formation in a Regional Context," *Industrial and Corporate Change* 10 (2001): 861–91.

[16] H. Rocha and R. Sternberg, "Entrepreneurship: The Role of Clusters; Theoretical Perspectives and Empirical Evidence from Germany," *Small Business Economics* 24, no. 3 (2005): 267–92.

[17] For a comparison between Silicon Valley and Route 128 (in the Greater Boston area), see A. Saxenian, *Regional Advantage: Culture and Competition in Silicon Valley and Route 128* (Cambridge, MA: Harvard University Press, 1994).

[18] R. Sternberg and T. Litzenberger, "Regional Clusters in Germany—Their Geography and Their Relevance for Entrepreneurial Activities, *European Planning Studies* 12 (2004): 767–91.

[19] K. Koschatzky, K. Räumliche Aspekte im Innovationsprozess [Spatial aspects of innovation processes]: (Münster/Hamburg/London: Lit, 2001).

[20] R. Florida, "Toward the Learning Region," *Futures* 27 (1995): 527–36.

[21] C. Lawson, "Territorial Clustering and High-Technology Innovation: From Industrial Districts to Innovative Milieux," Economic and Social Research Council Working Paper Series, 54 (Cambridge: University of Cambridge, 1997).

[22] C. Lawson and E. Lorenz, "Collective Learning, Tacit Knowledge and Regional Innovation Capacity," *Regional Studies* 33 (1999): 305–17.

[23] See W. B. Arthur, "Urban Systems and Historical Path Dependence," in *Cities and their Vital Systems*, ed. J. Ausubel and R. Herman (Washington, D.C.: National Academy of Engineering, 1988), 85–97.

[24] Howells, "Tacit Knowledge."

[25] For an example, see A. Jaffe, "The Real Effects of Academic Research," *American Economic Review* 79 (1989): 957–70.

[26] For an example, see M. P. Feldman, "The New Economics of Innovation Spillovers and Agglomeration," *Economics of Innovation and New Technology* 8 (1999): 5–26.

[27] R. Florida, "The Economic Geography of Talent," *Annals of the Association of American Geographers* 92 (2002): 741–55.

[28] For an example, see L. G. Zucker, M. R. Darby, and M. B. Brewer, "Intellectual Human Capital and the Birth of U.S. Biotechnology Enterprises," *American Economic Review Change* 88 (1998): 290–306.

[29] See Oinas and Malecki, "Evolution of Technologies in Time and Space."

[30] Oinas and Malecki, "Evolution of Technologies in Time and Space," 123.

[31] Oinas and Malecki, "Evolution of Technologies in Time and Space."

[32] See J. Wagner and R. Sternberg, "Start-up Activities, Individual Characteristics, and the Regional Milieu: Lessons for Entrepreneurship Support Policies from German Micro Data," *The Annals of Regional Science* 38 (2004): 219–40.

[33] In 1933 Shanghai's contribution to world trade was larger than the contribution by Japan and Hong Kong together, according to M. Schüller and L. Diep, *Shanghai—Modell für Chinas Wirtschaftsentwicklung?* [Shanghai—A role model for China's economic development?] In *China Aktuell* 30, no. 10 (Hamburg: Institut für Asienkunde, 2001), 1102.

[34] Z. Lin, "Shanghai's Big Turnaround Since 1985: Leadership, Reform Strategy, and Resource Mobilization," in *Provincial Strategies of Economic Reform in Post-Mao China: Leadership, Politics, and Implementation*, ed. P. Cheung, J. H. Chung, and Z. Lin (New York: Armonk, 1998), 50.

[35] A. Segal, *Digital Dragon: High-Technology Enterprises in China* (Ithaca and London: Cornell University Press, 2003).

[36] Segal, *Digital Dragon*.

[37] S. S. Han, "Shanghai between State and Market in Urban Transformation," *Urban Studies* 37, no. 11 (2000): 2091–112.

[38] Segal, *Digital Dragon*.

[39] Segal, *Digital Dragon*.

[40] Interview with Wang Rong of the Shanghai Hi-Tech Business Incubator Network. See an appendix to this chapter for annotations concerning this citation. The appendix is available at <http://sprie.stanford.edu/publications/greater_chinas_quest_for_innovation>.

[41] D. Zweig, "Learning to Compete: China's Strategies to Create a 'Reverse Brain Drain'" (working paper no. 2: Hong Kong: Center on China's Transnational Relations, 2005).

[42] China Economic Information Service, "State-level Business Park to Attract More Returned Overseas Students," *Daily Newsletter of China Xinhua News Agency*, December 31, 2004).

[43] If a founder does not declare or register the company as a domestic corporation, it does not show up in the register. The fact that many of the companies visited have registered as a wholly foreign-owned or joint venture indicates that the real number of companies founded by returnees in Zhangjiang is higher.

[44] Chinese Academy of Sciences, CAS Institutes, <http://english.cas.ac.cn/eng2003/dmk01a/institutes.asp> (March 30, 2005).

[45] H. Liu and Y. Jiang, "Technology Transfer from Higher Education Institutions to Industry in China: Nature and Implications," *Technovation* 21, no. 3 (2001): 183.

[46] NBS, *China Statistical Yearbook* (Beijing: China Statistics Press, 2003), 769.

[47] Segal, *Digital Dragon*.

[48] W. Hong, "An Assessment of the Business Environment for High-Tech Industrial Development in Shanghai," *Environment and Planning C: Government and Policy* 21, Issue 1 (2003): 107–37.

[49] The total number of interviews is larger. However, because this paper follows an industry-specific approach and also is limited in space, entrepreneurs in other industries are not considered. For details, see C. Müller (forthcoming), "Zur Bedeutung von Remigranten für Innovationsprozesse in China. Eine theoretische und empirische Analyse" (On the relevance of re-immigrants for innovation processes in China. A theoretical and empirical analysis), in *Europäische Hochschulschriften*, Reihe 5: Volks- und Betriebswirtschaft, Bd. 3235 (Frankfurt a.M. et al.: Peter Lang).

[50] China Semiconductor Industry Association, *A Report on Development Status of Semiconductor Industry in China* (Shanghai: China Semiconductor Industry Association, 2004).

[51] Interviews with Xue Zi of the Shanghai Semiconductor Association, various dates in 2004.

[52] "China to Become Worldwide IC Market Leader in 2005," *IC Insights*, <http://www.icinsights.com/news/releases/press20050106.html> (January 30, 2005).

[53] China Semiconductor Industry, 2004.

[54] T. Tschang and L. Xue, "The Chinese Software Industry: A Strategy of Creating Products for the Domestic Market" (ADB Institute Working Paper, 2003).
[55] Interview R35, various dates in 2003.
[56] Interview R35.
[57] Interview R61.
[58] The creative "buzz" has become a topic of research lately; interestingly, this expression was used by one of the respondents himself. See, for example, H. Bathelt, A. Malmberg, and P. Maskell, "Clusters and Knowledge: Local Buzz, Global Pipelines and the Process of Knowledge Creation," *Progress in Human Geography* 28, no. 1 (2004): 31–56.
[59] Interview R114.
[60] Interview R46.
[61] Interview R66.
[62] Interview R44.
[63] Interview R40.
[64] Interview R109.
[65] Interviews R66, R46, and R36.
[66] Interview R109.
[67] Interview R107, IC.
[68] Interview R35, IC.
[69] Interview R114, IC.
[70] Interview R111, IC.
[71] Interview R107, IC.
[72] Interview R114, IC.
[73] Interview R46.
[74] Interview R14.
[75] Interview R35.
[76] The founder here alludes to the common practice of inviting officials to dinner and other kinds of amusement.
[77] Interview R109.

Statistical Indicators: Patents and Journals

WHAT DO THEY PATENT IN CHINA, AND WHY?

Albert Guangzhou Hu[*]

China is experiencing an explosion in patents. An important aspect of the explosion is that foreigners accounted for two-thirds of all invention patents granted by China's State Intellectual Property Office (SIPO) in 2004. In contrast, the share of foreign applicants of all patents granted in the United States is about 50 percent, and in Japan it is 10 percent. Also, Chinese patents granted to foreign applicants have been growing at a higher rate than those granted to domestic applicants.

What is behind such aggressive acquisition of intellectual property rights (IPRs) in China when their enforcement is weak at best? To what extent is this surge driven by the need of foreigners to protect their proprietary technologies against Chinese imitators? And how much of it is driven by competition among foreign investors in the Chinese market? Such interactions imply a strategic value in patent protection there.

Hu and Jefferson[1] were the first to investigate the forces behind China's patent explosion. They found several economic forces contributing to it, including foreign direct investment (FDI), propatent legislation, and ownership reform. The data for large- and medium-size Chinese enterprises that they used covered only patents granted to such enterprises, an important but small portion of all SIPO patents. The enterprise data did not allow them to differentiate between invention and utility model patents either. Utility model patents, which dominate in numbers, arguably carry much less technology than invention patents but receive equal legal protection during their shorter statutory lives.

I have constructed a data set that provides applicant and technological information on all 1.37 million patents that the SIPO granted from 1985 to 2004. I then use the Organisation for Economic Co-operation and Development (OECD) Technology Concordance (OTC) to assign them to International Standard Industrial Classification (ISIC) industries. With this data set and

---

[*] I would like to thank Gary Jefferson and participants in the CISTP (Tsinghua/SPRIE [Stanford]) workshop, Tsinghua University (Beijing, China, May 20–21, 2006), for their useful comments. Li Jia provided superb research assistance. I also gratefully acknowledge the financial support of the National University of Singapore Academic Research Grant (R-122-000-091-112).

similar data from the U.S. Patent and Trademark Office (USPTO) as a reference, I investigate a number of questions:

- In what industries are Chinese and foreign inventors taking out SIPO patents?
- Is foreign patenting in China driven principally by increased patenting by domestic Chinese inventors, or by that of inventors from other foreign countries?
- Does Chinese patenting behavior react to foreign patenting in China beyond what is predicted by technological opportunity and a propensity to patent?
- What is the role of utility model patents?
- To what extent is patenting driven by trade and FDI?

The Chinese Patent System and Chinese Patents

China reinstituted its patent law in 1985. The second amendment to it, in 2000, largely brought it in line with the international norm.[2] The SIPO grants three types of patents: invention, utility model, and design. Since design patents apply mostly to new, original, and ornamental design for an article or manufacture, I omit them here.

An invention patent must meet the criteria of usefulness, novelty, and creativity. The requirements for utility model patents are different; while invention patent applications are subject to "substantive examination," which requires the patent examiner to conduct a search of prior works and ensure that the three criteria are met, utility model applications do not receive substantive examination and are basically granted on a registration basis. As a result, the application cycle for utility model patents is much shorter than that of invention patents. Nevertheless, the same postgrant reexamination and opposition procedure that applies to invention patents also applies to utility model patents.

Once they are granted by the SIPO, invention patents receive a life of legal protection of twenty years from the date of application and are subject to annual renewal; utility model patents are protected for ten years.

Utility model patents, sometimes referred to as "petty" patents, are usually marginal improvements in technology. They tend to be justified on the grounds that such innovations can be imitated easily and that these patents are useful in protecting the intellectual property (IP) of small enterprises. The USPTO does not grant utility model patents, although many jurisdictions—including Japan, Korea, Taiwan, France, and Germany—do grant them.[3]

Another idiosyncratic feature of China's patent system is the high proportion of individual patents, as distinct from patents that are assigned to institutions. Those that remain unassigned or that are assigned to individuals at the time they are granted accounted for about 17 percent of all USPTO patents granted to U.S. inventors in 2004, but they accounted for 33 percent of SIPO patents granted to Chinese applicants. Part of the explanation is that the intellectual property rights (IPRs) granted to an employee's invention are not clearly defined and assigned in Chinese organizations; therefore, patents that could have belonged to the inventor's employers have been assigned to the inventors themselves or to other individuals.

Top Patenting Countries

Countries with the most patents granted by the SIPO and the USPTO are listed in Table 14.1, along with the number of patents granted in 2004 and growth rates from 1995 through 2004. It is not a coincidence that the same countries are on both lists. In both China and the United States, Japan, Germany, Korea, and Taiwan have been granted more patents than other foreign countries.

Table 14.1 Patenting in China and the United States

	China SIPO				USPTO	
	Invention		Utility model		Utility	
	Growth rate (%) 1995–2004	Count 2004	Growth rate (%) 1995–2004	Count 2004	Growth rate (%) 1995–2004	Count 2004
China	25.60	15,733	9.00	60,561	20.80	404
United States	28.20	6,572	20.80	218	4.60	84,271
Japan	36.60	12,439	12.40	217	5.40	35,350
Germany	34.90	3,043	15.40	16	5.50	10,779
Taiwan	39.90	1,773	9.90	7,424	14.40	5,938
Korea	58.20	2,267	3.30	70	14.90	4,428
All patents	30.10	49,054	9.10	68,889	5.40	164,293

Source: Author's calculation using SIPO and USPTO data.

Although the number of invention patents granted to Chinese applicants has been growing annually by almost 26 percent, inventors from the United States, Japan, Germany, Taiwan, and Korea have been taking out Chinese invention patents at even faster rates. The number of Korean invention patents in the SIPO, for example, has been growing by 58 percent a year.

Table 14.1 also shows that utility model patents still dominate in the SIPO, particularly from China, but that they have not been growing nearly as fast as invention patents; this suggests a rapid improvement in their average quality. Utility model patents are dominated by Chinese and Taiwanese applicants, who together are responsible for over 98 percent of the total. (Taiwan's share—11 percent of SIPO utility model patents—is an interesting phenomenon.)

Compared with China, the growth of USPTO patents has been modest. Those from the world's three top inventors—the United States, Japan, and Germany—grew about 5 percent a year from 1995 to 2004. Those granted to Korea, Taiwan, and particularly China, grew at three to four times the world average rate of 5.4 percent. The much faster rate of foreign patenting in China than in the United States suggests that it is unlikely to have been driven by faster knowledge production in those foreign countries. They must be patenting a larger proportion of their inventions in China.

Countries in the Technology Space

A country's research and development (R&D) outlay is spread over many technological fields, and the distribution is shaped by the technology opportunities in them. One would expect the existence of more such opportunities to lead to more R&D and to more patenting. Countries that do research in similar fields are therefore likely to see a concurrent increase in innovation outputs (measured by patents). To understand the rapid growth in SIPO patents granted to foreign entities, I first locate the relative position of countries in the technology space.

When a patent is granted, a SIPO patent examiner assigns it a primary technology class and one or more secondary technology classes, using the International Patent Classification (IPC) system, which is used by all national patent offices. The IPC is a hierarchical system in which all technologies are divided into eight primary sections, with each technology subdivided into patent classes and subclasses. At the lowest level of the hierarchy are technology groups.[4] Therefore, the IPC is a natural instrument to position countries in technology space.

I use an uncentered correlation between the technology class distributions of the patents of two countries to measure how close they are in technology space.[5] I define the technology space as a 164-dimension space corresponding to the 164 IPC patent classes. The innovation activity of each economy is then projected onto this technology space, as represented by a 164-element vector, with each element occupied by the share of SIPO patents that the economy receives in that year in the relevant patent class. The technology proximity between two economies, i and j, in year t, is defined as:

$$TP_{ijt} = \frac{V_{it}' V_{jt}}{\sqrt{V_{it}' V_{it}} \sqrt{V_{jt}' V_{jt}}} \quad \text{Equation (14.1)}$$

where V_{it} is a 164-element vector of patent class shares of country i's SIPO patents granted in year t. TP is bounded between 0 and 1 and is increasing in the similarity between the two countries' patent portfolios.

I calculate pair-wise technology proximity between countries using SIPO data and report the results in Table 14.2. The top panel of Table 14.2 is based on patents granted to China, Germany, Japan, Korea, Taiwan, and the United States from 1995 through 1997. The bottom panel uses the last three years of data. For China, I use both utility model patents ($China_U$) and invention patents ($China_I$) to compute technology proximity with others; only the invention patents are used among the other countries.

Table 14.2 Technology Proximity

	$China_U$	$China_I$	Germany	Japan	Korea	Taiwan	United States
			1995–1997				
$China_I$		1	0.59	0.63	0.33	0.46	0.62
Germany	0.34	0.59	1	0.87	0.32	0.51	0.85
Japan	0.81	0.63	0.87	1	0.48	0.58	0.85
Korea	0.84	0.33	0.32	0.48	1	0.42	0.46
Taiwan	0.86	0.46	0.51	0.58	0.42	1	0.43
United States	0.57	0.62	0.85	0.85	0.46	0.43	1
			2002–2004				
$China_I$		1	0.72	0.54	0.4	0.39	0.79
Germany	0.59	0.72	1	0.77	0.62	0.51	0.89
Japan	0.75	0.54	0.77	1	0.89	0.76	0.86
Korea	0.87	0.4	0.62	0.89	1	0.61	0.78
Taiwan	0.75	0.39	0.51	0.76	0.61	1	0.7
United States	0.49	0.79	0.89	0.86	0.78	0.7	1

Source: Author's calculation using SIPO and USPTO data.

Using utility model patents as an indicator for *imitation* and invention patents to represent *invention*, Table 14.2 shows that China's imitative efforts align with the technologies of Japan, Korea, and Taiwan, whereas China's inventions mostly track the technologies that Germany, Japan, and the United States patent in China. (Note that Japan appears in both categories.) Although the technology proximity measure does not tell us about causality, there is much overlap between what Chinese applicants imitate (utility patents) and what Japan, Korea, and Taiwan patent in China. In contrast, the pattern of Chinese inventions is highly correlated with patenting choices made by Germany, Japan, and the United States.

Table 14.2 shows that the correlation between the patent class distributions of China on the one hand, and Germany and the United States on the other hand, has increased over time. From the mid-1990s through 2004, the technology proximity between China and Germany measured by invention patents increased from 0.59 to 0.72, while it increased from 0.62 to 0.79 between China and the United States. China's imitation effort also moved closer to Germany's invention patenting over the same time period, from 0.34 to 0.59.

Among foreign patent applicants, the portfolios of Taiwan and Korea were closely aligned with the others—as well as with each other. For instance, in the mid-1990s, the degree of overlap between the portfolios of Korea and Japan was 0.48; toward the end of the sample period, the overlap increased to 0.89. The proximity between Korea and the United States increased from 0.46 to 0.78. The proximity of Germany, Japan, and the United States has been high from the beginning but has been stable or has even declined slightly over time.

Without attempting to draw causal inferences, the technology proximity inquiry reveals several interesting patterns. China's imitation is focused on fields where Japan, Korea, and Taiwan take out SIPO patents; the high-technology proximity of Germany, Japan, and the United States shows that the patenting strategies of these leading innovators are highly correlated; and finally, Korea and Taiwan have been increasingly patenting in fields previously dominated by Germany, Japan, and the United States.

Matching Patents to Industry

A difficulty in using data compiled by patent offices to study innovation is that it does not tell us which industries the applicants come from. It is difficult to use patent data to understand resource-allocation decisions. Economists have been trying to overcome this deficiency since the early efforts of Schmookler and Scherer.[6] Schmookler focused on patents related to capital goods inventions and aggregated patents from a number of patent subclasses that he determined were relevant to a given industry. Scherer went to the firm level and aggregated firm patent totals into industry totals. Both approaches have their limitations in that only a small portion of the vast universe of patent data has been used.[7]

There has been no effort to match Chinese SIPO data to firm- or industry-level economic variables. The data that Hu and Jefferson[8] used, perhaps the best available, contains a firm-level patent count, but it does not differentiate between invention and utility model patents. Only large and medium-size Chinese firms are in the database, which accounts for 8.7 percent of all patent applications at the SIPO in 2001.

The OECD Technology Concordance

Another approach was undertaken by a group of economists at Yale University.[9] They constructed the Yale Technology Concordance, using the practice of the

Canadian Patent Office, which assigned, in addition to IPC patent classes, an industry of manufacture (IOM) and a sector of use (SOU) for each of the over 300,000 patents granted between 1972 and 1995. This is a natural base for constructing a concordance between patent classes and the IOM and SOU of the patents. Assuming that such a concordance remains stable over time and across countries, one can use it to study patents granted by any national patent office that assigns IPC classes to the patents it grants.[10]

I aggregate SIPO patent data to three-digit ISIC industries, using the OTC, which assigns to each IPC patent class a probability that patents from this IPC class belong to a three-digit ISIC IOM and a different probability that these patents belong to an SOU, also at the three-digit ISIC level. With this matrix of probabilities, I can then assign most patents—the OTC does not cover all IPC classes—to three-digit ISIC industries. I apply the OTC to both SIPO and USPTO data, and use IOM to classify patents.[11]

Top Patenting Industries

In this section I concentrate on patents from manufacturing industries, since most R&D and patents are in those industries. Figures 14.1 and 14.2 provide histograms of the manufacturing industry distribution of patents.

Figure 14.1 contrasts the distributions of SIPO invention patents and USPTO patents for manufacturing industries. The two distributions look quite similar. For both, most patents go to the machinery industry (ISIC 290). For the SIPO, the next three major patenting industries are other chemicals (ISIC 242), other radio and TV (ISIC 320), and basic chemicals (ISIC 241). At the USPTO, other radio and TV, medical instruments (ISIC 331), and office equipment (ISIC 300) are the next three industries.

A comparison of the distributions of SIPO utility model patents and USPTO patents in Figure 14.2 shows more differences between Chinese and U.S. patents than Figure 14.1. SIPO utility model patents are much more concentrated than USPTO patents. The lone spike of the machinery industry highlights the difference; machinery accounts for nearly 40 percent of all utility model patents, whereas it accounts for less than 30 percent of invention patents. Chemical industries account for nearly a quarter of invention patents, but less than 10 percent of utility model patents.

Figure 14.1 Industry Distribution of SIPO Invention Patents and USPTO Patents

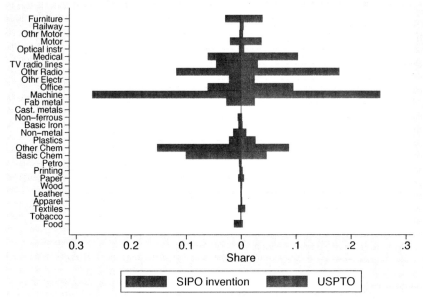

Source: SIPO and USPTO data.

Figure 14.2 Industry Distribution of SIPO Utility Model Patents and USPTO Patents

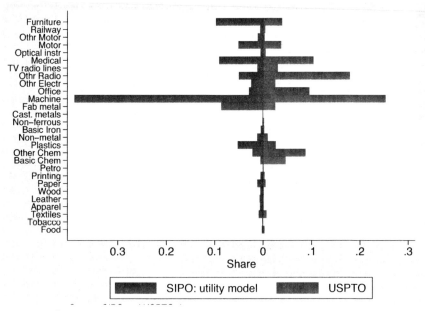

Source: SIPO and USPTO data.

ALBERT GUANGZHOU HU

Why Do They Patent So Much?

The number of patents that a country applies for and is granted in China and its technological areas are determined by these factors:

- The country's R&D and the new knowledge produced
- The propensity to patent
- The strength of IPR protection
- The extent of market contact between the foreign country's firms and Chinese firms and other foreign firms
- Strategic motivation: preemptive entry, limiting the scope of IP

I use the following model in examining the causes of the rapidly rising number of Chinese SIPO patents:

$$P^C_{kijt} = \sum_{\substack{n=1 \\ n \neq k}}^{6} \alpha_n P^C_{nijt} + \sum_{\substack{n=1 \\ n \neq k}}^{6} \beta_n P^U_{nijt} + \gamma \hat{P}^C_{cijt} + f(D_i, D_t) + v_{kijt} \quad \text{Equation (14.2)}$$

where P is the number of invention patents, and superscripts C and U denote China SIPO patents and USPTO patents, respectively. $\hat{P}$ is the number of utility model patents taken out by Chinese inventors. The letter k denotes the country of the patent applicant—that is, k = China, United States, Japan, Germany, Korea, and Taiwan. The industry (three-digit ISIC), the IPC class (three-digit) within the industry, and the year the patent is granted are represented by i, j, and t, respectively. The last term on the right side of equation 14.2, $f(D_i, D_t)$, captures industry-specific and grant-year-specific fixed effects. In estimating equation (14.2), I enter all patent count variables into a natural logarithm.

The inclusion of the USPTO patents in equation (14.2) accounts for the effects of knowledge production of a country, knowledge spillover between countries, and the propensity to patent in a technological field. The industry and grant-year controls account for time-invariant, industry-specific characteristics, such as propensity to patent and openness in terms of FDI and trade, the changing/strengthening of the patent-protection regime in China, and other economy-wide macroeconomic shocks.

After controlling for these effects, the estimation of parameters of primary interest, α's and γ, allow me to gauge the impact of the change over time of market contact among firms from various countries on their patenting decisions and the extent to which the intensity of patenting is related to strategic interaction among these firms.

Results and Discussion

The regression results are reported in Table 14.3. Columns 1 and 3 through 7 show the estimates of equation (14.2) for China, the United States, Japan, Germany, Korea, and Taiwan. The dependent variable of column 2 is the number of China SIPO utility model patents granted to Chinese applicants. Full sets of year and industry (three-digit) dummies are included in each regression.

China's SIPO Patents

There is almost no relationship between China's invention patents and utility model patents. Columns 1 and 2 of Table 14.3 show that the partial correlation between the two types of patents is around 0.03, with a significance level of 5 percent, which is much lower than the correlation between China's invention patents and those of the United States, Japan, Germany, Korea, and Taiwan, which range from 0.049 to 0.356 (all significant at the 1 percent level). This suggests that Chinese invention and utility model patents serve different purposes. It suggests also that Chinese patenting responds to foreign patenting in China beyond what the propensity to patent and knowledge spillovers would predict.

The first two columns of Table 14.3 show also that Chinese invention patents and utility model patents interact differently with the invention patents of other countries. While invention patents are all positively correlated with each other, China takes out more utility model patents in fields where Japan, Korea, and Taiwan receive more invention patents. The fact that utility model patents are imitative rather than innovative suggests that China actively imitates the technologies of Japan, Korea, and Taiwan while conducting more inventions in the areas where the United States and Germany patent intensively.

Another interesting observation can be summarized by the following inequalities:

$$Corr(P^C_{China}, P^C_j) > Corr(P^C_{China}, P^U_j), \quad j = Japan, Korea, Taiwan \quad \text{Equation (14.3)}$$

While China's SIPO invention patents are positively correlated with the SIPO invention patents of Japan, Korea, and Taiwan, they are negatively or uncorrelated with these economies' USPTO patents in the same technological fields. In other words, companies in these respective economies adopt different patenting strategies in China and the United States: where they patent actively in China, they patent little in the United States, and vice versa.

This pattern is consistent with the strategic interaction hypothesis and the market contact hypothesis. China conducts the greatest amount of trade with Japan, Korea, and Taiwan; firms from these economies have significant contact with Chinese firms and therefore want to protect their proprietary technologies in China.[12] However, comparative advantage dictates that Chinese firms not compete with Japanese,

Table 14.3 Determinants of SIPO Patenting

	China^I	China^U	USA	Japan	Germany	Korea	Taiwan
P^C_{China}		0.038*	0.057**	0.131**	0.061**	0.04**	0.125**
		(0.02)	(0.01)	(0.01)	(0.02)	(0.02)	(0.02)
P^C_{China}	0.03*		-.148**	0.052**	0.009	0.184**	0.09**
	(0.01)		(0.01)	(0.01)	(0.01)	(0.01)	(0.01)
P^C_{USA}	0.124**	-.409**		0.225**	0.399**	0.153**	0.084**
	(0.03)	(0.03)		(0.02)	(0.02)	(0.02)	(0.03)
P^C_{Japan}	0.356**	0.182**	0.282**		0.333**	0.496**	0.242**
	(0.03)	(0.03)	(0.02)		(0.03)	(0.03)	(0.03)
$P^C_{Germany}$	0.082**	0.016	0.247**	0.164**		0.041	0.044*
	(0.02)	(0.02)	(0.01)	(0.02)		(0.02)	(0.02)
P^C_{Korea}	0.049**	0.285**	0.086**	0.221**	0.037		0.185**
	(0.02)	(0.02)	(0.01)	(0.01)	(0.02)		(0.02)
P^C_{Taiwan}	0.133**	0.121**	0.041**	0.095**	0.034*	0.162**	
	(0.02)	(0.02)	(0.01)	(0.01)	(0.02)	(0.02)	
P^U_{USA}	0.812**	1.092**	0.668**	-.323**	-.391**	-.131**	0.118*
	(0.04)	(0.05)	(0.03)	(0.03)	(0.04)	(0.04)	(0.05)
P^U_{Japan}	-.519**	-.848**	-.106**	0.475**	-.191**	-.084**	0.098**
	(0.03)	(0.03)	(0.02)	(0.02)	(0.03)	(0.03)	(0.03)
$P^U_{Germany}$	0.367**	0.296**	0.022	-.096**	0.714**	-.460**	-.148**
	(0.03)	(0.03)	(0.02)	(0.02)	(0.03)	(0.03)	(0.03)
P^U_{Korea}	0.005	-.274**	-.060**	-0.015	-.079**	0.577**	-.128**
	(0.03)	(0.03)	(0.02)	(0.02)	(0.02)	(0.02)	(0.03)
P^U_{Taiwan}	-.489**	0.482**	-.104**	0.07**	0.042*	-0.025	0.233**
	(0.02)	(0.03)	(0.02)	(0.01)	(0.02)	(0.02)	(0.02)
Obs.	3,976	3,976	3,976	3,976	3,976	3,976	3,976
R^2	0.837	0.833	0.916	0.935	0.871	0.873	0.842

Source: Author's regression analysis using China SIPO and USPTO data.
Note: All regressions include full sets of industry and year dummies.
* significant at 95-percent level
** significant at 99-percent level

Korean, and Taiwanese firms in the U.S. market. Hence the negative—or lack of—correlation between Chinese SIPO invention patents and these economies' USPTO patents.

Foreign Patenting in China

All foreign patent applicants react positively to China's invention patents. This result is particularly prominent for Japan, Korea, and Taiwan. Chinese patents—invention and utility model patents combined—have a partial correlation with Japan's SIPO patents of 0.183, which is greater than Japan's correlation with Germany and Taiwan. For Korea and Taiwan, such correlation reaches 0.22, making Taiwan and Korea more closely related to China than they are to Japan. Korea's SIPO invention patents are particularly highly correlated with Chinese utility model patents.

However, the fact that SIPO patents from Korea and Taiwan are even more highly correlated with Japan's SIPO patents indicates that competition among foreign firms in China is one of the driving forces of the foreign-patenting surge in China. Taiwan's SIPO patents, for instance, are more correlated with Japan's SIPO patents than even with their own USPTO patents.

The United States and Germany, in contrast, are closer to each other and to Japan than they are to China, Korea, and Taiwan. It is interesting that patenting by the United States and Germany is either negatively correlated or uncorrelated with Chinese utility model patents. Either Chinese imitation eschews areas where the United States and Germany patent intensively, or it does not create a competitive threat to the interests of American and German firms in China.

While foreign firms generally patent in similar fields in China, this is not true in the United States. For example, Germany's SIPO invention patents are significantly correlated with those of the United States and Japan, but they are significantly negatively correlated with the USPTO patents of U.S. and Japan firms. This result applies to almost every other foreign country. One possibility is that foreign firms compete against each other in a smaller range of products and markets than they do in the U.S. market. This leads them to patent in similar technological fields.

Conclusion

The patents granted by China's SIPO to foreign applicants have been growing at a higher rate than those granted to domestic Chinese applicants. The fact that foreign patenting in China has been growing at three to five times its rate in the United States suggests that this explosion cannot be simply a result of a higher level of invention in the foreign countries.

I find that even after allowing for determinants of patenting such as knowledge production, propensity to patent, and the strength of China's IPR protection regime, there is a significant correlation between patents granted to foreign countries and those granted to domestic Chinese applicants. The degree of interaction is particularly strong between Chinese patenting and East Asian patenting in China—that is, patents granted to applicants from Japan, Korea, and Taiwan, and among these East Asian economies.

The results are consistent with the bilateral patterns of China's trade with these countries. Competition between foreign firms, and between foreign firms and domestic Chinese firms, is likely to be an important driving force behind China's patent explosion and the rapid growth of foreign patenting in China.

Notes

[1] Albert Guangzhou Hu and Gary H. Jefferson, "A Great Wall of Patents: What Is Behind China's Recent Patent Explosion?" unpublished, 2005, <http://courses.nus.edu.sg/course/ ecshua/.alberthu.htm.>

[2] For a more detailed discussion of the evolution of China's patent system, see Hu and Jefferson, "A Great Wall of Patents."

[3] The following countries and organizations grant and protect utility model patents: Argentina, African Regional Intellectual Property Organization (ARIPO), Armenia, Australia, Austria, Belarus, Belgium, Brazil, Bulgaria, China, Colombia, Costa Rica, Czech Republic, Denmark, Estonia, Ethiopia, Finland, France, Georgia, Germany, Greece, Guatemala, Hungary, Ireland, Italy, Japan, Kazakhstan, Kenya, Kyrgyzstan, Malaysia, Mexico, Netherlands, African Intellectual property Organization (OAPI), Peru, Philippines, Poland, Portugal, Republic of Korea, Republic of Moldova, Republic of Trinidad and Tobago, Russian Federation, Slovakia, Spain, Tajikistan, Turkey, Ukraine, Uruguay, and Uzbekistan. See the World Intellectual Property Organization (WIPO) website at <http://www.wipo.int/>.

[4] The eight primary sections are coded A to H. For example, Section C is titled "Chemistry; Metallurgy." "Fertilisers; Manufacture thereof" is a class under Section C, coded C05. Under C05 are several subclasses, one of which is C05C, defined as "Nitrogenous fertilizers." Going one level down, one group under subclass C05C is "Ammonium nitrate fertilizers," with the code C05C 1/00. For a detailed description of IPC, see the WIPO website at <http://www.wipo.int/classifications/en/>.

[5] Adam B. Jaffe used uncentered correlations as weights to construct a measure of knowledge pool that he found to be highly correlated with firm performance. See Adam B. Jaffe, "Technological Opportunity and Spillovers of R&D: Evidence from Firms' Patents, Profits, and Market Value," *American Economic Review* 76, no. 5 (December 1986): 984–1001.

[6] Jacob Schmookler, *Invention and Economic Growth* (Cambridge: Harvard University Press, 1966); F. M. Scherer, "Corporate Inventive Output, Profits, and Growth," *Journal of Political Economy* 73, no. 3 (1965): 290–97; and F. M. Scherer, "Firm Size, Market Structure, Opportunity, and the Output of Patented Inventions," *American Economic Review* 55 (1965): 1097–1125.

For an early survey and discussion of the data and conceptual issues involved in studying patent statistics, see Zvi Griliches, "Patent Statistics as Economic Indicators: A Survey," *Journal of Economic Literature* 28, no. 4 (December 1990): 1661–1707.

[7] In constructing the National Bureau of Economic Research (NBER) patent and patent citations database, Bronwyn H. Hall et al. undertook the task of matching the name of a patent assignee to the name of a public, listed company, so that patent data could be matched with company balance-sheet and income-statement data. This is a big step forward, but still only publicly listed companies are matched. See Bronwyn H. Hall, Adam B. Jaffe, and Manuel Trajtenberg, "The NBER Patent Citation Data File: Lessons, Insights and Methodological Tools" (NBER working paper #8498, 2001).

[8] Hu and Jefferson, "A Great Wall of Patents."

[9] Robert E. Evenson and Daniel Johnson, "Introduction: Invention Input-Output Analysis," *Economic Systems Research* 9, no. 2 (June 1997): 149–60; and Samuel Kortum and Jonathan Putnam, "Assigning Patents to Industries: Tests of the Yale Technology Concordance," *Economic Systems Research* 9, no. 2 (June 1997): 161–75.

[10] The original industry assignments that the Canadian Patent Office used and that the Yale Technology Concordance adopted were based on the Canadian Standard Industrial Classification (SIC) system, which is different from the International Standard Industrial Classification (ISIC) system. ISIC is used internationally to define economic sectors. Daniel K. N. Johnson added another layer of translation from the SIC to the ISIC to construct the OECD Technology Concordance (OTC). See Daniel K. N. Johnson, "The OECD Technology Concordance (OTC): Patents by Industry of Manufacture and Sector of Use" (OECD STI Working Paper Series, March 2002).

[11] The USPTO started IPC patent class assignment in 1975, although the USPTO patent data goes back to 1963.

[12] In 2003 China ran a trade surplus of $58.6 billion with the United States and trade deficits of $40.3, $23, and $14.7 billion, respectively, with Taiwan, Korea, and Japan, according to the statistics of China's NBS, <www.stats.gov.cn/english>.

A COMPARATIVE ANALYSIS OF CHINESE AND INTERNATIONAL AGGREGATED CITATION RELATIONS AMONG JOURNALS

Ping Zhou and Loet Leydesdorff*

For more than a decade, the scientific production of China has shown exponential growth.[1] In terms of the world's share of publications, China advanced from seventeenth place in 1993 to second in 2007. Figure 15.1 shows that both Chinese publications and citations have grown exponentially.

At the Annual Science and Technology Conference of China held in January 2006, a new Medium- and Long-Term Science and Technology Development Plan was made public. According to this plan, China is striving to become an innovative country by the year 2020, and the role of its scientific publications is crucial. The number of citations received by publications is considered a more important indicator of innovation by scientometricians, since citations show the visibility or impact of scientific output. Although the number of citations received by Chinese publications continues to increase, it is not growing as rapidly as that of publications. China's position advanced from seventeenth place in 1993 to fifth in 2004 in the world's share of publications, but it ranked only fourteenth in the share of citations. Why is this? Can Chinese science and technology (S&T) institutions make a larger contribution to knowledge?

The visibility of Chinese journals and the citations of Chinese publications are important for Chinese scientists' communications with domestic and international counterparts. It works in two ways. On the one hand, articles in journals with high visibility have better chances of being read, therefore bringing more citations. On the other hand, high-quality articles raise the visibility of a journal. Exploring citation patterns as indicators of the domestic and international visibility of Chinese journals helps us to understand the relatively low citation impact of Chinese papers.

This study is based on two databases: the China Scientific and Technical Papers and Citations Database (CSTPCD) and the *Science Citation Index (SCI)*. Among the issues we examine are a comparison of average citation patterns of Chinese journals with their international counterparts, and citation patterns of

---

* We are grateful to the statistics team of the Institute of Scientific and Technical Information of China (ISTIC) for supplying relevant data, especially Ma Zheng for providing us with the aggregated journal-to-journal citation data.

Chinese journals in general science. We use the *Chinese Science Bulletin* (*CSB*) as our example, because it is considered a leading journal in general science in China, and because it has editions in both Chinese and English, the former included in the CSTPCD and the latter in the *SCI*.

Figure 15.1 The Exponential Growth of Chinese Publications and Citations

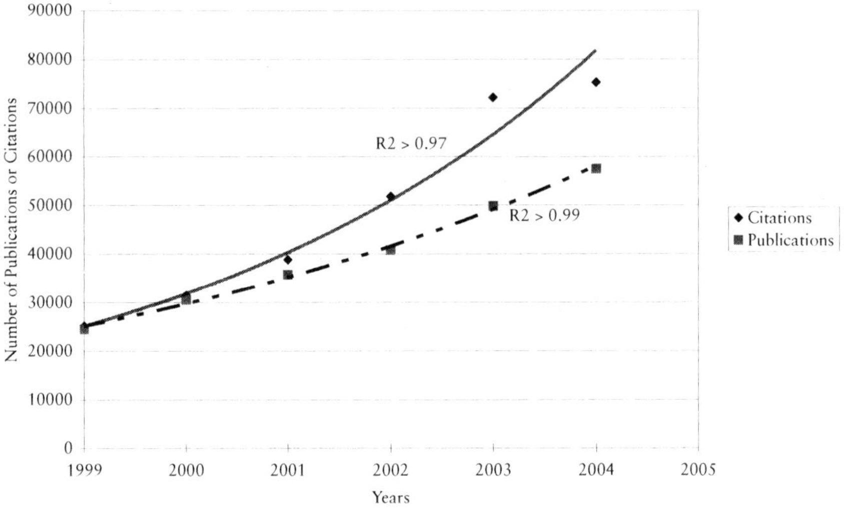

Source: Science Citation Index (SCI).

Methods and Materials

Leydesdorff and Cozzens developed a series of routines that generate aggregated journal-to-journal citation matrices on the basis of feeding a seed journal, or a set of seed journals, into the data of the *Journal Citation Reports* of the *SCI*.[2] Here we apply these routines to both Chinese and international data sets. These two environments can be very asymmetrical for the same journal in the two databases (the CSTPCD and the *SCI*). We show in this chapter that some journals are heavily cited domestically but cite only internationally.

The routines can generate an aggregated journal-to-journal *citing* network that includes only journals that are cited by the seed journal above a certain threshold (for example, 1 percent of its total citing), while a *cited* network covers journals that cite the seed journal above the threshold (that is, 1 percent of its total cited). The *cited* networks can be considered relevant impact environments, and the *citing* networks as indicators of the intellectual community of authors publishing in these journals.

The citation matrices are imported into the Statistical Package for the Social Sciences (SPSS) for factor analysis and read into Pajek for visualization. SPSS is a computer application that provides statistical analysis of data. It allows for in-depth data access and preparation, analytical reporting, graphics, and modeling. Pajek—meaning "spider" in Slovenian—is an open-source Windows program for analyzing and visualizing large networks, with thousands or even millions of vertices. Data are harvested from aggregated journal-to-journal citations, as reported in the CSTPCD and the corresponding *Journal Citation Report* of the *SCI 2003* and *SCI 2004*. SPSS is used for the descriptive statistics.

Results

We conducted the study at both macro and micro levels. At the macro level, we compared the differences and similarities between the CSTPCD and the SCI. For microlevel analysis, we used a specific Chinese journal that is covered by both the CSTPCD and the *SCI* in order to analyze citation patterns of Chinese journals in the domestic and international environments.

The Average Citation Patterns of Chinese and International Journals

Based on the CSTPCD, we derived the average citation patterns of Chinese journals. Such information can also be retrieved from the *SCI* for the international journals. Data are from 2003 and 2004 in the two databases; this makes it possible to conduct both horizontal and vertical comparisons. (This chapter updates a more extensive study that used only data for 2003 but analyzed more journals.[3])

Table 15.1 provides general information about the two databases. Both increased their coverage of journals in 2004; the CSTPCD covered 32 more journals, while the *SCI* increased by 62. Inclusion of Chinese journals in the SCI keeps growing. In 2003, 67 journals in the *SCI* were from China, and one year later the number had increased to 76. As shown in the first column of Table 15.1, both the Chinese and international data sets increased.

From Table 15.1 we can also obtain average journal-to-journal citation information about both Chinese and international journals. Longitudinal information about three indicators is shown in Table 15.2. From 2003 to 2004, the three indicators of both Chinese and international journals increased. However, the net value of the increase is different: the former is less than the latter. The average journal-to-journal citation relations per Chinese journal increased by 21.8 percent, while the number of international citations (*SCI*) increased by 6.5 percent. The average number of references per Chinese journal increased by 14.1 percent, while this number increased by 6.6 percent in international journals. The average number of citations received by a Chinese journal increased by 19.9 percent, while that of an international journal increased by 6.1 percent.

Table 15.1 Comparison of the Data for the CSTPCD and the *SCI*

	Years	CSTPCD	Percentage increase	*SCI*	Percentage increase
Number of source journals processed	2003	1,576	+ 2.0	5,907	+ 1.0
	2004	1,608		5,969	
Unique journal-to-journal relations	2003	157,659	+ 9.1	971,502	+ 6.9
	2004	172,010		1,038,268	
Sum of journal-to-journal relations	2003	573,543	+ 24.2	17,604,594	+ 7.6
	2004	712,194		18,943,827	
Total "citing"	2003	2,233,524	+ 16.3	23,953,246	+ 7.7
	2004	2,599,778		25,798,965	
Total "cited"	2003	570,384	+ 22.2	19,497,302	+ 7.2
	2004	697,242		20,909,401	

Sources: CSTPCD and *SCI*.

Citation relations among Chinese journals are much lower than among international journals. In 2004 the average number of journal-to-journal citation relations for a Chinese journal was 443, while that of an international journal was 3,174; the number of references provided by a Chinese journal was 1,617, while that of an international journal was 4,322; the number of citations received by a Chinese journal was 434, while that of an international journal was 3,503.

In terms of the net increase of the three indicators given in Table 15.2, Chinese journals increased less than international journals. However, the Chinese journals increased in percentage terms by double digits, while those of international journals increased by only one digit. Authors in Chinese journals make fewer references than their international counterparts. This causes the three levels of indicator to be much lower for Chinese journals than for international journals. In 2004 average references in a Chinese journal were about three times fewer than those in an international journal. In the 2004 data, for example, a Chinese journal provided 1,617 references to journals on average, but a Chinese journal received only 434 citations. Where has the difference of 1,183 (= 1,617 – 434) references gone? The answer is that authors in Chinese journals cite more international journals than domestic journals. Nevertheless, the gap narrowed in 2004. In 2003 journal-to-journal citation relations of Chinese journals were fewer than eight times those of international journals, up from seven times less in 2004. Similar observations hold for the other two indicators: the average number of references per journal and the average number of citations per journal.

Table 15.2 Citation Relations among Chinese or International Journals in 2003 and 2004

	Year	Chinese journals	Percentage increase	International journals	Percentage increase
Average journal-to-journal citation relations per journal	2003	364	+ 21.8	2,980	+ 6.5
	2004	443		3,174	
Average number of references per journal	2003	1,417	+ 14.1	4,055	+ 6.6
	2004	1,617		4,322	
Average number of citations per journal	2003	362	+ 19.9	3,301	+ 6.1
	2004	434		3,503	

Sources: CSTPCD and SCI.

The net values of the indicators of Chinese journals shown in Tables 15.1 and 15.2 are far lower than those of international journals. However, the increase in the Chinese journals is higher than the international journals. This implies that Chinese authors are improving their citation habits by providing more references to other journals.

Citation Patterns of Chinese Journals

The journals selected are in general science, and the one analyzed here in detail is the *CSB*, which is published in two independent editions, Chinese and English. We label the Chinese edition as *CSB-C* and the English edition as *CSB-E*. The *CSB-C* is covered by the CSTPCD, while the *CSB-E* is included in the SCI. According to data provided by the Institute of Scientific and Technical Information of China (ISTIC), the Chinese edition ranked first in the general science class—with an impact factor of 0.935—in the CSTPCD in 2004.[4] The English edition of the *CSB* had an impact factor of 0.683 in the *SCI* in 2004.[5]

273

Citation Patterns of the Chinese Science Bulletin (Chinese edition) in the CSTPCD

Cited Pattern

In 2004, 785 journals included in the CSTPCD cited the CSB-C, providing 4,511 citations. Among the 1,608 total journals included in the CSTPCD, nearly half (49 percent) had cited the CSB-C. This means that the CSB-C has high visibility among Chinese scientific and technological journals. This visibility matches its reputation.

To analyze which fields have close citation relations with the CSB-C, we collected the cited environment of the CSB-C by setting the threshold at 1 percent (Figure 15.2). Each of the twelve journals included comprised a number of citations of more than 1 percent of the total number of citations of the CSB-C—that is, 1,105 citations to CSB-C. Two journals are in general science (including the CSB-C itself); the other ten are classified by the ISTIC in the geosciences (Figure 15.2).

Although the CSB-C is a general science journal, its citation reception is mainly in the geosciences. There are two possibilities for this high visibility in the geosciences. First, articles in the geosciences in the CSB-C have higher citation impact than articles in other disciplines published in the journal. Second, the CSB-C mainly publishes articles in the geosciences. However, when we checked articles in the CSB-C, we found various disciplines. This implies that the visibility of this multidisciplinary journal is not evenly distributed among journals in various fields.

Citing Pattern

The CSB-C contained a total of 10,982 references in 2004, among which 1,612 were provided to 281 journals included in the CSTPCD; authors in the CSB-C gave only 15 percent of their references to journals covered by the CSTPCD. When the threshold is set at the convenient value of 1 percent (given the expectation of a highly skewed distribution), only one journal other than the CSB-C is included. In other words, only one journal other than the CSB-C included in the CSTPCD received 1 percent of the CSB-C's total citations. Where have the rest of the 9,370 (= 10,982 − 1,612) references gone? We conjecture that most of them have been given to international journals: when making citations, authors in the CSB-C favor international journals. International journals would then account for approximately 85 percent of the total references of the CSB-C.

Citation Environment of the Chinese Science Bulletin (English edition) in the SCI

Chinese scientific authors favor citations in international journals when publishing in the CSB-C (further demonstrated in the citing pattern of the CSB-E, discussed later). Do their publications receive the same return from their international counterparts?

Figure 15.2 Cited Environment of the *Chinese Science Bulletin* (Chinese edition, CSTPCD, 2004; threshold = 1%; cosine ≥ 0.2)

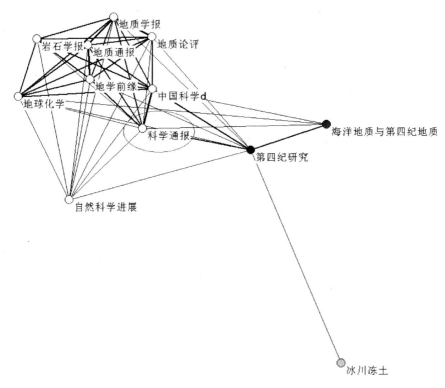

CSTPCD is a database. CSTPCD 2004 means the data in 2004 in the database.

地质学报: Acta Geologica Sinica
岩石学报: Acta Petrologica Sinica
地质通报: Geological Bulletin of China
地质论评: Geological Review
地学前缘: Earth Science Frontiers
地球化学: Geochimica
中国科学d: Science in China—Series D
科学通报: Chinese Science Bulletin
海洋地质与第四纪地质: Marine Geology and Quaternary Geology
第四纪研究: Quaternary Sciences
自然科学进展: Progress in Natural Science
冰川冻土: Journal of Glaciology and Geocryology

Citing Pattern

The *CSB-E* cited a total of 958 journals two or more times in 2004, and cited another 2,041 journals only once, generating 10,633 citations in total.[6] Among these journals, 662 are included in the *SCI* and are cited two or more times (8,210 citations total). This means that at least 68 percent of the references by *CSB-E* are given to journals included in the *SCI*. International journals have a very significant citation impact on Chinese authors in *CSB-E*. Among the journals cited more frequently by these authors, leading international journals in general science and the geosciences prevail (Figure 15.3).

Figure 15.3 Citing Network of the *Chinese Science Bulletin* (English edition; *SCI*, 2004; threshold = 1%; cosine ≥ 0.2)

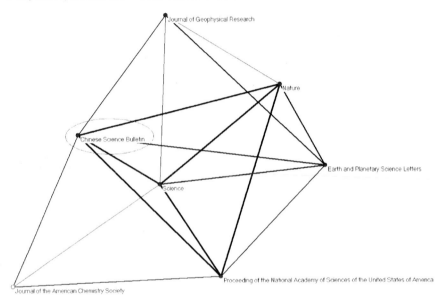

Cited Pattern

The *CSB-E* was cited 2,739 times in 2004 by journals in various disciplines. However, most of the journals citing the *CSB-E* are from China. When the threshold is set at 1 percent, all the journals included in the cited environment of the *CSB-E* are from China (Figure 15.4). Figure 15.4 shows that the *CSB-E* has visibility in more fields than the *CSB-C*.

Figure 15.4 Cited Pattern of the *Chinese Science Bulletin* (English edition; *SCI*, 2004; threshold = 1%; cosine ≥ 0.2)

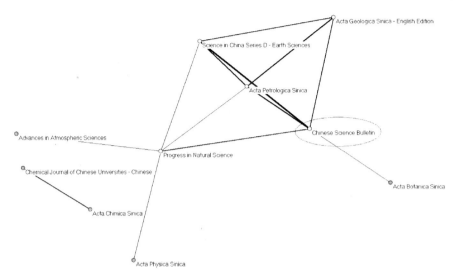

In summary, the *CSB-C* is an important journal in China. It had the highest impact factor (0.935) among journals included in the CSTPCD in the general science category in 2004. However, the international visibility of Chinese journals in this category is low. With an impact factor of 0.683, the *CSB-E* ranked only seventeenth among the 45 journals in the multidisciplinary category in the *SCI 2004*. Journals with top impact factors in this category are *Nature* and *Science*. Their impact factors are 32.182 and 31.853, respectively—more than forty times that of the *CSB-E*.

Although the two editions of the *CSB* contain the same articles, they seem to have separate disciplinary impacts in terms of their citations. The impact of the Chinese edition is mainly focused on the geosciences, while the English edition is more like a multidisciplinary journal when evaluated in terms of its citation impact.

Conclusion

Authors in Chinese journals make fewer references than their international counterparts. However, the situation is improving: the number of citation relations among Chinese journals increased between 2003 and 2004, with a greater change than those among international journals.

The visibility of Chinese journals in the multidisciplinary category is not evenly distributed among fields. For example, the Chinese edition of the *CSB*, a multidisciplinary journal, is highly visible among journals in the geosciences. The English edition of the CSB has visibility in more fields, but its visibility is also not evenly distributed among various fields.

Authors who publish in high-quality Chinese journals prefer to cite articles in international journals instead of domestic ones, and few domestic journals appear in the citation graph of some high-quality journals when the threshold is set at 1 percent of their citing environments. However, this pattern does not affect the domestic visibility of these journals, which for high-quality journals in terms of citations is high.

Although authors in Chinese journals prefer to cite papers in international journals rather than in domestic journals, their international counterparts do not provide the same return; the international visibility of Chinese journals is low.[7]

Policy Implications

Because Chinese journals are important publication channels for Chinese scientists, increasing their visibility may help raise the impact of Chinese papers. Because of the *SCI*'s influence in evaluating scientific output, inclusion in the *SCI* has become a major objective among editorial boards of Chinese journals. However, our analysis shows that inclusion in the *SCI* does not necessarily lead to an increase in visibility. More needs to be done to increase this visibility, especially in terms of efforts on the part of research institutes, scientific authors, and editorial boards.

The present evaluation system of some research institutes overemphasizes counts of scientific publications, which leads Chinese authors to focus more on the number rather than on the quality of their publications. The high share of publications and relatively low share of citations are direct results. Chinese authors need to focus more on original and innovative research. And producing a paper in English is a plus for increasing visibility.

Chinese editorial boards may be able to increase their international visibility in two ways. First, they could improve accessibility to international readers by cooperating with international publishers and online journal database providers, as well as by adopting open access. And second, they could publish an English-language edition. Our analysis in another study shows that Chinese journals in English have higher international visibility than journals in Chinese in the same field.[8]

Notes

[1] B. H. Jin and R. Rousseau, "Evaluation of Research Performance and Scientometric Indicators in China," in *Handbook of Quantitative Science and Technology Research*, ed. H. F. Moed, W. Glanzel, and U. Schmoch (Dordrecht, Netherlands: Kluwer Academic Publishers, 2004), 497–514; and P. Zhou and

L. Leydesdorff, "The Emergence of China As a Leading Nation in Science," *Research Policy* 35, no. 1 (2006): 83–104.
    [2] L. Leydesdorff and S. E. Cozzens, "The Delineation of Specialties in Term of Journals Using the Dynamic Journal Set of the Science Citation Index," *Scientometrics* 26 (1993): 133–54.
    [3] P. Zhou and L. Leydesdorff, "A Comparison between the China Scientific and Technical Papers and Citations Database and the Science Citation Index in Terms of Journal Hierarchies and Inter-journal Citation Relations," *Journal of the American Society for Information Science and Technology* 58, no. 2 (2007): 223–36.
    [4] Institute of Scientific and Technical Information of China (ISTIC), 2004 年度中国科技论文统计结果 (Statistics of Chinese Publications in 2004), (Beijing: Scientific and Technical Documents Publishing House, 2005).
    [5] *Journal Citations Report of the Science Citation Index 2004* (Philadelphia, 2004: Thomson Scientific, 2004). Hereafter *JCR 2004*.
    [6] *JCR 2004*.
    [7] S. L. Ren and R. Rousseau, "International Visibility of Chinese Scientific Journals," *Scientometrics* 53, no. 3 (2002): 389–405; and M. Liu, "A Study of Citing Motivation of Chinese Scientists," *Journal of Information Science* 19 (1993): 13–23.
    [8] P. Zhou and L. Leydesdorff, "China Ranks Second in Scientific Publications since 2006," *ISSI Newsletter*, no. 13 (March 2008): 7–9.

THE ROLE OF GLOBAL MULTINATIONAL CORPORATIONS VERSUS INDIGENOUS FIRMS IN THE RAPID GROWTH OF EAST ASIAN INNOVATION: EVIDENCE FROM U.S. PATENT DATA

Poh Kam Wong

East Asia (excluding Japan) has experienced very rapid growth in innovative capacities in recent years. In this chapter I investigate this phenomenon through the analytical lens of U.S. patents granted to inventions in East Asia, with a focus on the four Asian newly industrialized economies (NIEs)—Taiwan, Korea, Hong Kong, and Singapore—plus China and India. While patent data is an imperfect measure of innovation capabilities, and data on patents granted in the U.S. Patent and Trademark Office (USPTO) carry systematic biases, prior research has nonetheless found that patent data provide a valuable, if limited, window on innovation outputs by countries outside the United States.[1]

This chapter also investigates the relative contribution of indigenous organizations versus foreign firms to the growth of patents in Asia. In particular, we trace the timing and geographical diversification pattern of patenting activities in Asia by the five hundred largest patent-owning companies in the world from 1976 through 2004, to assess the extent to which they influence the direction and growth of innovation activities in Asia (excluding Japan). We also provide several case studies of the patenting patterns of leading global multinational corporations (MNCs). We believe that such analysis provides valuable insights into where MNCs are doing their innovations, and that these insights are not otherwise obtainable even through major surveys of the corporations' research and development (R&D) activities.[2]

Patents as Innovation Indicators

While some of the data reported in this paper are extracted from the USPTO website,[3] most are drawn from a database[4] that we have developed based on the raw-text annual U.S. patent data set provided by the USPTO and covering the period 1976 through 2004. While the exclusion of patents granted before 1976 results in an underestimation of patenting by Japan, it should have little effect on the rest of Asia, since there were very few patents granted to Asia-based inventors outside Japan prior to 1976. These data allow for several conceptual measures to be constructed, as described in the following sections.

Location of Invention

The location of a patented invention can be classified by the residential address of the inventor stated in the patent. This is difficult in the case of more than one inventor. A common method is to use the residential address of the first-named inventor as an indicator of the source country. While widely used, this method introduces some bias in the case of multiple inventors. An alternative approach is to count all the patents with at least one inventor residing in a given country. The drawback of this multiple-count method is that the sum of patent counts by country exceeds the total tally. A more refined approach is to assign fractional counts of patents to each coinventor country. However, because assigning patents in this way is much more time-intensive, in this chapter we use the multiple-count approach.

Location of Assignee

A patent invented in a particular country is treated as having a local or foreign assignee according to whether the address of the first-named assignee is inside or outside a country. While patents with foreign assignees are probably foreign-owned, patents with a local assignee might be owned by a subsidiary or branch of a foreign firm. Consequently, calculating the proportion of patents that are invented locally but are owned by foreign assignees is likely to underestimate the actual extent of foreign ownership. The decision to assign patents to corporate headquarters or to the subsidiary unit depends on corporate intellectual propety (IP) management strategy and hence may vary across firms and across countries. Notwithstanding this, the location of the assignee is widely used as a conservative proxy measure of the foreign ownership of local inventions.

Patent Citations

Patent citations are the number of forward citations attracted by a patent. Analagous to scientific citations, they are used as a measure of patent quality—although they suffer from well-documented problems, including the truncation of more recent patents.[5] Also, limitations of backward and forward citation links of a patent to other patents, as a measure of the flows of knowledge between inventions, have also been well documented, particularly the fact that many patent citations were inserted by the patent examiners rather than by the inventors themselves.

Top Patent-owning Companies in the World

Based on our database,[6] we constructed another database comprising the cumulative number of patents owned by the five hundred largest corporations in the world from 1976 through 2004, and we called this new database the Global IP 500. Because we excluded patents granted before 1976, there may be some undercounting of patents by older corporations or by those that started patenting substantially before 1976. However, for the purpose of this chapter, such a bias is not a serious issue, because, as we have noted, patenting in Asia (excluding Japan) was relatively rare before 1976.

Overview of Patenting Trends in Asia

In this section, we first present a broad overview of the pattern of patenting across Asian countries, before focusing on the specific patterns for the four Asian NIEs, China, and India. All subsequent references to patenting pertain to U.S. patents.

Overall Growth Trends of Patenting in Asia

Table 16.1 shows the growth trend of cumulative patent counts in Asian economies. While the growth rate of total patenting in the world steadily declined from 15.1 percent per annum (p.a.) over 1980–1990 to 7.0 percent p.a. over 2000–2004, the increase in the cumulative number of patents granted to inventors in Asia has been relatively faster, resulting in a steady rise of the Asian share from below 11 percent at the end of 1980 to 22 percent by the end of 2004. However, patenting in Asia has been overwhelmingly dominated by Japan, which accounted for 84 percent of the Asian total over the entire period, although its share declined from 96 percent in the period 1980–1989 to 74 percent in the period 2000–2004.

Behind Japan, the four Asian NIEs—Taiwan, Korea, Hong Kong, and Singapore—and China and India occupy the next six rungs, in that order. Collectively, their total patents amounted to less than one-fifth (18.6 percent) of Japan's over 1977–2004, but because they had increased their patenting rate much faster than Japan, their patenting exceeded one-third of Japan's in the most recent period, 2000–2004. While Japan's patent growth rate decreased from 20 percent p.a. over 1980–1989 to 8 percent p.a. over 2000–2004, Taiwan and Korea saw their patent portfolio growing at 34 to 39 percent p.a. over 1980–1989, before decreasing to around 17 percent over 2000–2004, still more than twice the rate of Japan's. Hong Kong's patent growth rate was uniformly lower than that of Taiwan and Korea, and similarly showed a declining trend from 1980–1989 to 2000–2004 (23 percent to 13 percent p.a.), whereas Singapore, starting from a much smaller base, consistently maintained a high growth rate of around 26 to 27 percent p.a., although its cumulative total remained the smallest among the four Asian NIEs due to its small initial base rate. In contrast, both China and India exhibited an increase in their patent growth rate from 1990–1999 to 2000–2004, with China registering the fastest growth rate among all Asian countries in 2000–2004 (30 percent), followed closely by India at 27 percent.

There is a considerable gap between the aforementioned economies and the Association of Southeast Asian Nations (ASEAN) economies excluding Singapore. The combined patenting by the ASEAN-4 of Malaysia, Thailand, the Philippines, and Indonesia over 1977–2004 amounted to only 1.3 percent

Table 16.1 Number of Patents in Selected Asian Countries, 1980–2004 (cumulative from 1977)

	Number of patents				Growth rate p.a. (%)		
	As of 1980	As of 1990	As of 2000	As of 2004	1980– 1990	1990– 2000	2000– 2004
Japan	26,563	168,184	429,352	574,865	20.3	9.8	7.6
Taiwan	196	3,535	30,448	57,606	33.5	24.0	17.3
South Korea	34	940	19,098	35,673	39.4	35.1	16.9
Hong Kong	138	1,057	3,917	6,449	22.6	14.0	13.3
Singapore	10	98	1,049	2,719	25.6	26.8	26.9
China	3	196	917	2,593	51.9	16.7	29.7
India	46	175	743	1,921	14.3	15.6	26.8
Malaysia	8	37	273	547	16.5	22.1	19.0
Thailand	5	24	165	348	17.0	21.3	20.5
Philippines	24	76	178	258	12.2	8.9	9.7
Indonesia	4	26	102	162	20.6	14.6	12.3
Other Asia	4	12	49	99	11.6	14.9	19.5
Asia total (incl. Japan)	27,035	174,360	486,291	683,240	20.5	10.8	8.9
Asia total (excl. Japan)	472	6,176	56,939	108,375	29.3	24.9	17.5
World total	259,207	1,058,586	2,364,872	3,101,716	15.1	8.4	7.0
Asia patents (incl. Japan) as % of world total	10.4	16.5	20.6	22.0			
Asia patents (excl. Japan) as % of world total	0.2	0.6	2.4	3.5			

	% Share (including Japan)				% Share (excluding Japan)		
	1980– 1989	1990– 1999	2000– 2004		1980– 1989	1990– 1999	2000– 2004
Japan	96.1	83.7	73.9		NA	NA	NA
Taiwan	2.3	8.6	13.8		58.5	53.0	52.8
South Korea	0.6	5.8	8.4		15.9	35.8	32.2

Hong Kong	0.6	0.9	1.3	16.1	5.6	4.9
Singapore	0.1	0.3	0.8	1.5	1.9	3.2
China	0.1	0.2	0.9	3.4	1.4	3.3
India	0.1	0.2	0.6	2.3	1.1	2.3
Malaysia	0.0	0.1	0.1	0.5	0.5	0.5
Thailand	0.0	0.0	0.1	0.3	0.3	0.4
Philippines	0.0	0.0	0.0	0.9	0.2	0.2
Indonesia	0.0	0.0	0.0	0.4	0.1	0.1
Other Asia	0.0	0.0	0.0	0.1	0.1	0.1
Asia total (incl. Japan)	100.0	100.0	100.0	–	–	–
Asia total (excl. Japan)	–	–	–	100.0	100.0	100.0

Source: "Historic Patents by Country, State, and Year—All Patent Types" (December 2004), USPTO, <http://www.uspto.gov/web/offices/ac/ido/oeip/taf/cst_utlh.htm>.
Note: Patents are allocated to countries according to the residence of the first inventor.

of the total of the four Asian NIEs, and while it experienced patent growth rates exceeding 15 percent p.a. over the years, collectively the group's share in Asian patenting (excluding Japan) shrank between 1980–1989 and 2000–2004.

The differences in patenting performance among the Asian economies reflect more than just differences in levels of economic development. Table 16.2 compares the patenting profile across Asian countries relative to their population and their gross domestic product (GDP). A regression analysis of one log (patent output in 2004 per capita) versus another log (GDP 2004 per capita) indicates an elasticity coefficient of 1.16, suggesting that patenting intensity increased faster than income level. Thus, just as there appears to be a digital divide among the Asian economies in terms of information technology (IT) diffusion and adoption,[7] there is an evident "IP divide" between the more advanced Asian economies of Japan and the four Asian NIEs on the one hand, and the less advanced ASEAN (excluding Singapore) and South Asian regions (excluding India) on the other hand. Moreover, the divide appears to be increasing. China and India are the exceptions, exhibiting increasing rates of patent growth in recent years and catching up with the Asian NIEs in terms of total patent counts.

Notwithstanding the rapid increase of patenting in the four Asian NIEs plus China and India over the last fifteen years, their share remains small—only about 4 percent of the cumulative global patent stock during 1977–2004. Nevertheless, it is noteworthy that Taiwan has already become the third-largest patenting economy in the world behind the United States and Japan.

Table 16.2 Overall Profile of Patenting Among Asian Countries, 2004

	Nominal GDP, 2004 (US$bn at PPP)	GDP per capita, 2004 (US$bn at PPP)	Cumulative patents (all patents, 1st inventor), 1977–2004	All patents, 1st inventor, granted in 2004	Population, 2004 (millions)	Patents per million population, 2004	Patents/GDP, 2004 (patents per US$bn PPP)
Japan	3,722	29,231	574,865	37,034	127.3	290.84	9.95
East Asian NIEs							
Hong Kong	206	30,120	6,449	641	6.9	93.51	3.11
Singapore	135	31,880	2,719	485	4.4	111.39	3.59
South Korea	1,006	20,911	35,673	4,671	48.6	96.03	4.64
Taiwan	615	27,300	57,606	7,207	22.7	316.79	11.72
China	7,283	5,600	2,593	597	1,298.8	0.46	0.08
India	3,428	3,170	1,921	376	1,065.1	0.35	0.11
ASEAN (excl. Singapore)							
Indonesia	778	3,260	162	23	238.5	0.10	0.03
Malaysia	259	10,180	547	93	23.5	3.95	0.36
Philippines	383	4,440	258	21	86.2	0.24	0.05
Thailand	513	7,890	348	28	63.7	0.44	0.05
Brunei	N.A.	N.A.	2	0	0.4	0.00	na
Myanmar	38	769	3	0	46.5	0.00	0.00
Vietnam	224	2,710	9	1	82.7	0.01	0.00
Other Asian countries							
Bangladesh	241	1,730	1	0	141.3	0.00	0.00
Pakistan	340	2,210	26	4	159.2	0.03	0.01
Sri Lanka	70	3,390	46	3	19.9	0.15	0.04

Source: U.S. Census Bureau, International Database, <http://www.census.gov/ipc/www/idb>; and "Historic Patents by Country, State, and Year—All Patent Types" (December 2004), USPTO, <hhttp://www.uspto.gov/web/offices/ac/ido/oeip/taf/cst_utlh.htm>; EIU Country Data.

Quality-adjusted Asian Patenting Trends

Because the preceding analysis draws on raw patent counts, it does not take quality into account. Notwithstanding the limitations highlighted earlier, it may still be useful to examine the quality-adjusted trend in Asian patenting using patent citations. This can be done in two ways. The first method is to examine the changing share of citation counts of Asian patents instead of just the patent counts. By taking the ratio of the two, a relative citation impact index can be computed, with a value bigger than 1 indicating a higher share of citations versus patent counts. The second method is to examine the changing shares of Asia among the most highly cited patents (for example, the top 5 percent of all patents in terms of number of forward citations attracted). By taking the ratio of a country's share of the most highly cited patents to its share of all patents, a relative technological influence index can be computed, with a value bigger than 1 indicating a higher relative influence than share of patent count alone would suggest.

Tables 16.3 through 16.6 summarize the relevant data computed from our database.[8] As shown in Tables 16.3 and 16.4, the four Asian NIEs plus China and India have increased both their share of patents and patent citations over the periods of 1976–1985, 1986–1995, and 1996–2004, with the ratio between the two more or less constant for Taiwan and Hong Kong, declining slightly for Singapore, and declining more for Korea, China, and India. However, if we bear in mind their much higher growth rate of patenting in the most recent period (1996–2004), the truncation effect has probably biased their citation shares downward. It is also important to note that the relative citation ratios of the four East Asian NIEs in the most recent period (1996–2004) are actually higher than those of many advanced countries, suggesting that the average quality of Asian patents (excluding Japan) is not worse off than that of many advanced countries, at least as measured by forward citations attracted.

By focusing on the share of the most influential patents, Tables 16.5 and 16.6 are more revealing about the change in Asia's ability to compete closer to the technological frontier. Not only did the four Asian NIEs rapidly increase their share of the top 1 percent and top 5 percent of the most highly cited patents in the world over the three time periods (Table 16.5), but their relative technological influence index also increased substantially over the last two periods; that is, their share of the top 5 percent of the most influential patents increased, relative to their share of all patents. Hong Kong and Korea did experience a drop between the periods 1976–1985 and 1986–1995, but their patent

Table 16.3 Citations Received by Utility Patents,* Selected Countries

	Utility patents (at least one inventor) in each country as share of world total			Citations received by utility patents in each country as share of world total		
	1976–1985	1986–1995	1996–2004	1976–1985	1986–1995	1996–2004
Large OECD**						
U.S.	60.01	54.01	54.48	65.75	63.25	65.1
France	3.4	3.24	2.67	2.72	2.32	1.82
Germany	6.85	8.25	6.97	5.3	5.37	4.14
U.K.	3.87	3.03	2.66	3.48	2.54	2.12
Japan	12.91	21.5	20.96	12.88	19.38	18.28
East Asian NIEs						
Singapore	0.01	0.04	0.22	0.01	0.04	0.2
Taiwan	0.1	0.95	2.85	0.08	0.76	2.32
Korea	0.02	0.5	2.22	0.02	0.38	1.58
Hong Kong	0.04	0.07	0.14	0.04	0.06	0.12
Small European economies						
Ireland	0.03	0.06	0.1	0.03	0.08	0.09
Israel	0.21	0.36	0.63	0.22	0.4	0.66
Finland	0.17	0.33	0.51	0.13	0.23	0.39
Netherlands	1.12	1.05	0.98	0.92	0.79	0.65
Sweden	0.89	0.89	1.02	0.74	0.69	0.79
Switzerland	2.1	1.5	1.1	1.54	1.06	0.69
ASEAN NIEs						
Malaysia	0	0.01	0.04	0	0.01	0.03
Thailand	0	0.01	0.02	0	0	0.01
Emerging economies						
China	0	0.06	0.19	0	0.05	0.08
India	0.03	0.04	0.16	0.02	0.03	0.07

Source: P. K. Wong and Y. P. Ho, "Knowledge Sources of Innovation in a Small Open Economy: The Case of Singapore," Scientometrics 70, no. 2 (2006): 223-49. Computed from the database of the USPTO (various years).
Note: *Patents where at least one inventor resides in the specified nation.
**Organisation for Economic Co-operation and Development.

Table 16.4 Relative Citation Index (Ratio of Share of Citations to Share of Patents)*

	Share of citations received/ share of utility patents		
	1976–1985	1986–1995	1996–2004
Large OECD			
U.S.	1.096	1.171	1.195
France	0.8	0.716	0.682
Germany	0.774	0.651	0.594
U.K.	0.899	0.838	0.797
Japan	0.998	0.901	0.872
East Asian NIEs			
Singapore	1	1	0.909
Taiwan	0.8	0.8	0.814
Korea	1	0.76	0.712
Hong Kong	1	0.857	0.857
Small European economies			
Ireland	1	1.333	0.9
Israel	1.048	1.111	1.048
Finland	0.765	0.697	0.765
Netherlands	0.821	0.752	0.663
Sweden	0.831	0.775	0.775
Switzerland	0.733	0.707	0.627
ASEAN NIEs			
Malaysia	–	1	0.75
Thailand	–	0	0.5
Emerging economies			
China	–	0.833	0.421
India	0.667	0.75	0.438

Source: P. K. Wong and Y. P. Ho, "Knowledge Sources of Innovation in a Small Open Economy: The Case of Singapore," *Scientometrics* 70, no. 2 (2006): 223–49. Computed from the database of the USPTO (various years).

Note: *Patents where at least one inventor resides in the specified nation.

base in the first period (1976–1985) was rather small. Moreover, the ratios achieved by these economies recently are comparable to those observed in advanced countries; Singapore's ratio of 1.6 is particularly noteworthy, being higher than even that of the United States. In terms of global share of the top 5 percent of the most highly cited patents in 1996–2004, Taiwan had overtaken the U.K. and approached Germany, while Korea had overtaken France and approached Switzerland.

While there are likely sectoral biases that skew in favor of countries with more patents in certain sectors (for example, semiconductors have higher and faster citation propensities) that need to be adjusted for in a more refined analysis, the NIEs have not only expanded their patenting activities in general, but have also become more competitive in the league of the most highly cited patents in recent years.

Table 16.5 Share in the World's Total of Highly Cited Patents,* Selected Countries

	Top 1% most highly cited			Top 5% most highly cited		
	1976–1985	1986–1995	1996–2004	1976–1985	1986–1995	1996–2004
Large OECD						
U.S.	77.53	80.95	79.53	72.88	73.57	72.83
France	1.9	1.93	1.09	2.14	1.57	1.19
Germany	2.71	1.95	1.83	3.39	2.92	2.45
U.K.	2.98	2.03	1.66	3.15	2.14	1.86
Japan	8.96	9.22	10.4	11.26	15.12	14.63
East Asian NIEs						
Singapore	0	0.03	0.33	0	0.05	0.36
Taiwan	0.03	0.23	1.8	0.04	0.49	2.43
Korea	0.03	0.16	0.95	0.03	0.25	1.38
Hong Kong	0.06	0.01	0.19	0.06	0.05	0.13
Small European economies						
Ireland	0.02	0.09	0.04	0.03	0.09	0.1
Israel	0.34	0.55	0.84	0.25	0.45	0.8
Finland	0.58	0.43	0.67	2.7	2.41	3.94
Netherlands	0.56	0.45	0.35	3.59	2.78	2.13
Sweden	0.54	0.66	0.55	3.05	3.25	3.25
Switzerland	0.62	0.54	0.25	2.95	2.68	1.43
ASEAN NIEs						
Malaysia	0	0	0.03	0	0.01	0.03
Thailand	0	0	0	0	0	0.01
Emerging economies						
China	0	0.1	0.04	0	0.04	0.07
India	0.02	0.04	0.07	0.03	0.02	0.04

Source: P. K. Wong and Y. P. Ho, "Knowledge Sources of Innovation in a Small Open Economy: The Case of Singapore," *Scientometrics* 70, no. 2 (2006): 223–49. Computed from the database of the USPTO (various years).
Note: *Patents where at least one inventor resides in the specified nation.

Table 16.6 Relative Technological Influence Index (Share of Each Country in the World's Top 5 Percent Most Cited Patents Relative to Its Share in the World's Total Patents)*

	1976–1985	1986–1995	1996–2004
Large OECD			
U.S.	1.21	1.36	1.34
France	0.63	0.48	0.45
Germany	0.49	0.35	0.35
U.K.	0.81	0.71	0.7
Japan	0.87	0.7	0.7
East Asian NIEs			
Singapore	0	1.25	1.64
Taiwan	0.4	0.52	0.85
Korea	1.5	0.5	0.62
Hong Kong	1.5	0.71	0.93
Small European economies			
Ireland	1	1.5	1
Israel	1.19	1.25	1.27
Finland	15.88	7.3	7.73
Netherlands	3.21	2.65	2.17
Sweden	3.43	3.65	3.19
Switzerland	1.4	1.79	1.3
ASEAN NIEs			
Malaysia	0	1	0.75
Thailand	0	0	0.5
Emerging economies			
China	0	0.61	0.39
India	0.84	0.58	0.27

Source: P. K. Wong and Y. P. Ho, "Knowledge Sources of Innovation in a Small Open Economy: The Case of Singapore," *Scientometrics* 70, no. 2 (2006): 223–49. Computed from the National University of Singapore (NUS) database of U.S. patents.
Note: *Patents where at least one inventor resides in the specified nation.

Foreign versus Indigenous Ownership of Patenting in Asia

How much patenting in Asia in recent years has been driven by indigenous organizations versus foreign firms? Here we can use the proportion of assignees with foreign addresses (remembering that some of the local assignees could be subsidiaries of foreign firms). Based on data for Singapore, where we were able to check the ownership status of all local assignees, about 4.3 percent of local assignees of patents granted over 1976–2004 were subsidiaries of foreign firms. As foreign assignees accounted for about 50 percent of all patents granted, this implies that foreign ownership was underestimated by around 8 percent. We expect the underestimation to be lower for most Asian economies, given that Singapore is one of the most open economies, with a proportionately larger share of foreign firms that have innovating activities in their subsidiary operations.

Notwithstanding the preceding caveat, Table 16.7 shows that the overall extent of foreign ownership of patenting in Asia (excluding Japan) is 13.7 percent for all patents granted during 1976–2004 to the four Asian NIEs, China, India, and the ASEAN-4 combined. However, there is significant variation across economies and over time. Foreign ownership is highest for the ASEAN-4 as a group (over 70 percent), followed by China (62 percent), Singapore (48 percent), India (43 percent), and Hong Kong (29 percent); Taiwan (7.3 percent) and Korea (3.6 percent) have the lowest foreign ownership. It declined in several economies, with Korea, Taiwan, India, and Singapore, showing the growing importance of indigenous organizations in patenting. Foreign ownership increased in China and stayed about constant in Hong Kong over 1990–2004 (after registering an increase from the earlier period of 1980–1999). With the exception of Indonesia, where foreign share declined, the other three ASEAN economies registered no clear trend.

Asian Patenting Trend of the Largest Patent-owning Firms

In this section we will explore the Asian patenting trends of the Global IP 500. We begin with an overall profile of these companies before examining in more detail their patenting activity in Asia by location and by technology field.

Profile of the Global IP 500[9]

As Table 16.8 shows, North America had the largest share (57.2 percent) of patents in terms of nationality of corporate headquarters, followed by Japan (29.0 percent) and Europe (11.2 percent), with Asia (excluding Japan) accounting for 2.4 percent, or twelve firms, of which seven are based in Korea, four in Taiwan, and one in Singapore. The one remaining is based in Australia. These five hundred corporations owned a total of 1.22 million patents (cumulative between 1976 and 2004).

Despite their global nature, their inventive activities remain highly localized within their national borders. As shown in Table 16.9, just one out of ten patents owned by those corporations headquartered in North America was invented in a foreign country, with the percentage even lower (3.2 percent) for Japanese corporations. European firms show a slightly higher patenting propensity outside their home base, with close to one-third of their patenting being done in a foreign country (which is understandable, given the much smaller size of most European economies). Among the small sample of Asian firms (excluding Japan), the Singaporean firm did more than one-quarter of its patenting outside its home base, versus 17.8 percent for the four Taiwanese firms and only 3.8

percent for the seven Korean firms. The sole Australian firm had a miniscule 0.2 percent of its patents invented in a foreign country. The average share of patenting outside the home base for all Global IP 500 firms is 10.3 percent.

Table 16.7 Percentage of Asian-invented Patents with Foreign Ownership, 1976–2004

	1976–1989	1990–1999	2000–2004	1976–2004
Asian NIEs				
Hong Kong	22.1	30.8	29.1	28.8
Singapore	61.5	57.1	43.6	48.0
South Korea	16.7	4.1	3.0	3.6
Taiwan	31.6	10.4	5.4	7.3
NIE total	25.2	10.0	7.0	8.6
China	47.1	59.5	63.2	61.7
India	74.3	57.0	32.9	43.1
ASEAN-4				
Indonesia	94.1	60.6	33.3	53.1
Malaysia	66.7	76.5	67.6	70.4
Philippines	62.5	98.0	96.4	93.3
Thailand	84.6	47.8	65.8	60.2
ASEAN-4 total	74.4	72.7	67.7	70.1
Total (average)	23.2	14.7	12.1	13.7

Source: NUS patents database.
Note: Asian-invented patents are so classified based on the residence of the first inventor. Patents are classified as belonging to foreign assignees based on the first-named assignee.

Asian Patenting by the Global IP 500

While almost two in every three corporations in North America and Europe had patents invented in Asia (excluding Japan)—63.3 percent and 64.3 percent, respectively—the proportion was much lower for Japan (42.8 percent) (see Table 16.8). On patenting counts, however, the cumulative total of 34,190 patents in Asia (excluding Japan) was a very small fraction of the total portfolio of Global IP 500 firms headquartered in the triad—only 1 out of 200 patents (about 0.5 percent) for North American and European firms, and 1 out of 1,000 for Japanese firms. However, for the twelve firms headquartered within Asia (excluding Japan), 95.5 percent of their patents are for inventions in Asia (excluding Japan), resulting in an average of 2.8 percent across all the Global IP 500 firms.

Table 16.8 Global IP 500 Companies by Location (Region) of Headquarters

	No.	Percentage	% with patents invented in Asia (excluding Japan)	% of patents invented in Asia (excluding Japan)
North America	286	57.2	63.3	0.5
Japan	145	29.0	42.8	0.1
Europe	56	11.2	64.3	0.4
Asia (excl. Japan)	12	2.4	100.0	95.5
Australia	1	0.2	0.0	0.0
Total	500	100	58.2	2.8

Source: Global IP 500 patents database.
Note: The location of a patent's invention is determined by whether any of the inventors resided in that region/country.

Table 16.9 Percentage of Foreign Patenting among the Global IP 500 Companies by 2004

Location of headquarters	Percentage of patents with any inventor residing in foreign countries
North America	9.7
Japan	3.2
Europe	32.2
Asia (excl. Japan)	8.3
• Korea	3.8
• Taiwan	17.8
• Singapore	28.7
Australia	0.2
Total	10.3

Source: Global IP 500 patents database.

While its share remains small, the good news is that Asia (excluding Japan) recorded the fastest growth rate of all the regions where these firms are patenting. Table 16.10 shows that their cumulative patent counts in Asia (excluding Japan) grew at 41.2 percent p.a. over 1990–2000 and 17.9 percent p.a. over 2000–2004, much faster than their overall patent portfolio growth rate of 8.4 percent and 7.2 percent p.a. over the same periods.

Table 16.10 Cumulative Number of Patents of the Global IP 500 Companies by Residence of Any One Inventor, 1990–2004

Residence of any one inventor	Cumulative no. of patents			Growth per year (%)	
	As of 1990	As of 2000	As of 2004	1990–2000	2000–2004
North America	229,251	473,659	610,376	7.5	6.5
Japan	115,520	312,242	422,782	10.5	7.9
OECD Europe	70,484	138,680	178,232	7.0	6.5
Asia (excl. Japan)	561	17,687	34,190	41.2	17.9
Other	748	2,804	5,246	14.1	17.0
Total no. of Global IP 500 patents	412,961	928,930	1,224,671	8.4	7.2
	Percentage of total no. of Global IP 500 patents				
North America	55.5	51.0	49.8		
Japan	28.0	33.6	34.5		
OECD Europe	17.1	14.9	14.6		
Asia (excl. Japan)	0.1	1.9	2.8		
Other	0.2	0.3	0.4		

Source: Global IP 500 patents database.

Table 16.11 shows that, cumulatively to the end of 2004, Korea was by far the largest location, with 1.8 percent of all Global IP 500 patents, followed by Taiwan with 0.7 percent. However, these figures are skewed by the four Taiwanese firms and Korean firms included in the Global IP 500 that do most of their patenting at their home base. If we exclude the eleven Asian firms (excluding Japan) from the Global IP 500, then Singapore emerges as the largest location for patenting, followed by India, Taiwan, and Korea. Although, as recently as 2000, China had less than two-thirds of the patent counts for Hong Kong, it has since overtaken Hong Kong to rank fifth in 2004. It was followed by the ASEAN-4, together with Hong Kong.

Table 16.11 Cumulative Number of Patents in Asia (excluding Japan) by Global IP 500 Companies

Residence of any one inventor	Cumulative no. of patents			Growth per year (%)	
	As of 1990	As of 2000	As of 2004	1990–2000	2000–2004
South Korea	297	12,588	21,673	45.5	14.5
Taiwan	40	3,520	8,503	56.5	24.7
Singapore	41	716	1,800	33.1	25.9
China	10	144	921	30.6	59.0
India	83	323	720	14.6	22.2
Malaysia	18	166	348	24.9	20.3
Hong Kong	46	163	275	13.5	14.0
Thailand	3	30	105	25.9	36.8
Philippines	8	40	76	17.5	17.4
Indonesia	9	31	44	13.2	9.1
Other Asia	11	24	33	8.1	8.3
Total no. of Global IP 500 patents with Asian inventors (excl. Japan)	561	17,687	34,190	41.2	17.9
	Percentage of total no. of Global IP 500 patents			Growth per year (%) 1990–2004	
South Korea	0.072	1.355	1.770	35.9	
Taiwan	0.010	0.379	0.694	46.6	
Singapore	0.010	0.077	0.147	31.0	
China	0.002	0.016	0.075	38.1	
India	0.020	0.035	0.059	16.7	
Malaysia	0.004	0.018	0.028	23.6	
Hong Kong	0.011	0.018	0.022	13.6	
Thailand	0.001	0.003	0.009	28.9	
Philippines	0.002	0.004	0.006	17.4	
Indonesia	0.002	0.003	0.004	12.0	
Other Asia	0.003	0.003	0.003	8.2	
Total no. of Global IP 500 patents with Asian inventors (excl. Japan)	0.136	1.904	2.792	34.1	

Source: Global IP 500 patents database.

Global IP 500 Asian Patenting by Technology Field

The three largest technological fields in which the Global IP 500 have been patenting are electrical/electronic, computers/communications, and chemicals (see Table 16.12). Compared to total patenting in the world, their patenting was relatively more concentrated in these three fields (67.4 percent versus 53.8 percent). In contrast, their share of the pharmaceuticals/medical technology field appears to be lower than that of total world patenting, suggesting that a proportionately larger share of such patents may be held by smaller biotech firms.

Compared to their North American counterparts, Global IP 500 firms with European headquarters appear to have a relatively higher patenting propensity in chemicals and in the pharmaceuticals/medical technology field, slightly less in electrical/electronic, and much less in computer/communications. In contrast, firms from Japan and other Asian countries exhibit greater concentration in electrical/electronic and in computers/communications, and much less in chemicals and the pharmaceuticals/medical technology fields. Japan also shows greater patenting propensity in the mechanical field.

The relative specialization of Asia (excluding Japan) in electrical/electronic and computers/communications is apparent when we examine the patenting of the Global IP 500 in Asia (excluding Japan) versus that in other parts of the world (Table 16.13). Cumulatively over 1976–2004, these two fields alone accounted for 76.2 percent of all their patenting in Asia (excluding Japan), versus just 45.8 percent of their global patenting. Conversely, only 8.6 percent of Global IP 500 patenting in Asia (excluding Japan) is in chemicals and pharmaceuticals/medical technology, versus 27.6 percent globally. Mechanical innovations are also significantly underrepresented in Asia (excluding Japan), measuring 9.0 percent versus 17.7 percent globally.

Overall, as measured by the Herfindahl Index,[10] patenting by technological fields in Asia (excluding Japan) exhibits not only a much higher concentration than other regions (0.334 versus 0.194 globally for cumulative patents until the end of 2004), but the degree of concentration has increased over time (0.211 at the end of 1990, 0.313 at the end of 2000), even though on a global scale there has been little change in the degree of technological specialization.

A detailed breakdown (Table 16.14) shows a consistent emphasis on electrical/electronic and computers/communications in Taiwan, Korea, and Singapore since the late 1980s, while the shift toward these two fields was more recent in China. India presents an interesting contrast: while there is also a shift toward these two fields, computers/communications appear to be the much more important of the two, reflecting India's strength in software. Moreover, India still has a sizable (albeit declining) share of patenting in chemicals and pharmaceuticals/medical technology; indeed, it is of a magnitude comparable to the global share of these fields in Global IP 500 patenting.

Overall, the Global IP 500 are expanding their patenting activities in Asia (excluding Japan) to exploit the region's growing technological advantages in the

Table 16.12 Technological Classification of Patents of Global IP 500 Companies by Location of Headquarters

	North America	Japan	Europe	Asia (excl. Japan)	Australia	Total	World total for all U.S. patents
	No. of patents						
Chemicals	137,780	60,397	50,353	2,069	49	250,648	525,540
Computers/ communications	120,300	100,246	24,024	8,034	363	252,967	402,182
Pharmaceuticals/ medical technology	45,276	9,770	14,516	29	1	69,592	270,540
Electrical/electronic	123,161	107,854	31,870	15,742	45	278,672	530,373
Mechanical	86,703	94,418	22,034	2,588	75	205,818	588,060
Others	61,702	30,624	10,866	1,717	6	104,915	557,442
Total	574,922	403,309	153,663	30,179	539	1,162,612	2,874,137
	Percentage						
Chemicals	24.0	15.0	32.8	6.9	9.1	21.6	21.4
Computers/ communications	20.9	24.9	15.6	26.6	67.3	21.8	13.8
Pharmaceuticals/ medical technology	7.9	2.4	9.4	0.1	0.2	6.0	9.8
Electrical/electronic	21.4	26.7	20.7	52.2	8.3	24.0	18.6
Mechanical	15.1	23.4	14.3	8.6	13.9	17.7	19.4
Others	10.7	7.6	7.1	5.7	1.1	9.0	17.1
Total	100.0	100.0	100.0	100.0	100.0	100.0	100.0

Source: Global IP 500 patents database.

Table 16.13 Technological Classification of Patents of Global IP 500 Companies by Residence of Any One Inventor (%)

As of 1990	North America	Japan	Europe (OECD members only)	Asia (excl. Japan)	Other countries	All countries
Chemicals	32.8	18.1	38.9	23.9	34.6	29.7
Computers/communications	10.8	17.1	9.1	16.0	13.4	12.2
Pharmaceuticals/medical technology	6.1	2.9	9.8	9.2	7.1	5.8
Electrical/electronic	20.2	23.4	18.8	31.8	15.8	20.9
Mechanical	17.2	29.1	15.4	8.8	15.1	20.2
Others	13.0	9.3	8.1	10.3	14.0	11.1
Total	100.0	100.0	100.0	100.0	100.0	100.0
Herfindahl Index	0.210	0.211	0.235	0.211	0.210	0.204

As of 2000	North America	Japan	Europe (OECD members only)	Asia (excl. Japan)	Other countries	All countries
Chemicals	27.1	16.4	35.8	8.2	21.4	24.3
Computers/communications	17.5	22.9	12.5	27.2	36.4	18.7
Pharmaceuticals/medical technology	7.6	2.8	10.8	1.0	8.5	6.2
Electrical/electronic	20.9	25.3	18.3	46.6	15.6	22.6
Mechanical	15.7	24.7	14.8	9.5	9.3	18.6
Others	11.3	7.9	7.8	7.4	8.8	9.6
Total	100.0	100.0	100.0	100.0	100.0	100.0
Herfindahl Index	0.191	0.212	0.217	0.313	0.226	0.193

As of 2004	North America	Japan	Europe (OECD members only)	Asia (excl. Japan)	Other countries	All countries
Chemicals	24.0	15.1	32.2	7.9	16.2	21.6
Computers/communications	21.2	24.5	15.9	26.8	45.9	21.8
Pharmaceuticals/medical technology	7.6	2.6	10.3	0.7	7.7	6.0
Electrical/electronic	21.8	26.8	19.3	49.4	13.7	24.0
Mechanical	14.9	23.4	14.5	9.0	9.1	17.7
Others	10.5	7.6	7.7	6.2	7.4	9.0
Total	100.0	100.0	100.0	100.0	100.0	100.0
Herfindahl Index	0.189	0.216	0.204	0.334	0.276	0.194

Source: Global IP 500 patents database.

Table 16.14 Technological Classification of Patents of Global IP 500 Companies by Residence of Any One Inventor in Asia (excluding Japan), %

As of 1990	Korea	Taiwan	Singapore	China	India	Other Asia	Asia Total
Chemicals	10.1	35.9	6.7	50.0	53.0	29.3	23.9
Computers/communications	21.7	17.9	23.3	0.0	3.6	11.0	16.0
Pharmaceuticals/medical technology	1.8	0.0	3.3	50.0	28.9	12.2	9.2
Electrical/electronic	47.5	35.9	33.3	0.0	4.8	18.3	31.8
Mechanical	10.1	2.6	20.0	0.0	2.4	11.0	8.8
Others	8.8	7.7	13.3	0.0	7.2	18.3	10.3
Total	100.0	100.0	100.0	100.0	100.0	100.0	100.0
Herfindahl Index	0.301	0.297	0.229	0.500	0.374	0.192	0.211
As of 2000	Korea	Taiwan	Singapore	China	India	Other Asia	Asia Total
Chemicals	7.1	8.4	8.3	25.4	32.3	16.1	8.2
Computers/communications	32.8	5.0	19.5	24.6	23.3	23.8	27.2
Pharmaceuticals/medical technology	0.4	0.1	1.1	15.6	23.6	5.4	1.0
Electrical/electronic	39.8	75.4	52.3	25.4	14.1	25.6	46.6
Mechanical	11.6	2.8	9.6	7.4	2.2	11.9	9.5
Others	8.4	3.2	9.3	1.6	4.5	17.1	7.4
Total	100.0	100.0	100.0	100.0	100.0	100.0	100.0
Herfindahl Index	0.292	0.601	0.336	0.220	0.237	0.195	0.313
As of 2004	Korea	Taiwan	Singapore	China	India	Other Asia	Asia Total
Chemicals	7.3	7.9	7.0	9.8	20.5	11.9	7.9
Computers/communications	34.0	3.6	24.4	18.1	37.4	20.2	26.8
Pharmaceuticals/medical technology	0.3	0.1	1.0	5.6	13.4	3.3	0.7
Electrical/electronic	40.2	76.0	52.8	55.5	21.3	37.4	49.4
Mechanical	11.2	3.6	8.8	6.2	1.9	14.4	9.0
Others	7.0	3.7	5.9	4.8	5.5	12.8	6.2
Total	100.0	100.0	100.0	100.0	100.0	100.0	100.0
Herfindahl Index	0.300	0.594	0.355	0.359	0.248	0.233	0.334

Source: Global IP 500 patents database.
Note: "Other Asia" includes Hong Kong, Malaysia, Thailand, Indonesia, Philippines, Vietnam, Pakistan, Bangladesh, Sri Lanka, Brunei, Burma, Laos, Macao, and North Korea.

electrical/electronic and computers/communications fields, with the chemicals and pharmaceuticals/medical technology fields a second area of focus in India. The recent public policy thrust to promote biotechnology in the four NIEs (particularly Singapore) did not have a visible impact in terms of patenting outputs, at least as of 2004.

Pattern of Asian Patenting by Global MNCs: Some Examples

Focusing on MNCs in the IT/electronics industry, we now compare the extent and pattern of Asian patenting by the largest Japanese, American, and European IT/electronics firms, respectively: Matsushita Electric, Texas Instruments, and Philips Electronics. There are several interesting differences among them. First, reflecting the generally lower propensity of Japanese firms to move R&D activities abroad, Matsushita has a smaller amount of patenting in Asia outside Japan, both in absolute number of patents (75) and in relative share of the company's total portfolio (0.4 percent) than Philips (164, or 0.8 percent) and Texas Instruments (310, or 2.7 percent) (see Tables 16.15 through 16.17).

Second, Matsushita concentrates its patenting activities in Asia outside Japan, with Singapore as the primary location (about 70 percent of all patents generated) and Taiwan as a secondary location (almost one-fifth). Only 8 of its 75 patents (10.7 percent) in Asia were generated in other locations. In contrast, both Philips and Texas Instruments were more geographically diversified. Philips started patenting in Hong Kong in the early 1980s, expanded to include Singapore and Taiwan from the mid- to late 1980s, and expanded into China from the late 1990s. By the end of 2004, Singapore and China had overtaken Hong Kong in terms of patenting outputs, with Taiwan catching up to Hong Kong as well. Texas Instruments used Singapore as the first base for patenting activity in Asia (excluding Japan) during 1980; however, in the 1990s, it rapidly expanded its patenting activities by diversifying into Taiwan, India, Korea, and Malaysia. By the end of 2004, Texas Instruments' patenting portfolio in Asia (excluding Japan) was distributed across India, Singapore, and Taiwan—in that order—followed by smaller outputs from Korea and Malaysia.

The companies also appear to differ in terms of the extent of knowledge linkages with the host countries in particular or with the Asian region (excluding Japan) in general. Using the proportion of backward citations to, and forward citations from, other patents invented in the region as a measure of the strength of knowledge linkages within a region, Texas Instruments' patents in the region exhibited the weakest backward linkages with the region itself. Only 6.7 percent of its backward citations were to other patents invented in the region, compared with 9.6 percent for Matsushita and 8.0 percent for Philips. Conversely, Matsushita patents in the region have the weakest forward linkages with Asia (excluding Japan), with only 12.7 percent of the forward citations they attracted coming from the region, compared with 15.5 percent for Texas Instruments and 20.9 percent for Philips. This seems to indicate that Texas Instruments' R&D activities in Asia (excluding Japan) are less dependent on locally generated knowledge than those of Philips, which is in turn less dependent on the region than Matsushita. However, knowledge generated by Philips' R&D activities in the region is more intensively used by other inventors in the region than knowledge generated by the R&D activities of Texas Instruments and Matsushita.

Table 16.15 Citation Profile of Matsushita Electric, Patents Invented in Asia (excluding Japan), as of 2004

	No. of patents
Patents assigned to Matsushita	19,698
Matsushita's patents invented in Asia (excluding Japan)	75
Backward citations	
Matsushita's patents invented in Asia (excluding Japan) with citations to other U.S. patents	75
Unique patents cited by Matsushita's Asian (excluding Japan) patents	428
Backward citations among Matsushita's Asian (excluding Japan) patents	588
Of backward citations, no. of Japan-invented patents	182 (31.7%)[1]
Of backward citations, no. of patents invented in Asia (excluding Japan)	55 (9.6%)[1]
Average no. of backward citations per patent (for all Matsushita's Asian [excluding Japan] patents)	7.8
Forward citations	
Matsushita's Asian (excluding Japan) patents that have been cited by other U.S. patents	49
Unique patents citing Matsushita's Asian (excluding Japan) patents	368
Citations to Matsushita's Asian (excluding Japan) patents	385
Of forward citations, no. of citing patents from Japan	113 (29.4%)[2]
Of forward citations, no. of citing patents from Asia (excluding Japan)	49 (12.7%)[2]
Average forward citations per patent (for all Matsushita's Asian [excluding Japan] patents)	5.1

Source: Global IP 500 patents database.
Notes: [1] Number in parentheses refers to percentage of total backward citations. Percentage is calculated based on a total of 574 backward citations, that is, excluding 14 citations to patents which have no inventor information.
[2] Number in parentheses refers to percentage of total forward citations.

Table 16.16 Citation Profile of Texas Instruments Patents Invented in Asia (excluding Japan), as of 2004

	No. of patents
Patents assigned to Texas Instruments	11,476
Texas Instruments' patents invented in Asia (excluding Japan)	310
Backward citations	
Texas Instruments' patents invented in Asia (excluding Japan) with citations to other U.S. patents	308
Unique patents cited by Texas Instruments' Asian (excluding Japan) patents	1,654
Backward citations among Texas Instruments' Asian (excluding Japan) patents	2,097
Of backward citations, no. of U.S.-invented patents	1,492 (76.5%)[1]
Of backward citations, no. of patents invented in Asia (excluding Japan)	131 (6.7%)[1]
Average no. of backward citations per patent (for all Texas Instruments' Asia [excluding Japan] patents)	6.8
Forward citations	
Texas Instruments' Asian (excluding Japan) patents that have been cited by other U.S. patents	201
Unique patents citing Texas Instruments' Asian (excluding Japan) patents	1,519
Citations to Texas Instruments' Asian (excluding Japan) patents	1,765
Of forward citations, no. of citing patents from U.S.	1,174 (66.5%)[2]
Of forward citations, no. of citing patents from Asia (excluding Japan)	273 (15.5%)[2]
Average no. of forward citations per patent (for all Texas Instruments' Asian [excluding Japan] patents)	5.7

Source: Global IP 500 patents database.

Notes: [1] Number in parentheses refers to percentage of total backward citations. Percentage is calculated based on a total of 1951 backward citations, that is, excluding 146 citations to patents which have no inventor information.
[2] Number in parentheses refers to percentage of total forward citations.

Table 16.17 Citation Profile of Philips Electronics Patents Invented in Asia (excluding Japan), as of 2004

	No. of patents
Patents assigned to Philips	19,626
Philips' patents invented in Asia (excluding Japan)	164
Backward citations	
Philips' patents invented in Asia (excluding Japan) with citations to other U.S. patents	165
Unique patents cited by Philips' Asian (excluding Japan) patents	853
Backward citations among Philips' Asian (excluding Japan) patents	996
Of backward citations, no. of U.S.-invented patents	52 (5.9%)[1]
Of backward citations, no. of patents invented in Asia (excluding Japan)	70 (8.0%)[1]
Average no. of backward citations per patent (for all Philips' Asian [excluding Japan] patents)	6.1
Forward citations	
Philips' Asian (excluding Japan) patents that have been cited by other U.S. patents	103
Unique patents citing Philips' Asian (excluding Japan) patents	554
No. of citations to Philips' Asian (excluding Japan) patents	599
Of forward citations, no. of citing patents from U.S.	28 (4.7%)[2]
Of forward citations, no. of citing patents from Asia (excluding Japan)	125 (20.9%)[2]
Average no. of forward citations per patent (for all Philips' Asian [excluding Japan] patents)	3.7

Source: Global IP 500 patents database.
Notes: [1] Number in parentheses refers to percentage of total backward citations. Percentage is calculated based on a total of 877 backward citations, that is, excluding 119 citations to patents which have no inventor information.
[2] Number in parentheses refers to percentage of total forward citations.

Internationalization of Patenting by the Global IP 500

Globally, there has been an increase in the internationalization of inventions among the Global IP 500. The percentage of patents granted to inventors from more than one country has increased from 1.3 percent in 1980 to 2.9 percent in 2004 (Table 16.18). Thus, although most patents still represent inventions in a single country, the numbers reflect the increasing internationalization of R&D, which has been well documented.[11]

Table 16.18 Percentage of Patents with More than One Inventor Country, by Location of Headquarters

	Cumulative to 1980	Cumulative to 1990	Cumulative to 2000	Cumulative to 2004
North America	1.0	1.3	2.6	3.3
Japan	0.2	0.4	0.8	1.0
Europe	3.3	4.0	5.9	6.9
Asia (excl. Japan)	NA	0.0	1.7	2.6
Australia	NA	NA	0.0	0.0
All Global IP 500 patents	1.3	1.5	2.4	2.9

Source: Global IP 500 patents database.

The internationalization dynamics are remarkably stable for North America, Japan, and Europe (hereafter, the triad regions), which have increasing shares of patents granted to inventors from more than one country, viewed by assignee or by inventor (see Tables 16.18 and 16.19). Japan is the least internationalized; 1.0 percent of patents from companies headquartered in Japan are granted to inventors from more than one country, while 1.6 percent of patents granted to Japanese inventors include inventors from other countries. Europe is the most internationalized, again reflecting the smaller average size of the European countries; 6.9 percent of patents belonging to European companies are granted to inventors from more than one country, while 13.6 percent of patents granted to European inventors include inventors from other countries.

Asia (excluding Japan) is different from the triad. Asian companies began their internationalization much later; as of 1990, no Asian company had patents granted to inventors from more than one country. However, once they began, their internationalization grew fairly rapidly, so that by 2004, 2.6 percent of patents belonging to Asian companies (excluding Japan) were granted to inventors from multiple countries.

If one looks at Asian (excluding Japan) patents by region of inventor, a different picture emerges. In this case, Asian patents have exceptionally high rates of cross-border inventorship. In 1980, 58 percent of patents granted to Asian inventors (excluding those from Japan) included inventors from other countries. This percentage decreased rapidly over time, however, so that by 1990 the share had fallen by almost half, to 29.8 percent. The next decade saw another dramatic fall in internationalization of inventorship among Asian-invented patents, with only 8.3 percent granted to inventors from more than one country in 2000, although this percentage rose to 8.8 in 2004. Thus, R&D activities by Asian scientists working for the Global IP 500 companies, in contrast to scientists in the more advanced countries, have started as highly internationalized and have become more localized over time. This reflects the need to depend more heavily on foreign knowledge sources when the local innovation systems were less developed. As the local knowledge base has strengthened, this dependence has become less necessary. It remains to be seen whether the slight increase in the internationalization of Asian patents from 2004 is the beginning of a trend in which Asia (excluding Japan) once again engages more intensively in the internationalization of R&D.

Table 16.19 Percentage of Patents with More than One Inventor Country, by Location of Any One Inventor

	Cumulative to 1980	Cumulative to 1990	Cumulative to 2000	Cumulative to 2004
North America	1.2	1.6	3.4	4.2
Japan	0.6	0.7	1.4	1.6
Europe	5.4	6.6	11.3	13.6
Asia (excl. Japan)	58.0	29.8	8.3	8.8
Other countries	45.1	51.3	52.9	45.7
All Global IP 500 patents	1.3	1.5	2.4	2.9

Source: Global IP 500 patents database.

Conclusion

These findings are preliminary, and more work is needed. Nevertheless, we can make several relevant inferences. First and foremost, while the rapid growth in innovation activities in Asia (excluding Japan) as a whole is not really a surprise, we have also shown that the four Asian NIEs have increased their share of the most influential patents in the world relative to their share of all patents over the last two decades or so. This shows an advance not only in quantity but also in quality of innovation. It is clear that Asia is becoming competitive as a location for innovative activities not just in terms of cost; significant clusters of highly advanced innovative capabilities already exist in Asia, particularly in the electrical/electronic and computers/communications fields, consistent with other recent findings in the literature.[12]

Second, despite starting relatively late, the rapid rise in the innovation capabilities of China and India is reflected in IP outputs in the form of patents. Moreover, foreign MNCs have been a major driver for the growth of patenting activities in these two economies. The competitive pressure they exert on other Asian economies, particularly the developing ASEAN economies, will only intensify. Unless these economies accelerate their investment in innovation, they risk falling further behind.

These findings show that patenting activities in the region have grown high enough to make the use of patent data analysis statistically meaningful, at least for the four Asian NIEs, China, and India. While there is increasing literature on Asian innovations using patent analysis, it has tended to focus on individual countries,[13] and we believe that there is a scope for more research that takes a cross-country comparison or a regionwide focus.

The role of global MNCs in knowledge transfer, and the pattern of local spillovers of innovation by subsidiaries of foreign firms, have also begun to be studied using patent data, particularly patent citations data.[14] We believe that the growing innovation links between Asia and the United States, and the emerging trend of regional integration of innovation activities across Asian economies by global MNCs, can be usefully researched using patent citations data. In particular, data on the most active patenting firms in Asia, such as the Global IP 500 database that we have constructed, can provide useful insights on behavior that presages broader development trends.

Notes

[1] A. B. Jaffe and M. Trajtenberg, *Patents, Citations & Innovations: A Window on the Knowledge Economy* (Cambridge, MA: MIT Press, 2002).

[2] R. Pearce and S. Singh, *Globalizing Research and Development* (London: Macmillan, 1992); and W. Kuemmerle, "Foreign Direct Investment in Industrial Research in the Pharmaceutical and Electronics Industries: Results from a Survey of Multinational Firms," *Research Policy* 28 (1999): 179–193.

[3] See United States Patent and Trademark Office (USPTO), U.S. Department of Comerce, <http://www.uspto.gov>.

[4] See <http://patents.nus.edu.sg>.

[5] Jaffe and Trajtenberg, *Patents, Citations & Innovations*.

[6] See <http://patents.nus.edu.sg>.

[7] P. K. Wong, "ICT Production and Diffusion in Asia: Digital Dividends or Digital Divide?" *Information Economics and Policy* 14, no. 2 (June 2002): 167–87.

[8] See <http://patents.nus.edu.sg>.

[9] A complete listing of all five hundred firms is available on request from the authors.

[10] The Herfindahl Index is typically used as a measure of market concentration; however, it has also been used to measure countries' specialization in manufacturing (see D. Weinhold, and J. E. Rauch, "Openness, Specialization

and Productivity Growth in Less Developed Countries," National Bureau of Economic Research (NBER) working paper no. 6131, 1997). In this paper, we have modified it to measure patent specialization.

[11] See, for example, P. Criscuolo and P. Patel, "Large Firms and Internationalisation of R&D: 'Hollowing Out' of National Technological Capacity?" (paper prepared for SETI (Sustainable Growth, Employment Creation and Technological Integration in the European Knowledge-based Economy) workshop, Rome, May 15–16, 2003), <http://www.seti.coleurop.be/LARGE%20FIRMS%20AND%20INTERNATIONALISATION%20OF%20R&D.PDF>; and C. Le Bas and P. Patel, "Does Internationalisation of Technology Determine Technological Diversification in Large Firms?" (SPRU electronic working paper series, no. 128, January 2005).

[12] See, for example, Henry S. Rowen, William F. Miller, and Marguerite Gong Hancock, eds., *Making IT: The Rise of Asia in High Tech* (Stanford, CA: Stanford University Press, 2006).

[13] For Singapore and Taiwan, see, for example, P. K. Wong and Y. P. Ho, "Knowledge Sources of Innovation in a Small Open Economy: The Case of Singapore," *Scientometrics* 70, no. 2 (2006): 223–49.

[14] See, for example, C. Le Bas and C. Sierra, "Location vs. Country Advantages in R&D Activities: Some Further Results on Multinationals' Locational Strategies," *Research Policy* 31, no. 4 (2002): 589–609; P. Almeida, "Knowledge Sourcing by Foreign Multinationals: Patent Citation Analysis in the U.S. Semiconductor Industry," *Strategic Management Journal* 7 (1996): 155–65; J. Singh, "Multinational Firms and Knowledge Diffusion: Evidence from Patent Citations Data," in *Best Paper Proceedings of the 2004 Meeting of the Academy of Management*, ed. D. H. Nagao, (New Orleans, LA: Academy of Management, 2004); and A. Hu "Multinational Corporations, Patenting and Knowledge Flow: The Case of Singapore," *Economic Development and Cultural Change* 52, no. 4 (2004): 781–800.

HIGH-TECH REGIONS

KNOWLEDGE SPILLOVERS AND GROWTH IN THE HSINCHU SCIENCE-BASED INDUSTRIAL PARK

Yih-Luan Chyi and Yee-Man Lai

This chapter investigates the mechanisms underlying growth and changes in the Hsinchu Science-based Industrial Park (HSIP) in Taiwan. The Taiwanese government established the park in 1980 to attract high-technology firms, including start-ups, and thereby to create a potential Silicon Valley of the East. The HSIP is located near the Industrial Technology Research Institute (ITRI) and two major research universities, National Tsing Hua University and National Chiao Tung University. To entice overseas Taiwanese engineers to return home, the HSIP administration, as well as ITRI, opened branch offices in Silicon Valley to provide information and local contacts. From the founding of the HSIP through the end of 2004, the Taiwanese government invested US$1.679 billion in its infrastructure. There were 384 park tenants by the end of 2004, and the HSIP had grown at an annual rate of 12 percent over the previous two decades. By then, returnees owned almost one-third of these firms. Total sales were NT$11 trillion (US$31.25 million), with an annual growth rate of 38 percent. The number of employees had increased more than ten times, from 8,275 in 1986 to 113,000 in 2004.[1]

Agglomeration economies and institutional networks may explain much of the HSIP's success. Economists define *agglomerations* ("clusters" for short) as a concentration of firms in a given industry within a region.[2] Hence, proximity and specialization could be key sources of efficiency in the HSIP. However, researchers in economic geography focus on the roles of cooperative networks.[3] The competitive advantages of firms in the HSIP—backward and forward linkages, labor pooling, and knowledge spillovers by means of interfirm or interindustry linkages—may be due to both cooperative networks and proximity. Although knowledge spills over in both cases, the mechanisms are different. "Cooperative networks" involve direct dealing among companies, which results in knowledge transfers. "Proximity" originally meant geographical closeness, but the great British economist Alfred Marshall was the first to observe that "ideas are in the air" in certain places with the necessity of direct dealings among companies.

Technology spillovers can occur when, without paying any costs, a firm can benefit from the research of other firms.[4] According to Jun Koo, "Research performed in one firm can stimulate the creation of new knowledge or the fruition of previous ideas in another firm. In this case, new knowledge is disembodied from new goods and becomes part of a general pool of knowledge (i.e., public goods)."[5] This type of knowledge spillover does not require direct input-output connections among firms or industries. Jaffe uses the term *technological closeness* for this phenomenon.[6] If two firms or industries use similar technology, innovations by one firm or industry can be useful to the other firm or industry.

In empirical studies surveyed by Nadiri,[7] spillover effects outside the firm (through whatever mechanisms) were shown to have significant positive influences on productivity at both the industry and firm level. However, the magnitude of these spillover effects varies substantially, depending on the means used to measure them. Among the methods used to investigate knowledge spillovers, spatial aspects have recently been addressed.[8] Using corporate profits or patent counts as dependent variables, Jaffe[9] finds significantly positive effects on innovations from technologically dependent research and development (R&D) spillovers. Furthermore, Jaffe[10] innovatively modifies the standard knowledge production function by introducing space. His results support the importance of geographic proximity for university and industry research. Adams and Jaffe, using the Cobb-Douglas production framework to measure intraindustry spillovers, find strong effects of R&D on productivity when firms or industries are dense in geographical and technological spaces.[11]

The standard method of measuring knowledge spillovers is limited in several aspects, as Cincera illustrates.[12] Firms may not apply for a patent, either because they have modest innovation capacity or because they deliberately decide to keep their R&D discoveries secret. The "zero" issue could be the main weakness of using patent statistics to postulate technological closeness among firms or industries. Therefore, we examined all firms with fewer than five patent applications by inspecting the consistency between their technological categories in patent statistics and industrial sectors.

Another distortion may occur when firms are technologically diversified; it is not reasonable to argue that such firms will benefit little from all stocks of technological spillovers. To avoid this problem, Jaffe[13] suggests that we distinguish local stocks of spillovers from external stocks by employing a clustering method. On the basis of the locations of firms in the technological space, we classified the firms into several technological clusters, using the k-means-clustering technique.[14] Based on this method, a local, indirect knowledge stocks index can be generated by a weighted sum of R&D activities of firms within the same cluster. This enabled us to construct measures of both intercluster (between different clusters in the HSIP), and intracluster (within the same cluster in the HSIP) knowledge spillovers. We measured total, local, and external R&D spillover stock by a weighted sum of R&D activities of all firms, intracluster firms, and intercluster firms in the HSIP, respectively. Here we applied Jaffe's concept of technological closeness, focusing on technological proximity because all the firms in the study are tenants of the HSIP. Thereafter, we used the production function approach for investigating impacts of knowledge spillovers on performances of HSIP firms during the period 2000–2004.

Data Description and Empirical Results

For the purposes of this chapter, we used data on the HSIP's firms spanning the period 2000 through 2004. There were 218 firms in 2003, but due to limited data availability, our sample consists of fewer than half the total. In addition, firms with fewer than five patent applications during a five-year period were excluded. Finally, we used data on 92 publicly listed firms in the Taiwan Stock Exchange, over-the-counter market, or emerging stock market. The final sample is panel data on 92 HSIP firms over the five-year period 2000–2004. *Taiwan Economic Journal (TEJ)* reports listed companies' profiles and financial statements, including variables such as net sales (SALES), number of employees, net property, plants and equipment, and annual R&D expenses. Data on the numbers of employees were not available for some of the firms in our sample. We used firm's age (AGE) as a proxy for number of employees. Net property and plans and equipment were combined to represent stocks of physical capital stock (K). All variables were deflated by the consumer price index (CPI, base year: 2001), as reported by the Taiwan Directorate General of Budget, Accounting and Statistics (DGBAS). The R&D capital stocks (RD) were imputed by the permanent inventory method.[15]

Patents assigned across various technologies allowed us to determine firms' positions in technological space. Each of the patent applications (totaling 15,032) that HSIP firms filed at the Taiwan Patent Office between 2001 and 2005 was attached to one of twenty-three categories. Based on these allocations, we classified firms into technological clusters by the popular k-means clustering technique. After having established total (TOTAL), local (LOCAL), and external (EXTERNAL) knowledge pools of R&D spillovers relevant for advancing the performances of firms, we estimated an augmented Cobb-Douglas production function, as in the Griliches-Jaffe methodology,[16] in order to investigate the knowledge spillover impact that firms' performances had on net sales. Table 17.1 shows summary statistics of the variables we consider in the empirical study.[17]

The empirical results suggest that the HSIP high-technology clusters show evidence of important knowledge spillovers within each cluster. Total R&D spillover stocks, the local R&D spillover stocks, and the external R&D spillover stocks are all statistically significant in relation to net sales. In particular, the estimated coefficients of total R&D spillover and external R&D spillover are higher than those of firms' own R&D stocks. Knowledge spillovers, among the factors contributing to agglomeration benefits, have attracted most attention in recent literature on new growth theory and industrial geography. Our study implies that the strong performance of firms or industries may well link to knowledge spillovers via interfirm or interindustry linkages in the HSIP.

Table 17.1 Summary Statistics

Variables	No. of observations	Mean	Median	Maximum	Minimum	Std. dev.
lnSALES	442	15.14	15.12	19.37	2.76	1.87
lnAGE	456	2.16	2.30	3.33	0.69	0.66
lnK	442	13.83	13.73	19.36	7.48	2.100
lnRD	441	13.77	13.68	17.86	9.35	1.50
lnTOTAL	460	18.47	18.59	19.37	15.30	0.66
lnLOCAL	460	17.21	17.30	19.10	13.18	1.36
lnEXTERNAL	460	17.76	17.77	19.10	15.02	0.71

Notes

[1] All statistics quoted here are from the HSIP website <http://service.sipa.gov.tw/WEB/Jsp/Page/index.jsp?thisRootID=115>. Most growth rates were calculated by the authors.

[2] P. Krugman, "Increasing Returns and Economic Geography," *Journal of Political Economy* 99 (1991): 483–99.

[3] See Annalee Saxenian, *Regional Networks: Industrial Adaptation in Silicon Valley and Route 128* (Cambridge, MA: Harvard University Press, 1994).

[4] The meaning of *technology spillover* is comprehensively reviewed in Jun Koo, "Technology Spillovers, Agglomeration, and Regional Economic Development" (manuscript prepared for the *Journal of Planning Literature*, Maxine Goodman Levin College of Urban Affairs, Cleveland State University, 2004).

[5] Koo, "Technology Spillovers," 6.

[6] A. Jaffe, "Technological Opportunity and Spillover of R&D: Evidence from Firms' Patents, Profits, and Market Value," *American Economic Review* 76 (1986): 984–1001.

[7] M. Ishaq Nadiri, "Innovations and Technological Spillovers" (NBER working paper series, no. 4423, 1993), <http://www.nber.org/papers/w4423.pdf>.

[8] See, among others, A. Jaffe, "Real Effects of Academic Research," *American Economic Review* 79 (1989): 957–70; M. P. Feldman, *The Geography of Innovation* (Dordrecht/Boston/London: Kluwer Academic Publishers, 1994); M. P. Feldman, "The New Economics of Innovation, Spillovers and Agglomeration: A Review of Empirical Studies," *Economics of Innovation and New Technology* 8 (1999): 5–25; J. Adams and A. Jaffe, "Bounding the Effects of R&D: An Investigation Using Matched Establishment-Firm Data" (NBER working paper series, no. 5544, 1996); L. Anselin, A. Varga, and Z. Acs, "Local Geographic Spillovers between University Research and High Technology Innovations," *Journal of Urban Economics* 42 (1997): 422–88;

and L. Anselin, A. Varga, and Z. Acs, "Geographic Spillover and University Research: A Spatial Econometric Perspective," in "Endogenous Growth: Models and Regional Policy," ed. P. Nijkamp and R. Stough, special issue, *Growth and Change* 31 (2000): 501–16.

[9] Jaffe, "Technological Opportunity and Spillover of R&D."

[10] Jaffe, "Real Effects of Academic Research."

[11] See Adams and Jaffe, "Bounding the Effects of R&D." Empirical findings in related literature are comprehensively surveyed in Gavin Cameron, "Innovation and Economic Growth" (discussion paper no. 277, London School of Economics, Centre for Economic Performance, 1996), <http://cep.lse.ac.uk/pubs/download/dp0277.pdf>.

[12] Michele Cincera, "Firms' Productivity Growth and R&D Spillovers: An Analysis of Alternative Technological Proximity Measures" (working paper #4894, London, Centre for Economic Policy Research, February 2005).

[13] Jaffe, "Technological Opportunity and Spillover of R&D."

[14] B. S. Everitt, *Cluster Analysis* (London: Edward Arnold, 1993).

[15] The depreciation rate of R&D capital stock is assumed to be 15 percent. The initial R&D capital stock is computed by supposing a 5 percent growth rate of R&D expenditure.

[16] Z. Griliches, "Issues in Assessing the Contribution of Research and Development to Productivity Slowdown," *Bell Journal of Economics* 10 (1979): 92–116.

[17] Details of the full empirical analysis are presented in Y. L. Chyi, and Y. M. Lai, "Knowledge Spillover and Growth in the Hsin-Chu Science Industry Park" (Workshop: Greater China's Innovative Capacities: Progress and Challenges, Stanford University, May 20–21, 2006).

THE DEVELOPMENT OF INNOVATIVE CAPACITY IN HONG KONG

Erik Baark

Since its return to Chinese sovereignty in 1997, the government of the Hong Kong Special Administrative Region (hereafter HKSAR or "Hong Kong") has been embarking on a new range of policies aimed at increasing the role of technology and innovation in the economy. The Asian financial crisis contributed to a dramatic bursting of the property-asset bubble that underlay much of Hong Kong's economy and triggered the territory's most acute economic recession. As a result, Hong Kong was no longer able to rely on its traditional methods of accruing wealth, such as serving as a niche for the trading of Chinese goods. In response to these conditions, in March 2008 the Hong Kong government formed the Chief Executive's Commission on Innovation and Technology (CIT) under the chairmanship of Prof. Chang-lin Tien, a former chancellor of the University of California, Berkeley.

In its first report, the CIT outlined a vision that "innovation and technology are vital to the future prosperity of Hong Kong."[1] Based on the commission's recommendations, the government increased the amount spent on research and development (R&D) beginning in the year 2000. Most notable was the establishment of the Innovation and Technology Fund (ITF), with HK$5 billion for projects that would contribute to innovation and technology upgrading in industry. The government also sought to facilitate high-technology innovation in Hong Kong by supplying venture capital and by creating new infrastructure such as the Cyberport, the Hong Kong Science and Technology Park, and the Applied Science and Technology Research Institute (ASTRI).

The new vision has encouraged investments in R&D. During most of the 1990s, R&D expenditures by the public and private sectors were estimated to be less than 0.30 percent of the gross domestic product (GDP). However, the proportion of R&D expenditures as a percentage of GDP increased gradually from 0.43 percent in 1998 to 0.79 percent in 2005, as illustrated in Figure 18.1. The government finances 44 percent of this expenditure, and private business spending has increased significantly since 1998, contributing more than 53 percent of total R&D expenditures in Hong Kong in 2005.

Moreover, the economic integration of Hong Kong and the Chinese Mainland has extended the innovative capacity of Hong Kong–based industries and services. Together with the strengthening of international R&D linkages that have emerged with the expansion of its research-based universities, the growth of overseas R&D activities by Hong Kong business has led to a new wave" of innovation networks. Hong Kong has also extended its traditional role as a financial center to include the capitalization of high-technology ventures in China. In this sense, Hong Kong has endeavored to create a new position as an innovation hub and intermediary between China and global markets.

Given the relative newness of policies explicitly designed to promote innovation, an analysis of the emerging innovative capacity requires a broad understanding that extends beyond conventional science and technology (S&T) indicators.

Figure 18.1 Gross Domestic Expenditure on R&D as a Percentage of GDP (1998–2005)

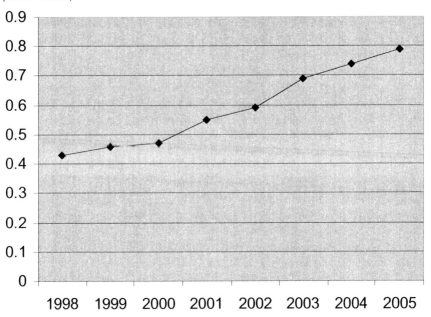

Source: HKSAR Census and Statistics Department. "Statistics on Research and Development in Hong Kong, 2001–2005," *Hong Kong Monthly Digest of Statistics* (for May 2004 and 2007).

Innovative Capacity: Conceptual Issues

Quantitative R&D statistics, patent statistics, and citations reflect only *formal* aspects of the processes of innovation. If there are few or no formal R&D expenditures, R&D statistics do not exist or, where they do exist, are of little use. Despite low levels portrayed in formal R&D statistics, there may still be extensive innovative activity taking place. In general, small economies especially need to focus on the benefits of acquiring and absorbing technologies from global sources, and need to emphasize the developing capabilities for doing this. The need to attract foreign direct investment (FDI) appears increasingly clear, especially together with policies that enforce competitive pressure on domestic firms in a relatively stable macroeconomic environment.[2] For late-industrializing countries, the diffusion and assimilation of technology have been essential to the upgrading of industrial facilities, as extensively documented by Amsden.[3] Nevertheless, efforts to promote R&D have been important complementary activities. L. Kim shows how Korean organizations have moved from an imitation stage to a more creative stage, and finally how some firms have reached the innovation stage.[4] It has been only in the latest innovation stage that a high intensity of R&D efforts toward generating new knowledge has been achieved.

The conventional emphasis on R&D expenditures and size of research organizations expresses a fundamentally mechanistic conceptualization of innovation systems. In this view, an innovation system is assessed in terms of indicators for key formal inputs, such as R&D expenditures or personnel, and key intermediate outputs, such as patents. This perspective emphasizes linear processes for transforming R&D inputs into innovative outputs. Sometimes, informal dimensions such as government support, the status of research organizations, or public perception of S&T supplement such assessments. In contrast, it is possible to develop more organic—or evolutionary—concepts that emphasize how learning and innovation take place under various conditions of market competition, absorptive capacity, and legal institutions. In this view, innovative processes are not merely determined by quantitative inputs, but are shaped in significant ways by informal cultural values and the formation of networks or clusters. Thus, I would propose a taxonomy of innovation system conceptualizations as shown in Table 18.1. One dimension is the extent to which formal or informal characteristics are the key focus, while the other dimension depicts an instrumental or organic perspective.

The most prominent study using the term *innovative capacity* in recent years is the comparative assessment of national innovative capacity published by Furman, Porter, and Stern.[5] They define it as "the ability of a country to produce and commercialize a flow of innovative technology over the long term." The study uses patent data as an indicator of innovative output, along with a range of indicators such as GDP per capita, aggregate S&T personnel, aggregate R&D expenditures, and share of GDP spent on higher education, as independent variables. In this sense, their work is firmly situated in the first quadrant of Table 18.1—emphasizing the quantitative and formal aspects of innovation inputs and outputs. The results of this analysis are in many ways unsurprising, showing a convergence of national innovative capacity among countries belonging to the Organisation for Economic Co-operation and Development (OECD). The results also suggest that the countries that had increased their innovative capacity had done more than increase the level of R&D resources by encouraging human capital investment and greater competition.[6]

Table 18.1 Taxonomy of Innovative System Concepts and Indicators

	Innovation system concepts	
	Mechanistic	Organic
Formal and/or quantitative indicators	(1) • R&D expenditures and funding • R&D personnel • Patents • Trade in high technology	(3) • Political economy structure • Absorption of technology • Educational system • Legal and regulatory framework
Informal and/or qualitative indicators	(2) • Environment for high-tech entrepreneurship • Government programs • Status of research organizations • Public perception of S&T	(4) • Linkages, clusters, and networks • Learning patterns • Cultural values and social cohesion • Labor market relations

A different approach, much more in line with the concept proposed here, is the study of regional innovative capacity of Lawson and Lorenz.[7] That study emphasizes collective learning among regionally clustered firms to encompass the generation and exploitation of tacit knowledge and the combining or recombining of diverse and complementary resources. Such an approach situates the concept of innovative capacity more closely to the literature on technological capabilities, competences, and resources of innovative firms. In this way, it draws upon the work of Lall, who explored the cumulative development of technological capabilities of individual firms, and linked these to aggregate national technological capabilities.[8] He noted that national technological capabilities were more than the sum of capabilities of individual firms in a country, being representative of an innovation system that includes externalities and synergy created by the learning process, ways of doing business, and the knowledge and skills residing in related institutions.[9] These concepts are representative of the fourth quadrant in Table 18.1.

My analysis of innovative capacity in Hong Kong will attempt to cover all quadrants in Table 18.1, in order to provide a more comprehensive understanding of various aspects of innovation in the territory. This method includes attention to formal R&D activities, but also recognizes the significance of the following attributes:

- Absorptive capacity for the exploitation of global knowledge sources
- Adaptation and modification of technology to new production assemblage
- Recombination of knowledge in new productive configurations; creative imitation
- Learning from advanced customers and markets
- Agile sensitivity to changing demands
- Organizational flexibility for orchestration of loosely coupled networks and business processes
- Systematic, stringent, and consistent quality control

This overview draws on quantitative statistical data, supplemented with a range of indicators compiled by the Hong Kong Census and Statistics Department since 2001.[10] I also draw on a few case studies to show the distinctive nature of much innovative activity.

Policies for the Promotion of Technology and Innovation in Hong Kong

The Commission on Innovation and Technology (CIT) was established on July 1, 2000 by the Hong Kong government. The CIT's mission was to spearhead Hong Kong's drive to build a solid foundation for innovation and technology development. The most important policy instrument in Hong Kong's effort to promote a new vision of knowledge-based development in the economy was the establishment of the ITF, with an endowment of HK$5 billion. The intention was to support R&D through cosponsored projects (with at least 10 percent of funding obtained from industrial partners) selected from applications submitted annually under the Innovation and Technology Support Programme, which has consumed three-quarters of the HK$2 billion during the first six years of the ITF's operation. However, an assessment of the outcome of these projects carried out in 2003 indicated that only about half of the results were considered useful by the industrial partners—thus adding less to the innovative capacity of Hong Kong industries than originally envisaged.

This assessment led to a new strategy, which the CIT launched in 2005. Accoding to this strategy, the major part of the future funding by the ITF would support R&D projects at five R&D centers, as well as a small number of major technological-breakthrough projects for specific sectors. There is a strong ambition to focus funding more precisely on sectors of particular relevance for Hong Kong's economic development.

Moreover, the importance of venture capital for high technology motivated the Hong Kong government in 1998 to expand the existing Applied Research Fund (ARF) into a publicly financed, but privately managed, venture capital fund. Hong Kong hosts an active community of venture capital companies, but typically these companies invest the vast majority of their funds outside Hong Kong. In the territory, there seems to be less fertile ground for such investments.

Although Hong Kong's entrepreneurs have historically been adept at exploiting available technology from the international market, they do not have a tradition of carrying out R&D for the purposes of creating proprietary technology on their own.[11] Thus, an unusually dynamic and adaptive entrepreneurial culture also fostered a short-term perspective of profit maximization with little long-term accumulation of technological knowledge or human resources.[12]

However, the government has undertaken several initiatives with the explicit purpose of supporting incubation of high-technology entrepreneurship. One of the earliest was the Hong Kong Institute of Biotechnology Ltd. (HKIB), founded in 1988 with a donation from the Hong Kong Jockey Club Charities Trust as a nonprofit but self-financing development center for biotechnology products. It has enjoyed only limited success so far.[13] The most successful venture in the field, CK Life Sciences, has not been connected to the HKIB. Most of the companies incubated by the HKIB either have disappeared or remain as small units lodged in the HKIB building.[14]

Another initiative was the Hong Kong Industrial Technology Centre Corporation (HKITCC), established in 1992. This publicly supported business innovation center was aimed at promoting technology development through three primary functions: technology-based business incubation and accommodation, technology transfer services, and product design and development and support services. The center hosted more than eighty new ventures in high-technology areas during several consecutive three-year incubation periods. In April 2002 the HKITCC was brought together with the new facilities at the Hong Kong Science and Technology Park to form an enhanced incubation scheme called the Incu-Tech Programme.

However, few high-technology firms have emerged from these incubation programs. Thus, among 201 companies listed on the Hong Kong Growth Enterprise Market (GEM), only three are graduates of the incubation program of the former HKITCC or the Incu-Tech Programme. Most high-technology companies launched on the GEM have originated in Mainland China. Among the few well-known GEM-listed firms established in Hong Kong are technology-related spin-off companies of large corporations, such as CK Life Sciences.

Another initiative came in March 1999, when the Hong Kong government announced its support for the Cyberport project, cooperating with the private sector to quickly create a strategic cluster of leading information technology (IT) service companies. The project provided advanced facilities and office space for 130 firms engaged in telecommunications, network and wireless communications, optical electronics, and Internet applications. However, in June 2002 the government reported that only seven companies had signed up to move into the newly completed Cyberport Phase I building, taking up about 80 percent of the 448,000 square feet of office space. The global economic slowdown and the bursting of the IT bubble also influenced the occupation rate for Phases II and III. In 2003 there were reports that Cyberport had problems attracting new IT firms and that one of the anchor tenants, Microsoft, had not brought in new investment in R&D, but merely had relocated its 250-strong marketing staff from downtown. The project has remained relatively isolated from other government-sponsored initiatives to support innovation in Hong Kong.[15]

An important innovation-related step was the establishment of ASTRI in January 2000. The intention was that ASTRI should perform R&D and the transfer of research results to industry for commercialization. It was intended to provide "midstream" and "downstream" research in order to facilitate the transfer of "upstream" research results from universities to industry. The model for ASTRI was the Industrial Technology Research Institute (ITRI) in Taiwan, which was seen as having contributed significantly to the development of Taiwan's high-technology industries. The first four years of ASTRI's existence were subject to considerable difficulties, due to both managerial problems and the uncertainties created by the crisis of IT industries. However, the arrival of a new management team in 2004, with a strong program of expansion and reorientation of ASTRI's R&D toward the need of customer industries, has strengthened the institute.

With these initiatives, the Hong Kong government embarked on policies very different from the policies of the past. While maintaining the overall philosophy of reliance on positive nonintervention, the government policies relating to innovation and technology in Hong Kong have been gradually moving toward an emphasis on market-driven and interactive knowledge flows as reflected in the concept of innovative clusters. Creating a new strategy was motivated partly by experience with schemes such as the ITF and the ARF, and also by the recognition of the need for a much closer integration of R&D with a select network of business enterprises—in other words, with existing or emerging clusters.

In 2004 the CIT prepared a consultation paper outlining a strategy that could better serve industry needs and raise the effectiveness of the government's support for innovation and technology. The paper reiterated the importance of technology upgrading in Hong Kong–related enterprises in the Pearl River Delta region, and the need to focus R&D and commercialization in selected areas.[16] Based on consultations with various stakeholders and industry representatives, in April 2006 the CIT officially established a range of R&D centers hosted by universities and/or public organizations in the following areas:

• Automotive parts and accessory systems
• Logistics and technology-enabling supply-chain management
• Textile and clothing
• Nanotechnology and advanced materials

Such activities will also be hosted by ASTRI in the following areas: communications technologies, consumer electronics, integrated circuit (IC) design, and optoelectronics. While this new strategy represents stronger commitment to steering technological innovation efforts—and thus "picking winners" rather than practicing nonintervention—there is no certainty that this approach will improve conditions for high-growth entrepreneurship. The selection of focus areas was clearly influenced by established industries or public organizations with vested interests and may not be representative of current or future high-growth firms or highly innovative technology-based industries. For example, the choice of the focus area "automotive parts and accessories systems" seems to reflect the ambitions of a small group of the Hong Kong Productivity Council's client base rather than any emerging high-technology industry.

It is clear that the Hong Kong government has embarked upon a transformation of the economy increasingly based on knowledge-based development and technology-intensive innovation. It is financing more R&D and has established various organizations to sponsor innovation and improve the environment for high-tech entrepreneurship. All this is focused on sectors where Hong Kong is believed to have potential competitive strength. With this perspective in mind, however, it is also necessary to understand the innovative activities that have been undertaken by organizations based in Hong Kong.

Innovation in Hong Kong: Status and Trends

As reported earlier, R&D intensity in Hong Kong grew from 0.43 percent in 1998 to 0.79 percent in 2005 but remains weak in comparison with other countries that have a similar GDP per capita; for example, Singapore invested 2.15 percent of GDP in R&D during 2003.

What is more important is that the business sector has increased its R&D activities. Figure 18.2 shows that while the higher-education sector was responsible for performing some 80 percent of R&D in 2000, it was around 47 percent in 2005, while the business sector's share grew from 18 percent in 2000 to 51 percent in 2005. Figure 18.2 also clearly shows that specialized government organizations still perform a very small part of R&D in Hong Kong—despite the setting up of ASTRI. At the same time, the universities in Hong Kong rely to a very large extent on government grants for research funding, and the higher-education sector in this sense represents a major public R&D commitment.

Together with the increase in R&D investments, Hong Kong's human resources for R&D expanded significantly, albeit from a modest level. Figure 18.3 shows that the human resources employed in R&D in Hong Kong more than doubled during the period 2000–2005. In terms of researchers (full-time equivalent) per 1,000 labor force workers, Hong Kong's proportion increased from 2.08 in 1998 to 5.09 in 2005.[17] Although this figure is not very high compared with Singapore's 9.3 researchers per 1,000 labor force workers in 2003, it brings Hong Kong close to the average figure of 5.7 researchers per 1,000 labor force workers in the European Union.

Figure 18.2 R&D Expenditure by Performing Sector (1998–2005)

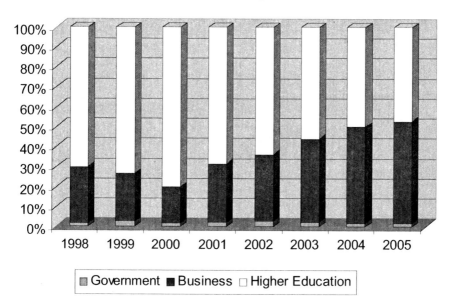

Source: HKSAR Census and Statistics Department, "Statistics on Research and Development in Hong Kong, 2001–2005," *Hong Kong Monthly Digest of Statistics* (for May 2004 and 2007).

It is likely that many of the new researchers employed by Hong Kong firms have been trained overseas. Moreover, Hong Kong firms have increasingly located their R&D activities on the Chinese Mainland, and it is therefore also likely that much of the increase in R&D personnel has occurred in units located there.[18]

With the increased R&D investments and an expansion of the human resources for R&D has come a stronger commitment to investments in technological innovation. Since 2001, when the Census and Statistics Department first conducted its survey of S&T indicators, there has been a steady increase in spending on technological innovation. As shown in Figure 18.4, such investments in the business sector more than doubled from 2001 to 2005, but declined somewhat in 2006. In fact, the percentage of technological innovation expenditure—in relation to total business receipt—grew from 0.09 percent to 0.27 percent during 2001–2004. The business sector's HK$18 billion expenditure in 2004 is also a large multiple of the HK$5 billion that the government pledged to commit to the ITF for expenditure over a longer period. In other words, government support of innovation may be more of a catalyst for further investments in the business sector, rather than a major source of funding in itself.

Figure 18.3 Human Resources Devoted to R&D, 1998–2005
(Number of full-time equivalent researchers)

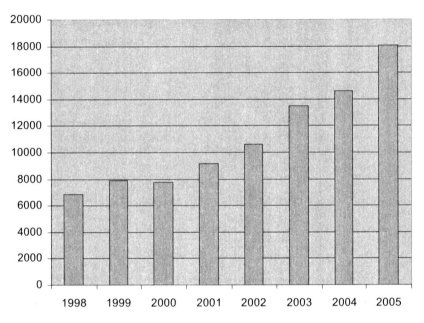

Source: HKSAR Census and Statistics Department, *Hong Kong as a Knowledge-based Economy: A Statistical Perspective*, 2007 Edition (Hong Kong: Census and Statistics Department, 2007).

The reported HK$18 billion that the business sector spent on technological innovation is likely to be a fraction of actual total business investments related to innovations. Businesses report more nontechnological innovations—that is, innovations concerned primarily with nontechnological aspects of the business. The data presented in Figure 18.5 indicate the number of firms that have reported technological product innovation, technological process innovation, and nontechnological innovation.

Figure 18.4 Business Sector Expenditure on Technological Innovation, 2001–2006

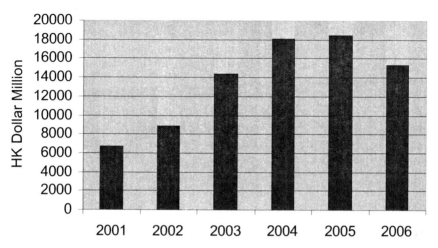

Source: HKSAR Census and Statistics Department, *Report on Annual Survey of Innovation Activities in the Business Sector*, various years, 2002–2006.

Figure 18.5 Technological and Nontechnological Innovations, 2002–2006 (Number of establishments undertaking each type of innovation)

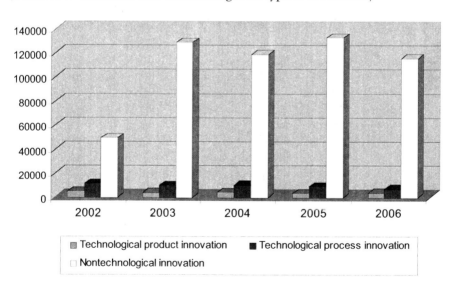

Source: HKSAR Census and Statistics Department, *Report on Annual Survey of Innovation Activities in the Business Sector*, various years, 2002–2006.

Table 18.2 Distribution of Firms That Have Undertaken Innovation in Hong Kong (No. of Firms)

Sector	Technological product innovation			Technological process innovation			Nontechnological innovation		
	2002	2003	2004	2002	2003	2004	2002	2003	2004
Manufacturing	779	393	160	1540	656	489	2,991	6,963	7,601
Electricity, gas, and water; construction*	437	11	18	771	71	57	2,355	6,410	4,104
Wholesale, retail, import/export trades; restaurants and hotels	2,434	2,117	2,618	6,237	4,402	4,607	30,437	79,189	73,902
Transport, storage, and communications*	42	76	44	918	1,104	666	2,487	8,883	6,561
Financing, insurance, real estate, and business services	814	775	1,013	1,313	2,094	3,220	8,364	21,482	19,072
Community, social, and personal services*	471	388	357	601	1,274	994	2,742	5,721	7,332
Total	4,977	3,760	4,210	11,380	9,574	10,033	49,376	128,648	118,571

Source: HKSAR Census and Statistics Department.

* Figures on these sectors have not been separately published in the reports of the Census and Statistics Department, because of considerations of precision. Care should therefore be taken in interpretation, as the statistics in these cells might be subject to large sampling errors. Besides, some industries—such as agriculture and fishing, mining and quarrying, and construction establishments, as well as those industries with relatively small economic contributions (e.g., taxis, public light buses, personal service)—are presumably not so involved in R&D and innovation activities, and are therefore not included in the survey.

The predominance of nontechnological activities by survey respondents reflects the dominant role of service sectors in the Hong Kong economy, with organizational marketing innovations usually playing a much more important role than the hardware-based new technologies that are predominant in much of manufacturing. Further details of the sectoral distribution of innovations reported are presented in Table 18.2.

This table shows that Hong Kong manufacturing firms have witnessed a declining output of technological product and process innovation. However, this does not mean that they are not innovative. The reason is that most Hong Kong firms with large-scale production facilities in the Pearl River Delta are registered under the category of "Import/export trade," since activities actually performed in Hong Kong tend to be focused on management, marketing, and development. It is noteworthy that these firms are deeply engaged in technological innovation and, in particular, in the kind of organizational streamlining and marketing associated with global production chains.

Table 18.2 also shows that business services have become increasingly innovative and contribute substantially to the continued competitiveness of this important sector.

Innovation Hong Kong Style: Continued Importance of Informal Capacity

Some of the most important Hong Kong industrial clusters were created in the wave of the expansion of manufacturing and services in the 1970s and 1980s. These include light manufacturing, transportation, tourism, financial and business services, and communication and media industries.[19] They have remained strong in specialized niches of global markets. Thus at the turn of the century, Hong Kong firms and their associated production bases in the Pearl River Delta remained the world's largest exporters of items such as toys and games, imitation jewelry, travel goods and handbags, fur clothing, telephone sets, and watches and clocks.[20] In such niche manufacturing and service industries, Hong Kong–based firms have been competitive due to a combination of low costs and very rapid turnaround in accordance with customers' demands.

The Consumer Electronics Industry: High Absorptive Capacity and Adaptive Capabilities

The Hong Kong electronics industry developed during the 1970s and 1980s, exporting around US$7.5 billion worth of electronics products in 1991—equivalent to 60 percent of the total exported by Taiwan at the time. Many overseas electronics firms first selected Hong Kong as a manufacturing base in Asia, and the electronics industry thrived on original equipment manufacturing (OEM) subcontracting arrangements.[21] Many small firms did labor-intensive assembly and later carved out niche positions in global commodity chains for products such as electronic toys, electronic watches, and chip packaging. They were aided by a free-trade framework, export orientation, and unrestrained labor markets. Organizations such as the Hong Kong Trade Development Council and the Hong Kong Productivity Council provided some government-sponsored assistance for linking up with foreign markets and technology transfer, respectively.

Speed to market has proven essential time and again. One example was provided during the Tamagotchi craze in 1996–1997. In 1996 the Japanese firm Bandai Company had invented a small electronic toy containing a relatively advanced IC and software, called a Tamagotchi (roughly translated as "a cute little egg"), which included a game with a small virtual pet animal that required regular feeding and attention to avoid "dying." This virtual pet toy developed into a market frenzy almost overnight, with sales of more than five million during the first six months. The craze came as a surprise for Bandai, which had only geared up production on a relatively small scale in Japan and was waiting to set up a manufacturing site in China. During this short interval, Hong Kong manufacturers capitalized on the overwhelming demand for Tamagotchis, which were in short supply in both Japan and the United States, and developed low-cost imitations or clones, such as Melko Chick or Pocket Dino, that were produced by a network of subsuppliers in southern China. During the short cycle of this craze, Hong Kong suppliers reaped substantial profits despite intense price competition, and when Bandai finally managed to get its new production facility in China under steam, the market had been swarmed by imitations—also from other production facilities in China—and the company finally faced losses from liquidating its overstock.[22]

Some of the Hong Kong companies that made Tamagotchis or that supplied components, such as LCD screens, later started to develop other products. For instance, the Hong Kong–based IDT International Ltd., a leading supplier of small LCD screens, developed a new range of electronic learning products based on its initial experience in the Tamagotchi market.

When electronics manufacturing moved to the Mainland in the 1980s, the low-cost Hong Kong production networks in South China had a tendency to capture the industries in the OEM patterns that had been successfully employed earlier. Several surveys of electronics firms in Hong Kong have found that 60 to 70 percent of these have copied or modified other products instead of doing independent product design.[23] The bulk of what little R&D expenditure there is by private firms is devoted to redesigning and improving products as well as making them easier and cheaper to produce. In other words, process innovation has often taken precedence over product innovation.

The example of the LCD industry in Hong Kong illustrates both the positive and negative aspects of the dominant strategy of its firms. They emphasize market intelligence and a keen market sense in order to be able to enter the right market niches at the right time. Coupled with high manufacturing flexibility and a recombination of knowledge in new designs or new productive configurations, they can shift production with astonishing speed. However, the absence of resources and government support for the continuous and extended upgrading of technologies tends to keep these firms locked in intensely competitive markets with a low (and often even declining) profit rate. They have become adept at surviving without the kind of support that LCD industries in, for instance, South Korea and Taiwan enjoy. Hong Kong LCD firms have moved into space left behind by more powerful competitors, and have avoided head-to-head competition. They have managed these moves on their own, without significant support from public institutions.[24]

This strategy has proven to be costly, risky, and time-consuming, yielding low returns. Chiu and Chung claim that if formal R&D investments had been channeled to the electronics sector (an existing industry in which Hong Kong firms had capabilities and skills), it is likely that the industry would have maintained its global position of leadership more effectively. They identify the predominance of the sales and marketing function in managerial strategies, and the bias toward a truncated OEM market, as the primary factors holding back R&D.[25] The weakness of product innovation and the low level of R&D investments has thus made it difficult for many firms to advance to original design manufacturing (ODM), although several firms in the garments and watch sectors have done this. Even more conspicuously, few electronics firms in Hong Kong have developed their own brand names (original brand manufacturing, or OBM). This is a persistent problem that has attracted increasing attention among policymakers in the government, and has resulted in several initiatives to improve the development and commercialization of new products. Therefore, if R&D were made a priority in traditional industries, such as electronics, the global performance of Hong Kong firms may be improved.

Service Innovation: A Hidden Strength?

The expansion of high-value-added clusters of service industries such as logistics, legal and financial services, or market research in Hong Kong in recent years has involved a range of light industries. In establishing and upgrading these networks, Hong Kong firms have exploited their traditional strategies of imitation and followership, while emphasizing organizational know-how rather than formal R&D.[26] In general, service industries do less formal R&D and rely on secrecy or copyright rather than patents to protect intellectual property (IP) and innovations. Given the relocation of manufacturing activities to the Chinese Mainland or Southeast Asian countries, services are now contributing more than 85 percent of the value-added in the Hong Kong economy. A study by Tao and Wong analyzing the emergence of producer services (such as business services directly associated with the production or distribution of goods) in Hong Kong indicates that their growth is closely related to the growth of FDI from Hong Kong to the Chinese Mainland, and with the growth rates of Hong Kong's reexports of goods from the Mainland.[27] Such producer services now contribute 50 percent of real GDP in Hong Kong. Producer networks from Hong Kong to the Pearl River Delta have supported the development of important services such as financial services, insurance, communications, and logistics in Hong Kong.

Consequently, the innovative character of Hong Kong's services is not adequately reflected in available R&D statistics. Producer services offered by Hong Kong firms have been constantly improved in terms of quality, reliability, and innovative solutions during the last two decades, and despite relatively high costs of qualified labor, the services of Hong Kong–based firms in logistics, telecommunications, and financial services have remained competitive. Transnational service firms making up a substantial portion of the service sector have often brought the latest technological advances to Hong Kong. But an increasing share of service firms originating in Hong Kong have been aggressively adopting IT-based systems to improve their offerings.[28]

One firm that has gained considerable fame in innovative services is Li & Fung, a Hong Kong trading company that was established in Canton in 1906 and that had sales amounting to US$4.2 billion in 2001. Li & Fung was one of the manufacturers that supported networking, subcontracting relationships, and an international search for markets before World War II. Today, Li & Fung has a specialized role as the orchestrator of loosely coupled supply-chain processes for a range of consumer products requiring labor-intensive manufacturing. Supplying well-known clients such as Levi Strauss, Reebok, and Disney, it uses a network of more than 7,500 suppliers in Asia and other continents. It handles the total chain of production and delivery of products to end-consumers, and the products are often packaged, marked with a price, and put directly on the shelf. This is achieved through a hybrid organization that includes a highly advanced and sophisticated electronic trading system linking five thousand people in the manufacturing process and various clients globally. The firm has also developed a number of customized extranets for major clients such as Coca-Cola.[29]

At the same time, Li & Fung utilizes more traditional networks of personal contacts and supervision for quality assessment and on-time deliveries. This extensive network of human resources coexists with the IT infrastructure to handle detailed design, production scheduling, logistics, final assembly, and customer relations. A dedicated team is engaged in extremely knowledge-intensive "disintegration" and optimization of supply chains, carrying out the design and planning of distributed manufacturing and coordination of the vast network. But few of these activities require formal R&D, and innovation is integrated into the development of new business processes and products. It is the specialized expertise in supply-chain management that provides Li & Fung with its unique competitiveness.

To be sure, innovations in services are not unique to Hong Kong. Many of the characteristics I have described are similar to those elsewhere. However, with the service industry occupying such a large proportion of the overall economy, innovation in services is doubly relevant and important to consider. Success is based on quick absorption of technology, organizational flexibility, learning from advanced customers and markets, skillful deployment of advanced management of productivity and quality control, appropriate levels of automation blended with labor-intensive procedures, and agile sensitivity to changing demands. Increasingly, these capabilities are being enhanced with greater abilities to produce and implement creative design and new product lines, and protect IP such as patents.

Conclusion: Innovative Capacity in Transition

It will continue to be important for Hong Kong's government to support innovation. However, intensified R&D efforts should enhance existing sectors rather than be directed at totally new ones. The recent push by the government to strengthen its innovation system should be even more closely aligned with existing strengths. The business sector's experience in more organic and informal types of innovation demonstrates the importance of exploiting existing knowledge and not being obsessed with the mechanistic creation and commercialization of new high technology. However important the creation of new knowledge—as reflected in R&D statistics—may be, it need not be the first step for a country seeking to strengthen its innovation system.

Nevertheless, an emphasis on new knowledge creation is likely to become more relevant at later stages, in order to strengthen a country's innovation system. This finding is acutely important for newly industrialized economies (NIEs) because there is more to a successful innovative strategy than simply pouring financial resources into formal R&D. Policymakers in these economies should take the more difficult route of identifying key organic strengths in their economies, and realize that it is possible to build on traditional strengths in order to attain innovative excellence.

Notes

[1] Hong Kong Special Administrative Region (HKSAR), "Commission on Innovation and Technology First Report" (Hong Kong, 1998), 13.

[2] David C. Mowery and Joanne Oxley, "Inward Technology Transfer and Competitiveness: The Role of National Innovation Systems," in *Technology, Globalisation and Economic Performance*, ed. Daniele Archibugi and Jonathan Michie (Cambridge: Cambridge University Press, 1997).

[3] Alice H. Amsden, *Asia's Next Giant: South Korea and Late Industrialization* (New York: Oxford University Press, 1989); and Alice H. Amsden, *The Rise of "The Rest": Challenges to the West from Late-Industrializing Economies* (Oxford and Hong Kong: Oxford University Press, 2001).

[4] Linsu Kim, *Imitation to Innovation: The Dynamics of Korea's Technological Learning*, The Management of Innovation and Change Series (Boston: Harvard Business School Press, 1997).

[5] Jeffrey L. Furman, Michael E. Porter, and Scott Stern, "The Determinants of National Innovative Capacity, *Research Policy* 31 (2002): 899–933.
[6] Furman et al., "The Determinants," 930–31.
[7] Clive Lawson and Edward Lorenz, "Collective Learning: Tacit Knowledge and Regional Innovative Capacity," *Regional Studies* 33, no. 4 (1999): 305–17.
[8] Sanjaya Lall, *Competitiveness, Technology and Skills* (Cheltenham, U.K., and Northampton, MA: Edward Elgar, 2001).
[9] Lall, *Competitiveness*, 266.
[10] I am grateful to Stephen Leung and Amy Yu at the Census and Statistics Department for their assistance in accessing and processing statistical data on innovation from the ASIA database.
[11] Howard Davies, "The Future Shape of Hong Kong's Economy: Why High-Technology Manufacturing Will Prove to Be a Myth," in *Hong Kong Management and Labour*, ed. Patricia Fosh et al. (London: Routledge, 1999).
[12] Tony Fu-Lai Yu, *Entrepreneurship and Economic Development in Hong Kong*, Routledge Advances in Asia-Pacific Business, vol. 5 (London and New York: Routledge, 1997).
[13] Lai Si Tsui-Auch, "Has the Hong Kong Model Worked? Industrial Policy in Retrospect and Prospect," *Development and Change* 29 (1998): 55–79.
[14] Erik Baark, "Innovation Policy Forensics: An Analysis of Biotechnology in Hong Kong" (paper presented at the DRUID Tenth Anniversary Summer Conference 2005, Copenhagen, Denmark, June 27–29, 2005).
[15] Erik Baark and Alvin Y. So, "The Political Economy of Hong Kong's Quest for High Technology Innovation," *Journal of Contemporary Asia* 36, no. 1 (2006): 102–20.
[16] HKSAR, Commission on Innovation and Technology, "New Strategy of Innovation and Technology Development" (Hong Kong, 2004).
[17] HKSAR Census and Statistics Department, *Hong Kong as a Knowledge-based Economy: A Statistical Perspective* 2007 Edition (Hong Kong: Census and Statistics Department, 2007).
[18] Federation of Hong Kong Industries, *Made in PRD: The Changing Face of Hong Kong Manufacturers* (Part II and Full Report, Hong Kong, 2003).
[19] Michael J. Enright, Edith Elizabeth Scott, and David Dodwell, *The Hong Kong Advantage* (Hong Kong and New York: Oxford University Press, 1997).
[20] Michael J. Enright, "Globalization, Regionalization, and the Knowledge-Based Economy in Hong Kong," in *Regions, Globalization, and the Knowledge-Based Economy*, ed. J. H. Dunning (Oxford and Hong Kong: Oxford University Press, 2000).
[21] J. W. Henderson, *The Globalisation of High Technology Production: Society, Space, and Semiconductors in the Restructuring of the Modern World* (London and New York: Routledge, 1989); and Michael Hobday, *Innovation in East Asia: The Challenge to Japan* (Aldershot, Hants, U.K., and Brookfield, VT: Edward Elgar, 1995).

[22] Naubahar Sharif and Erik Baark, "The Tamest of Tigers: Understanding Hong Kong's Innovation System and Innovation Policies," *International Journal of Technology and Globalization* 1, no. 3/4 (2005): 462–79.

[23] Tony F. Yu and Paul L. Robertson, "Technological Capabilities and the Strategies of Small Manufacturing Firms: The Case of Hong Kong," *Resources, Technology and Strategy*, ed. Nicolai J. Foss and Paul L. Robertson (London and New York: Routledge, 2000).

[24] Stephen W. K. Chiu and Wong Ka-Chung, "The Hong Kong LCD Industry: Surviving the Global Technology Race," *Industry and Innovation* 3, no. 1 (1998): 51–71.

[25] Stephen W. K. Chiu and Wong Ka-Chung, "Growth without Catching Up: Organizational Dynamics and Restructuring of the Electronics Industry in Hong Kong" (Hong Kong Institute of Asia-Pacific Studies Occasional Paper Series no.115, Hong Kong, 2001).

[26] Ishtiaq P. Mahmood and Jasjit Singh, "Technological Dynamism in Asia," *Research Policy* 32 (2003): 1031–54.

[27] Zigang Tao and Richard Y. C. Wong, "Hong Kong: From an Industrialized City to a Center of Manufacturing-Related Services, *Urban Studies* 39 (2002): 2345–58.

[28] Erik Baark, "New Modes of Learning in Services: A Study of Hong Kong's Consulting Engineers," *Industry and Innovation* 12, no. 2 (June 2005): 283–301.

[29] J. S. Brown, S. Durchslag, and J. Hagel III, "Loosening Up: How Process Networks Unlock the Power of Specialization," *The McKinsey Quarterly* (Special Edition: Risk and Resilience, 2002): 59–69.

CHINA AND THE EMERGING REGIONAL SYSTEM OF TECHNOLOGICAL ENTREPRENEURSHIP

Adam Segal

Since the late 1990s, China has been making what Richard Suttmeier calls a transition from a science and technology (S&T) policy to an innovation strategy. In Suttmeier's terms, policymakers increasingly complement the traditional tools of technology policy—top-down, state-directed programs that are often focused on specific sectors and government research institutes—with more bottom-up, multifaceted efforts to create a business environment supportive of innovation and entrepreneurship. Within the S&T policymaking community, there is an appreciation "that innovation, while involving science and technology, is a far more complex process than research and development (R&D) itself, and that while national R&D policy is an important element in a strategy, it must be nested in a socioeconomic context where risk-taking is rewarded and where social cooperation and inter-institutional coordination is highly developed."[1]

The Chinese strategy—reflected in 1999 and 2002 State Council decisions and further developed in the January 2006 National Medium- and Long-Term Science &Technology Development Plan—consists of three parts.[2] First, influenced by the experience of small, private innovative firms in the West, and in Silicon Valley in particular, policymakers put enterprises clearly at the center of the national innovation system (NIS). Moreover, support is now offered to all types of advanced enterprises, including nongovernmental, private, and small spin-offs, rather than just large state-owned enterprises (SOEs). Second, cuts in state agency personnel and oversight inhibit the government's efforts to select specific technologies for support. Now it provides broad support to all domestic enterprises designated as "high technology" and is making it easier to reward technologically inventive entrepreneurs. This is being done by defining and protecting property rights (including intellectual property, IP), using venture funds, and developing technology markets. The third part of the strategy encourages the less tangible software forms of technology transfer (such as licenses and consultancy) rather than hardware in the form of equipment imports.

The broad outlines of this strategy, and a similar desire to help foster a socioeconomic context for innovation, are visible throughout Asia. India has expanded financial support for small and medium-sized enterprises (SMEs), created incubators to foster university-industry linkages, and promoted the cooperation of state-run labs and multinationals. In Korea, the IT839 strategy—a government effort to introduce eight new information technology (IT) services, encourage investment in three key network infrastructures, and develop nine promising sectors—is complemented by new efforts to promote venture companies and "inno-biz."

Despite similarities, the genesis and trajectories of these projects are shaped by national industrial structures and political institutions. The legacy of the Soviet system of S&T is still felt in China, and the shadow cast by "mission-oriented" research institutes, especially in defense and nuclear energy, is long in India. Korean promotion of small technology enterprises, university-industry collaborations, and regional innovation systems (RISs) are tightly intertwined with efforts to reduce the gap between the *chaebol* (Korea's large business conglomerates) and small firms, and between Seoul and the rest of the country.

Yet there is also increasing interaction and competition among these systems. Policymakers are reacting to similar opportunities in the international environment. They are trying to exploit the opportunities created by the increasing globalization of R&D and the return flow of expatriates from Silicon Valley and other innovation clusters. There are also demonstration effects, as decision-makers evaluate and adopt policies from their neighbors. The Korean government is currently studying the Chinese experience with university-based technology enterprises. There are opportunities for cooperation among entrepreneurs in the region. Efforts to develop new standards in open software or home media bring firms in the region together. And there are competitive pressures as changes in one innovation system force a response in the others. The rise of China has created the concern among other actors that it will squeeze producers in both labor- and technology-intensive sectors; this adds to the impetus to promote local innovation. As a venture capitalist in Seoul put it, with only a little hyperbole, "All Koreans think about China all the time."[3]

This chapter looks at efforts to foster technological entrepreneurship in China, India, and Korea. It compares three policy arenas: university-industry collaboration, and university-related start-ups in particular; policy support for SMEs; and venture capital. Looking at the Chinese case in comparative context, there is clearly a degree of convergence in policies. Yet, many of the barriers to technological entrepreneurship that bedevil China—ineffective policy, dearth of early-stage capital, lack of scale, technological capabilities, management skills in small firms, and cultural barriers—are also present elsewhere. The question for policymakers throughout the region is how to fine-tune innovation strategy in order to promote innovation.

ADAM SEGAL

University-based Entrepreneurship

China

One of the greatest weaknesses of the pre-reform S&T system in China was separating R&D activities in universities and government research institutes from production and manufacturing. The reform period has seen efforts to reduce the barriers between research and production and to speed the commercialization of R&D. Within the ministries, there has been what Xue Lan calls a consistent focus on the "pull weakness of industrial R&D capabilities."[4] Most industrial firms lack innovative capabilities; in most firms, R&D expenses are only 0.56 percent of sales revenues, and in the large firms the percentage rises only to 0.71.[5] On the "push" side, university administrators, faced with reductions in support from the central government, have viewed university-industry collaboration as a way to supplement declining budgets.

University-industry collaboration occurs through several channels, including patent licensing, technology service contracts, joint research projects, university-based science parks, consulting agreements between individual faculty members and commercial firms, and university-affiliated enterprises.[6] University-affiliated enterprises received a big boost with the 1995 Decision on Accelerating Science & Technology Development, which called on universities and research institutes to establish firms using their own S&T capacity. These firms were often staffed and managed by the universities; they provided office space, graduate students and other staff, and IP. In 2002 Zhongguancun Science Park (ZGC) was home to more than 9,500 high-tech firms, among which more than 200 firms were university-affiliated.[7] Nationally, there were 2,216 technology-related spin-offs in 2002, with revenue of approximately $8.7 billion and net profits of $441.2 million.[8]

University-affiliated enterprises have attracted a great deal of attention in the rest of Asia, especially in Korea, and while there have been some success stories, such as Founder and Tongfang, the overall outcome has been mixed. The number of such firms has declined since the late 1990s, fewer have raised capital through initial public offerings (IPOs), and financial performance has worsened. A 2001 document, "Circular on the Experiment of Standardizing University-Run Enterprises Management at Peking University and Tsinghua University," actually called for the separation of universities and technology enterprises.[9]

A study by the Administrative Committee of ZGC found that, compared to other firms in the park, university-affiliated firms had lower labor productivity and profitability, fewer granted patents, and lower revenues from new products and technology transfer. Poor performance was the result of ill-defined property rights, high degrees of intervention from university administrators, a high level of bank debt and unnecessary risks, and a lack of management skills.[10]

There has also been a wider discussion about the possible negatives of blurring the lines between industry and academia, and in particular the impact of commercial activities on the academic environment and on university research directions.

India

The drivers to promote university-industry collaboration in India are similar to those in China: reductions in government funding, a desire to have universities play a greater role in the innovation process, and relatively low technological capabilities within Indian firms. Vijay Kelkar, former finance secretary, argues that one of the major weaknesses of the national S&T system has been "the feeble participation of universities in the innovation chain." Part of the problem has been funding. The central government provides the lion's share of R&D funds, and most of that money (60 percent) is funneled through the mission-oriented science agencies—the Department of Atomic Energy (DAE), the Indian Space Research Organization (ISRO), the Council for Scientific and Industrial Research (CSIR), the Defense Research and Development Organization (DRDO), the Department of Electronics (DOE), the Indian Council for Medical Research (ICMR), and the Indian Council of Agricultural Research (ICAR).[11] In 2003 higher-education institutions accounted for 4.2 percent of national R&D expenditures, significantly lower than the approximately 10 percent spent on university R&D in Korea and China.[12]

There has also been a mismatch between the orientation of the universities and the needs of industry.[13] Within universities, the incentives were for professors to publish, not to patent; there is also a sense that industry projects offered little intellectual challenge or pay for faculty. One professor at the Indian Institute of Technology, Mumbai, said the problems that industry needed help with seemed "trivial."[14] Academic institutes have been more likely to engage in join development with multinational corporations (MNCs) than with local companies.[15] Industry has seen little use for academic expertise and instead has preferred to look abroad for international collaborators.[16]

The IITs, Indian Institute of Science (IISC), and University Department of Chemical Technology, Mumbai, are exceptions to the general rule and have been more active in collaboration. The IITs in particular were a relatively better match with industry, placing stronger emphasis on fundamental principles over specialized knowledge, possessing a large number of postgraduate students, and encouraging interdisciplinary research and teaching.[17] The IITs were the earliest to develop mechanisms for working with industry—consultancy, midcareer work by industry, visiting faculty, joint student projects, and technology, development missions—and these expanded in the 1990s.[18] Still, the research capabilities of the IITs have tended to be limited—one professor at the IIT Mumbai claimed that the strength of research publication was not equal to a second-tier American university—and this has limited their overall impact on product or IP innovation.[19]

There has been a broad push to improve connections between industry and academia, led by the All India Council for Technical Education, Ministry of Human Resources, University Grants Committee, and Department of Science and Technology (DST). The National Research Development Corp., a department within the DST, licenses technologies that are developed in university- and government-run labs and invests in new ventures. The Home Grown Technologies program (HGT) and the Program Aimed at Technological Self-Reliance (PATSER) provide low-cost loans to industry to work with research institutes to adapt and commercialize technologies developed within the labs.

Within universities, the Technology Transfer Office (TTO) licenses technologies and arranges royalties for joint projects. The workings of the TTO have been criticized from both sides. Industry complains that the organization overvalues the technology that is transferred and understates how much more R&D is required before commercialization; academics say that licensing can be complex and bureaucratic. The biggest barrier has been a lack of trust between the two sides.[20]

The National Science and Technology Entrepreneurship Development Board (NSTDB), whose mission is to generate entrepreneurship among S&T personnel, has established industry-university cells, S&T entrepreneur parks, and technology business incubators. By 2005 there were some seventy parks and incubators, with the government adding another five a year, although government officials estimate that perhaps forty of them are successful.[21]

The structure and challenges faced by incubators are well illustrated by the case of the IIT Mumbai.[22] The incubator emerged out of a sense that more faculty members were involved in industrial consulting projects and were applying for patents. Since the terms of contracts were often unfavorable to individual professors, the prevailing thought was that creating companies would better protect the interests of the entrepreneurs and of the IIT. The IIT then had to decide how companies should be structured, what their relationship to the universities should be, and how much time founders could spend out of the department. Because professors are officially government employees, there was much sensitivity about questions of ownership and the social role of education.

The earliest solution was to designate a proxy as head of the company—usually a wife or son—and then take a leave to work one day a week. After the creation of the incubator in 2001, a pattern emerged: professors would build up the company over a year or two and then apply for an early sabbatical (they no longer had to wait for year 6). After the sabbatical year, they applied for a leave of absence, which allows them to keep their living quarters and seniority but decreases their salary. At the end of these two years, they made a decision: resign from the university and go back as an adjunct professor, or go back to their academic posting and work two days a week as a consultant to the company they founded.

For faculty members who started companies in the incubator, the university provided office space, phone lines, and other support in return for 3 percent of equity. When questions about the ownership of IP arose, a committee of outside consultants determined the value, setting a band between 2 and 7 percent—but no more since there was a fear that too large a share of university ownership is unattractive to outside funders and venture capitalists.

University-based entrepreneurship has had some notable successes. Prof. Ashok Jhunjhunwala and the Telecommunications and Computer Networks group at IIT Chennai have spun off or helped develop approximately twenty companies. Fifteen companies trace their roots back to IIT Mumbai. Younger faculty members are especially interested in starting their own companies, and continued pressure on academic budgets means that administrators will still see university-affiliated enterprises as an important source of revenue.

There are still significant barriers to the role of universities in technological entrepreneurship. Politically it remains a sensitive area, not only because of the large government subsidies to higher education but also because of the role of education as a public good. Within academic departments, there is resentment of outside commitments. The head of one department at an IIT complained that he did not like the ambiguity of where people's commitments lay—"they were here but always thinking of there"—and noted that while an "80-20 split between the department and the outside job might be workable, 50-50 would not be."[23]

As with university-affiliated firms throughout the world, management skills are often in short supply. The ability to manage a lab and commercialize new technologies does not often exist in the same person. "Awareness of IP"—the ability to value, market, and protect IP—is low, even among the scientific community.[24] And entrepreneurship remains a new concept. At least half of the graduates of the IITs leave the country; Indian Institute of Management (IIM) graduates go to work in big foreign companies with offices in India, such as McKinsey or the Hong Kong and Shanghai Banking Corporation (HSBC). The lack of a culture of networking (domestically, not among Indians abroad) and a fear of failure are significant barriers to technological entrepreneurship.

Korea

The Koreans, like the Chinese and Indians, believe there is a mismatch between what was supplied by the educational system and what industry needed during most of the 1980s and 1990s. Universities, focused primarily on undergraduate education, were weak in research capability. There were few incentives for professors to start their own companies, and since academics were considered public officials, there were social and legal restrictions to their receiving industry funds. At the same time, industry distrusted both the political leaning and practical skills of academics. Most institutes of higher education were not used to collaborating, and industry was not used to turning to academia.

As with the IITs and IISC in India, there were exceptions—the Korea Advanced Institute of Science and Technology (KAIST) and Pohong University of Science and Technology (POSTECH) in particular have been involved in industry collaboration—and there is now less distrust between the two sides. The Roh Moo Hyun government is promoting this collaboration by designating thirteen universities as industry-academia hubs and investing about $200 million in technology and human resource development. The Korea University of Technology and Education has initiated an "industry-oriented" curriculum, funded by Samsung, and President Roh has called for replication of the model throughout Korea. There have also been attempts to improve the university evaluation system. POSTECH, for example, now evaluates professors on two tracks, either on publications and citations or on contributions to industry and research money raised.[25]

To foster more university-related businesses, regulations now allow professors to form companies for a certain number of years, and universities can run businesses jointly with private firms. The Small and Medium Business Administration (SMBA) provides construction and operating-cost grants to incubators established by universities, public government R&D institutes, and local governments. The Ministry of Commerce, Industry, and Energy (MOCIE) has provided seed money and interest-free loans to start-ups—approximately $136 million to 1,500 companies from 1995 through 2003—through its Technology Business Incubation Program.

As with India and China, such efforts have raised questions about the roles of education and of university research. While improving the ties between research institutes and SMEs was expected to diminish the economic power of the *chaebol*, some have suggested that the influence of the large firms is increasing. The pressure on universities and government research institutes to commercialize new technologies makes them even more sensitive to the direction of the big companies, and more likely to move along technological trajectories determined by them.[26]

Policy Support for Small and Medium-sized Technology Enterprises

China

The last fifteen years could be seen as a great success for China's efforts to promote SMEs. From a few hundred firms in the late 1980s, there were over 100,000 nongovernmental technology enterprises in China in 2004, employing a staff of 6,444,300.[27] They had a total income of over $230 billion, profits of $14.4 billion, and R&D expenditures totaling $6.04 billion. The most important contributors to this success were the Torch Plan, introduced in 1988, and the high-technology development zones. Torch broadened sources of funds for nongovernmental enterprises. Although funds come directly from the plan, more importantly, designation as a Torch Plan recipient increases the likelihood of receiving local government support.[28] Advantages to industries in technology development zones include exemption from income taxes for the first three years after their establishment, as well as from construction taxes and licenses and duties on imported equipment.

Nevertheless, many barriers to SMEs remain in place.[29] They are much less likely to receive bank loans, and they do not receive the same preferential tax treatment as SOEs and foreign-invested enterprises (FIEs). Government procurement has generally been open only to government departments and SOEs. Local governments are slow to approve new enterprises; approval procedures, including registration and inspection, are burdensome and impose high transaction costs; and there are high barriers to entry in many fields, including high minimum capital requirements.[30]

As with previous policies, the 2006 National Guidelines trumpet the importance of SMEs to innovation and to the economy as a whole.[31] In a speech about the 2006 National Guidelines, Xu Guanhua noted that SMEs are responsible for 65 percent of patents and 80 percent of new products. They are the "birthing cribs" of big firms, and they are by definition innovative, for if they "did not have their own know-how they wouldn't survive and wouldn't develop."[32] Still, the guidelines state that "enterprises have not yet truly become technology innovators, and their independent innovative capabilities are not strong," mainly because of problems with technological capability, scale, ownership of property rights, and management skills.

The 2006 National Guidelines have at least eight provisions that directly or indirectly concern small and medium-sized technology businesses. As part of the broader process of shifting responsibility for R&D from government research labs and institutes of higher education to industrial enterprises, the guidelines state, "Enterprises take the lead in organizing and colleges and universities and science research institutes jointly participate in implementation." The guidelines reduce the income tax for high-tech firms that invest heavily in technology R&D, research expenses can be deducted from income tax for the year, and new enterprises registering in high-technology development zones will also enjoy two-year tax holidays and 85 percent tax reductions in subsequent years.[33]

Soft bank loans will also increase as commercial banks set up "bank-enterprise" relations with technology SMEs and building a credit-rating system for enterprises and individual entrepreneurs. Banks and other financial institutions are also called on to experiment with providing loans based on technology assets and to promote various kinds of credit-rating institutions.

India

The political and industrial structure has not been friendly to SMEs. There is, as in Korea, a polarization between the biggest family-owned companies (such as Tata, Reliance, and the Birla Group) and the tens of thousands of small enterprises whose support was seen as creating jobs and a more equitable regional distribution of industry. For the first four decades after independence, most policy interventions were protective, and there remains a political distinction between small-scale industries (SSIs) and SMEs. Through policies such as product reservation for SSIs, there was a mind-set not that small enterprises "could be an opportunity, but that they were also a vulnerability that needed to be protected."[34]

Even with the global success of the software industry, there is not yet a critical mass of small, dynamic technology companies capable of widespread innovation. In 2002 almost 90 percent of Indian companies did not report any R&D spending.[35] And even in the software sector, most companies spend only about 1 percent of sales on R&D. The incentives for the largest companies—such as Infosys and Wipro—to engage in risky R&D are not strong, given the high rates of return in outsourcing and IT services. A manager at a private equity firm argues that such companies have "growth rates of 15 to 20 percent in businesses they know. If you ask them to take their skills and apply [them] to new, untested technologies, they balk."[36] When these companies do conduct R&D, it is as a service to MNCs, and so benefits are captured by the foreign firm.[37]

There is a notable shift toward product innovation and a move in policy from protection to enabling competition, innovation, and globalization. As in Korea, policy initiatives have focused on raising the technological capabilities of SMEs through greater cooperation with research institutes, R&D grants, and incubating and supporting innovative technology firms.

The Technology Information, Forecasting and Assessment Council, through the Home Grown Technologies Program, provides technology-development-assistance funds for commercial enterprises at the pilot or demonstration stage of development; it also provides managerial assistance to start-ups. Grantees are expected to return funds in fixed installments. As of 2006, support had been granted to 77 projects, with the majority invested in SMEs; 21 projects had been commercialized, and 7 were entering the commercialization stage.[38] The Technology Development Board provides simple interest loans of 5 percent, to be repaid over five years to firms for commercialization. The Program Aimed at Technological Self-Reliance funds research, design, and engineering. The "Technopreneur Promotion" scheme, a joint effort of the NSTEDB and the Federation of Indian Chambers of Commerce and Industry, helps entrepreneurs find information about "technologies, projects, funding options and information about policy environment, incentive schemes and industrial infrastructure."[39]

The central and state governments also provide fiscal incentives and tax breaks for R&D expenditures. Companies with government-approved in-house R&D labs are allowed a weighted tax deduction of 150 percent of R&D expenditures. Companies whose principal objective is R&D are exempt from income tax for ten years from their inception. Accelerated depreciation is allowed for investment in plant and machinery equipment generated on the basis of indigenous technology. And companies are allowed customs and excise duty exemptions for capital equipment required for R&D.[40]

There are some innovative, product-oriented, small start-ups—for example, Sasken, Ittiam, and Tejas Networks—although IP/product still makes up only 5 percent of the market. These firms have some common characteristics: founders who came out of one of the MNCs (such as Texas Instruments), experience with more than one product cycle, and well-developed links to capital and global markets.[41]

Nevertheless, one analyst argues that the government sees the "need to foster SME growth, but does not know how to go about it."[42] Total spending on SME projects is approximately $45 million—compared to $6.3 billion in Korea. Most of the money is distributed through a very bureaucratic loan process; the government hesitates to use grants in fear that "industry will take the money and run."[43] Support comes from the DST and not from end-users such as the Ministry of Information Technology. As for procurement, the government has not viewed Indian producers as competitive. As one official describes it, the government is "midway in transition from acting as producer to acting as a regulator. We are not yet at the stage where we are thinking about the government as customer."[44]

Increasing government spending would only address part of the problem: the capacity of SMEs to absorb government support. Very few companies develop to scale. Unlike the founders of Ittiam or Sasken, the average engineer "is going to have much more difficulty in forming a team, getting capital, and understanding the market."[45]

Korea

In the early stages of Korea's development there were no explicit policies to promote SMEs. In contrast, many regulations kept SMEs from competing with the *chaebol*.[46] After the 1987 democratic transition and a corresponding attempt to prevent the abuse of economic power and ensure a more equitable distribution of income, the government began promoting SMEs.[47] When the 1997 Asian financial crisis created widespread unemployment, the government again looked to SMEs, this time to generate new jobs. The 1997 Venture Business Promotion Special Act enabled the government to certify and register "venture businesses"[48] and then supported new firms through tax benefits, credit loans for operating cash, public venture capital funds and matching funds for private investment, the lifting of the military service requirement for entrepreneurs and engineers, the lowering of financial requirements for initial public offerings, and the deregulation of the KOSDAQ. Venture businesses exploded—and then severely contracted. By 2001 there were approximately 11,000 venture companies, but some 60 to 70 percent failed within the next year or two.

In response to the controversy surrounding the rise and fall of venture businesses, the government introduced "inno-biz" companies—innovative businesses in traditional manufacturing sectors that were generally older and larger than "venture businesses." Inno-biz companies were also offered low-interest loans and preferential support, but such policies were as much about social policy and employment generation as they were about technology development. Success is clearer on the employment front; from 1998 to 2002 the largest firms shed 207,000 employees while employment in SMEs increased by 206,000.[49] In 2002 SME industrial output was $475 billion, 52 percent of total industry production.

The technological capabilities of small and medium-sized firms remain low, and there is a large gap between SMEs and the large firms in R&D spending; the twenty largest firms conduct more than half of the industrial R&D. In 2003 the R&D expenses of the largest firms totaled $115.7 billion versus $35.7 billion for SMEs (although the SMEs' share of total industrial R&D is increasing faster).[50] Of the over 110,000 SMEs in Korea, about 6,500 are officially certified as "venture business" and about 2,500 are "inno-biz." A 2004 study by the Science and Technology Policy Institute estimates that the ratio of "innovating" firms in Korea is lower than in other countries belonging to the Organisation for Economic Co-operation and Development (OECD): 37 percent in Korea, 67 percent in Germany, and 46 percent in France.

The Roh administration has redoubled efforts in SME support, integrating SME policy with that on regional innovation (regional technology service centers) and facilitating university-industry collaboration. Total government expenditures for SMEs equal $6.3 billion (5.5 percent of the total budget) with more than half as loans, the rest in subsidy or grant, and very little (1.2 percent) in equity investment. While the total number of venture businesses was expected to rise above 10,000 in 2006, there is the expectation that venture business and inno-biz policies will be integrated to reduce overlap and redundancy.

Efforts to raise the R&D capability and efficiency of SMEs exist in parallel to efforts to create new start-ups. In 2005 the government was expected to spend approximately $790 million on R&D in SMEs—12.7 percent of total government R&D expenditure. This money is disbursed through 14 government organizations and 230 separate programs, with the MOCIE and the SMBA taking the lead. Representative plans include the IT Commercial Technology Development Program, run by the Ministry of Information and Communication, and the Korea Core Industrial Technology Investment Association (KITIA) under the MOCIE.

The KITIA currently has investments in 319 companies, with part coming from the KITIA and the rest from venture capital companies, banks, and securities investment companies. The demand that applicants write their own business plans and raise part of the money on their own came from a shift in focus at the MOCIE. As the head of the management advisory group put it: "Before, the programs just gave a lot of money to many companies, and our aim was to develop technology. Now our aim is [to] develop technology and commercialize. Demanding they raise their own money creates partners of other companies that have a stake in their success."[51] Over the last six years, 20 to 40 percent of grant recipients have repaid their debt, twenty-two companies are now listed on the KOSDAQ, and four are bankrupt.

Critics question the positioning of SMEs in opposition to the *chaebol*, especially since many are suppliers to the larger firms. There are also questions about the continued use of industrial policy to support firms. The domestic market, and Korean society more broadly, are changing so rapidly that the government cannot keep up. "The real problem," according to an analyst at an S&T policy institute, "is that people believe too much in policy. Direct efforts to help SMEs have not been successful, so it is more important to create [an] environment in which SMEs can flourish."[52] While the newly created Office of Science and Technology is expected to coordinate all policy and controls the budget, the proliferation of SME policy is a sign of "too many organizations, too much overlap."[53]

In addition, there are questions about efficacy. A survey revealed that only 10 to 30 percent of government-funded projects resulted in commercialization, due to poor project selection and lack of resources for market entry. SMEs find it easy to say that they are undertaking R&D so that they receive government funds, and then fail to do any real development work. After explaining that R&D grants were given to firms to reduce risk, an official in the Industrial Policy Division of the MOCIE admitted, "I am not quite sure they wouldn't undertake the R&D on their own without our support."[54]

Venture Capital

China

Given the difficulties of small companies in accessing capital, there is much hope that venture capital and private equity, both within China and outside, can provide an additional, more efficient channel of funds for innovative firms. In a survey conducted by the State Council Development Research Center, 67 percent of the respondents listed "lack of funding" as the "biggest problem facing enterprise development."[55] According to Lu Kequn, chairman of the Beijing Securities Co. Ltd., ZGC needs a capital inflow of roughly RMB 200 billion ($29 billion) over the next ten years if it is to be successful.[56]

Initial funding for the first start-ups came from supervisory units (*zhuguan danwei*), personal savings, investment from family and friends, or some combination of these. Also, the Torch Plan provided direct investment as well, making it easier to approach banks for loans. High-technology parks and local governments also established their own investment funds for technology capital.

Although the China New Technology Venture Investment Corporation was formed in 1986, it was essentially an SOE, and it was not until the mid-1990s, according to White, Gao, and Zheng, that the "perception of venture capital shifted from its being a type of government funding to being a commercial activity necessary to support the commercialization of new technology."[57] Local government and universities established their own venture firms, and in 1998 the National People's Congress allowed corporations to establish their own venture capital firms. In 2005, according to the China Venture Capital Association, there were 175 firms in China, 35 of them foreign.

Despite the growth of the venture capital sector, domestic firms lack expertise in selecting, monitoring, and supporting new ventures. There is also a disjuncture between the internal structure and management systems of the venture capital companies and their requirements for investment managers.

A key issue for both domestic and foreign venture capital firms is the government's continued postponing of a second board for new venture listings. Perhaps most important, there are significant legal problems: Chinese law does not permit limited partnerships, minimum capitalization rules require large capitalization of corporations, rules for intangible assets make it difficult to price technology companies; requirements for pre-IPO investors to hold their shares for an extended "lockup" period (usually two years) after public listing increases risk and reduces potential returns to venture investors, and lack of transparency and frequent claims of improper use of proceeds have resulted in significant losses to investors.[58]

The 2006 National Guidelines address some of these concerns, calling for policies to "promote the development of venture capital investment in high technology."[59] Recent news reports suggest that venture capital firms will be eligible for a lower tax rate on investment income and possibly even be exempt from taxation. Venture funds will also be allowed to deduct the tax payable from their enterprise income tax for the year, based on a calculation related to investments. Local government departments will also be able to solicit public funds for investment, insurance companies will be able to invest in venture capital companies, and securities firms will be allowed to open venture capital investment businesses. The State Council is still considering launching a technology board patterned after the NASDAQ. A sub-board of the Shenzhen Stock Exchange, created to list small and mid-sized, tech-focused enterprises, was launched in the summer of 2004.

India

Venture capital developments in India mirror those in China: first government firms, then local, then foreign. In 1988 the Technology Development and Investment Corporation of India was created to administer a grant from the U.S. Agency for International Development (USAID) providing venture funding for the software industry. Soon afterward, the central government, and state governments, established approximately fifteen funds, with eight or nine surviving into the mid-1990s. In 1999 new regulations enabled various financial institutions to invest in venture capital. Perhaps the most important of these, in effect in April 1999, allowed banks to invest up to 5 percent of their new funds annually in venture capital.[60]

In the last half of the 1990s, foreign firms began arriving, and the first half of this decade saw the emergence of successful India-centric venture capitalists as well as increasing interest from U.S.-based firms.[61] Venture capital investments in India exceeded $2 billion in 2005—70 percent of which was funded by U.S. institutions—making India the largest destination for American venture capital outside the United States.

The firms have been investing in late-stage rather than early-stage ventures; "Even the existing venture capitalists would much rather fund late-stage ventures than take risks with the early-stage ventures."[62] The founder of one of India's most successful websites notes the widespread resistance to funding the young and inexperienced: "We [Indian capitalists] will not fund anyone under forty without ten years of experience."[63] While some successful entrepreneurs, such as N. S. Raghavan, one of the founders of Infosys, have supported new enterprises, there is a clear need for more angel capital.

Entrepreneurs complain that venture capitalists who are active in India have the mind-set of private equity investors (or worse, bankers), lack people on the ground, do too little mentoring, and lack technology experience. For investors, there are worries about transparency and management; companies are often seen as an extension of the founder. (This seems true of start-ups as well as of large companies.) "Until there is a clear separation of the person from the company and more visibility into the operations to an outsider, early-round investment will take a lot more work in India than it does in Silicon Valley."[64] The chief technology officer and the chief operating officer are hesitant to hand over authority to the chief financial officer; they "need to realize that ownership and control are not the same thing."[65] As a result, many venture capitalists prefer management teams that have global experience, with sales and marketing anchored in Silicon Valley and R&D located in India.

Korea

Venture capital officially dates back to the 1980s, but then it was a high-risk, low-return proposition. Funds were government sponsored with a very low rate of return. This all changed with the 1996 activation of KOSDAQ, which was designed to expand financing opportunities for small companies. The KOSDAQ exploded from 1998 to 2000, driven in part by the dot-com bubble. The KOSDAQ composite rose from 75.18 in December 1998 to 256.14 a year later; total capitalization went from Won 7.9 trillion in 1998 to 98.7 trillion in late 1999.[66]

Venture capital went along for the ride. The *chaebol* began setting up their own venture firms; Hyundai, Samsung, and the LG and SK groups set aside a combined $1.2 billion for technology investments.[67] By the trend's peak in 2000, there were 147 firms, 325 funds with total investment of Won 2 trillion. Since the bubble popped, funding has significantly decreased, and there is now greater reliance on government funding. In 2004 there were 105 companies and 422 funds, with a total of Won 527 billion; the government is responsible for 30 percent of the investment. To encourage stable growth, the SMBA is forming a "fund of funds" (Won 80 billion) for financing venture funds and wants to develop an active secondary market for venture capital (Won 800 billion).

The crisis of the *chaebol* in the 1990s convinced many in the government that Korea needed to diversify its industrial structure with more SMEs. One Ministry of Information and Communication official described the policy as "'small big companies'—create many small companies with the hope that maybe one, two, or three would grow to [a] large size."[68] Korea had just joined the World Trade Organization (WTO) and the OECD, so more interventionist policy tools were no longer available. There was also a desire to search for alternative funding avenues rather than bank loans. For SMEs, 72.7 percent of capital originates from bank loans, 19.8 percent from government funding, and only 0.3 percent from equity.[69] Many policymakers thought venture capital funds could fill this gap, and Korean officials visited Silicon Valley and invited American venture capitalists to come to Korea to lecture about how venture capital worked.

Korean officials believe that the role of state funds in venture capital will decline (and it has declined to about 15 percent)—except in such areas as content and cultural technologies. Currently, government funds do not "dominate the market, but all the other actors looking at government funds try and guess direction of investments."[70] As in India, there is a mismatch between the aspirations and expectations of entrepreneurs and venture capitalists. Korean entrepreneurs complain that there is a lot of money but little interest in early-stage investments. As one entrepreneur put it, venture capitalists "do not know how to evaluate technology, advise start-ups, or cultivate business contacts. In the United States, venture capitalists know the technology market as the innovator. Here [Korea] they do not know and do not spend enough time learning."[71] In the United States, venture capital partners do their own research on companies. In Korea, research staff do the work and then hand over the results to upper-level decision-makers, who have no intellectual stake in the companies. Only in 2005 were venture capital firms allowed to be listed as limited liability companies.

Venture capitalists complain that few entrepreneurs have "game-changing aspirations; they do not want to create a new Google but find a niche in a supply chain."[72] Governance is also complex. The founding CEOs of venture businesses tend to be excellent engineers but not as good managers. When venture capital firms want an outsider to step in, the venture business resists. Finally, exits are not easy. Cultural norms make it difficult to sell the company to another firm; selling is seen as a betrayal of the workers.

Conclusion

There is, throughout the region, a focus on the socioeconomic context of innovation, on creating market incentives to reward risk-taking, and on building horizontal cooperation and institutional coordination among stakeholders in academia, industry, and government. Fifteen years ago, few decision-makers saw the ability of small start-ups to access seed capital as critical to the country's innovative capability. They focused primarily on "big science" at state research institutes such as the Korea Institute for Science and Technology (KIST).

The issue, however, is how successful any policy—whether it governs S&T, innovation, or SMEs—can be in addressing the remaining barriers to technological entrepreneurship. More early-stage money can be made available through innovation funds, university technology liaisons can license to industry, and the registration process for SMEs can be simplified—but cultural, institutional, and political legacies shape these policies, creating unexpected outcomes.

In a sense, the shift from technology policy to innovation strategy has not gone far enough, because it still posits a central role for government actors. In the next stage, success will depend on the roles of civic and business associations. It is the growing number of successful Indian entrepreneurs who are involved in business-plan competitions, entrepreneurial clubs at the IITs, and angel capital that can effectively address cultural barriers to networking and risk-taking. It is the professor at POSTECH who introduces a course on computers and society who can address the distrust that many of the young in Korea feel for public institutions. Only through more active and independent involvement of these groups will a context of entrepreneurship be sustained.

Notes

[1] Richard Suttmeier, "Globalization, Structural Change, and the Role of Government in China's Search for a National Innovation Strategy" (unpublished manuscript, 2002).

[2] Barry Naughton and Adam Segal, "China in Search of a Workable Model: Technology Development in the New Millennium," in *Crisis and Innovation in Asian Technology*, ed. William Keller and Richard Samuels (New York: Cambridge University Press, 2003), 160–87.

[3] Interview, venture capitalist, LG Ventures, Seoul, March 8, 2006.

[4] Xue Lan, "University-Market Linkages in China: The Case of University-Affiliated Enterprises" (paper presented at the Conference on University, Research Institute, and Industry Relations in the U.S., Taiwan, and Mainland China, held at Stanford University, September 2004), <http://iis-db.stanford.edu/evnts/4097/LXue_University-Market_Linkages_in_China.pdf>.

[5] Xu Guanhau, "Regarding Several Big Problems in Independent Innovation," *Keji Ribao*, April 6, 2006.

[6] Hong Liu and Yunzhong Jiang, "Technology Transfer from Higher Education Institutions to Industry in China: Nature and Implications," *Technovation* 21 (2001): 175–88.

[7] Administrative Committee of Zhongguancun Science Park (ZGP), "A Research Report on the Performance and Problems of University-Owned Firms in the Zhongguancun Science Park" (paper presented at the Conference on University,

Research Institute, and Industry Relations in the U.S., Taiwan, and Mainland China, held at Stanford University, September 2004), <http://iis-db.stanford.edu/evnts/4097/MLZhao_University-Owned_Firms_in_Zhongguancun.pdf>, 2.

[8] Kun Chen and Martin Kenney, "Universities/Research Institutes and Regional Innovation Systems: The Cases of Beijing and Shenzhen" (BRIE working paper 168, September 2005), <http://brie.berkeley.edu/publications/wp168revised.pdf>.

[9] Jong-Hak Eun, Keun Lee, and Wu Guisheng, "Explaining the 'University-Run Enterprises' in China: A New Theoretical Framework and Applications" (working paper, January 2005), <http://www.ciber.gatech.edu/workingpaper/2005/019-05-06.pdf>.

[10] Administrative Committee of ZGP, "A Research Report," 9–10.

[11] And the first five agencies listed account for 86 percent of mission-oriented spending.

[12] Korea and China numbers are from the Organisation for Economic Cooperation and Development (OECD), *Main Science and Technology Indicators*, 2nd edition (Paris: OECD, 2007).

[13] The Ninth Five-Year Plan document notes that there "is organic inconsistency between skill requirement of the industry sector and the curriculum followed by the universities and technical institutions. The problem has [become] further compounded due to the inadequate linkage and interaction between universities and institutions of engineering and technology, government R&D organizations and user agencies such as industries." Planning Commission, Government of India, Ninth Five-Year Plan (1997–2002) Draft, vol. II, New Delhi (October 2000), <http://planningcommission.nic.in/plans/planrel/fiveyr/welcome.html>.

[14] Interview, research professor, IIT Mumbai, February 10, 2005.

[15] Interview, research professor, IIT Madras, February 2, 2005.

[16] P. V. Indiresan, "W(h)ither IITs?," <http://www.india-seminar.com/2000/494/494%20p.v.%20indiresan.htm>.

[17] Rakshat Hooja, "National Systems of Innovation and the Role of Academic Science Research in India," *Indian Journal of Education, Politics and Administration* 2, no. 1 (January–March 2004).

[18] Nimesh Chandra, "Knowledge Intermediaries in Globalizing India: The Case of the Technology Transfer Office at IIT Delhi" (paper presented at the 5th Triple Helix Conference, Turin, Italy, May 2005).

[19] Interview, research professor, IIT Mumbai, February 10, 2005.

[20] Chandra, "Knowledge Intermediaries."

[21] Interview, National Science and Technology Entrepreneurship Development Board (NSTEDB), DST, Delhi, February 14, 2005.

[22] Interview, research professor, IIT Mumbai, February 10, 2005.

[23] Interview, research professor, IIT Chennai, February 2, 2005.

[24] Interview, NSTEDB, DST, Delhi, February 14, 2005.

[25] Interview, research professor, POSTECH, March 7, 2006.

[26] Interview, economist, KDI, Seoul, March 6, 2006.

[27] 2001 *Niandu quanguo minying kejiqiye tongji baogao* [2001 Statistical

report on nongovernmental S&T enterprises], <http://www.mykj.gov.cn/manage/statistic/tj2001.htm>.

[28] Steven White, Jian Gao, and Wei Zhang, "China's Venture Capital Industry: Institutional Trajectories and System Structure" (prepared for the International Conference on Financial Systems, Corporate Investment in Innovation and Venture Capital, Brussels, November 7–8, 2002), 5.

[29] Adam Segal, *Digital Dragon: High-Technology Enterprises in China* (Ithaca, NY: Cornell University Press, 2002).

[30] OECD, *Economic Survey of China 2005: Improving the Productivity of the Business Sector* (Paris: OECD, 2005), 90–96; and Toshiki Kanamori and Zhijun Zhao, "Private Sector Development in the People's Republic of China" (Asian Development Bank Institute, October 2004), <http://www.adbi.org/book/2004/10/07/602.private.sector.prc>.

[31] The 1993 Decision on Several Problems Facing the Enthusiastic Promotion of Nongovernmental Technology Enterprises called nongovernmental firms a desirable outcome of the reform process, and the 1995 Decision on Accelerating S&T Development noted that nonstate companies were an important force in the high-tech field and were worthy of encouragement. The 2006 National Guidelines are available at <http://news.xinhuanet.com/english/2006-02/09/content_4155130.htm>.

[32] Xu Guanhau, "Regarding Several Big Problems in Independent Innovation," *Keji Ribao*, April 6, 2006.

[33] 2006 National Guidelines, <http://news.xinhuanet.com/english/2006-02/09/content_4155130.htm>.

[34] Interview, director of entrepreneur foundation, Mumbai, January 30, 2006.

[35] B. Bowonder, V. Kelkar, and N. G. Satish, "R&D in India" (ASCI Issue Paper No. 9, March 2003, Administrative Staff College of India, Hyderabad), 4.

[36] Interview, managing director, Warburg Pincus, Mumbai, January 30, 2005.

[37] Interview, vice president, Infosys, Bangalore, February 7, 2005.

[38] Nirmala Kaushik, TIFAC, "Home Grown Technologies Program" (paper presented to India-Israel Workshop on Technology Innovation & Finance, February 23–25, 2006, <ftp://132.68.13.3/events/INDIA-ISRAEL/Second%20Workshop%2002_2006/nirmala.pdf>.

[39] See <http://www.techno-preneur.net>.

[40] See Department of Scientific & Industrial Research (DSIR), Government of India, *Research and Development in Industry: An Overview* (New Delhi: DSIR, November 2000).

[41] Interview, manager, Ittiam, Bangalore, February 4, 2005.

[42] Interview, research professor, IIM Bangalore, February 5, 2005.

[43] Interview, research professor, IIM Bangalore, February 5, 2005.

[44] Interview, TIFAC, February 14, 2005.

[45] Interview, research professor, IIT Bangalore, February 5, 2005.

[46] Brian Lee, "Innovative SMEs and Promotion Policies in Korea," Trade and Investment Division, UNESCAP, January 20, 2006.

⁴⁷ Sung-Hee Jwa, "The Competitiveness Challenge to Korean Industry in a New Development Perspective," in *Industry Dynamism and Competitiveness in the East Asian Economies*, ed. Joonghae Suh (Seoul: KDI, 2004), 149.

⁴⁸ Venture business are defined as companies in which (1) R&D investment exceeds 5 percent of sales; (2) total R&D expenses are more than Won 50 million; or (3) for new technology firms, sales from tech/patent rights developed with government support and/or commercializing tech developed by a university or institute.

⁴⁹ Lee, "Innovative SMEs," 8.

⁵⁰ Joonghae Suh, "Enhancing Productivity through Innovation: Korea's Response to the Competitiveness Challenge," in *Industry Dynamism and Competitiveness in the East Asian Economies*, ed. Joonghae Suh (Seoul: KDI, 2004), 178.

⁵¹ Interview, Korea Core Industrial Technology Investment Association, March 8, 2006.

⁵² Interview, Science and Technology Policy Institute, Seoul, March 8, 2006.

⁵³ Interview, former director of science park, Seoul, March 5, 2006.

⁵⁴ Interview, Industrial Policy Division, MOCIE, March 6, 2006.

⁵⁵ *Wei Keji xing Zhong Xiao Qiye dailai Shenma* [What should be done for small and medium-sized enterprises?], *Keji Ribao*, July 25, 2002.

⁵⁶ Hou Mingjuan, "Zhongguancun Science Park Seeks $24 Billion," *China Daily*, October 7, 2000.

⁵⁷ White, Gao, Zheng, "China's Venture Capital Industry," 6.

⁵⁸ Feng Zeng, *Venture Capital Investments in China* (Pardee RAND Graduate School dissertation series, 2004), <http://www.rand.org/publications/RGSD/RGSD180>.

⁵⁹ 2006 National Guidelines, <http://news.xinhuanet.com/english/2006-02/09/content_4155130.htm>.

⁶⁰ Rafiq Dossani and Martin Kenney, "Creating an Environment: Developing Venture Capital in India" (BRIE working paper, 2001), <http://aparc.stanford.edu/publications/creating_an_environment_developing_venture_capital_in_india/>.

⁶¹ Sunil Mani, "Performance of India's Innovation System Since 1991" (paper presented at India-Israel Workshop on Technology Innovation and Finance, Bangalore, February 23, 2006).

⁶² Rajesh Jain, Emergic blog, (http://www.emergic.org), August 24, 2005.

⁶³ Interview, manager, Rediff.com, Mumbai, January 31, 2005.

⁶⁴ Abrar Hussain, "What's Holding Back the Indian Venture Capital Market?" January 16, 2006, <http://www.ventureintelligence.in/blog/2006_01_01_archive.html>.

⁶⁵ Interview, research professor, IIT Mumbai, February 20, 2005.

⁶⁶ Seong Somi, "Prospects of Korean Startups and Cooperation with Silicon Valley Firms," *Joint U.S.-Korea Academic Studies*, vol. 14 (Korea Economic Institute, 2004), 169.

⁶⁷ Moon Ihlwan, "The Latest Chaebol Game Plan: Stalking Cyber Startups," *BusinessWeek Online*, <http://www.businessweek.com/2000/00_14/b3675219.htm>.

[68] Interview, government official, MIC, Seoul, March 8, 2006.
[69] Lee, "Innovative SMEs," 24.
[70] Interview, fund manager, STIC Investment, March 8, 2006.
[71] Interview, founder, venture business, March 9, 2006.
[72] Interview, venture capitalist, LG Ventures, March 8, 2006.

Changing Industry Dynamics: Intellectual Property and the New Media

MANAGING INTELLECTUAL PROPERTY IN THE CHINESE SEMICONDUCTOR INDUSTRY

Xiaohong Quan, Henry Chesbrough, and Jihong Wu Sanderson[*]

Chinese semiconductor firms as a whole are still positioned at the low end of the value chain in the global integrated circuit (IC) landscape, although a few are emerging as world-class. Most of their technology is based on old-generation foreign technology.[1] Imitation instead of innovation dominates. While the growth rate of the IC industry almost doubles every year, it is still very hard for domestic firms to enter the high-end product market. In this industry, more than 80 percent of the patents registered in China are owned by multinational corporations (MNCs).[2] So most Chinese IC firms focus on peripheral products.

It is well known that intellectual property (IP) protection in China is very weak. As Shichang Zou, president of Shanghai IC Industry Association, reports, "In the centralized economy, we didn't have the concept of IP; we have just begun to study it."[3] Fei Teng, of the International Technology and Economy Institute (ITEI), the Development Research Center of the State Council, notes that the "IP issue is very complicated in China."[4] There are eight separate government sectors dealing with IP issues, and the unclear boundary of rights and obligations among various administrative units inevitably causes coordination problems and low efficiency. Accordingly, the cost of IP protection is very high.

Weak IP protection in China is not the result of a dearth of relevant regulations. A body of case law is forming in areas such as trademarks and copyrights and, to a much lesser degree, patents. Enforcement is the problem. "In China, IP laws are actually there, following international routines. The biggest problem is law enforcement. Our law firm has devoted a lot of time giving lectures to Chinese firms and government officials."[5] With the exception of the lawsuit between the Shanghai Manufacturing International Corporation (SMIC) and the Taiwan Semiconductor Manufacturing Company (TSMC), we heard no stories of effective enforcement of IP in the semiconductor industry. Instead, we

---

[*] This paper is part of a project sponsored by the Alfred P. Sloan Foundation, titled "The Globalization of R&D in the Chinese Semiconductor Industry."

heard stories on how to avoid IP issues or how to tolerate infringement. However, the government has started to realize the importance of IP in this industry. "We do need to protect other people's IP and our IP,"[6] as one industry leader put it. Indeed, domestic Chinese firms have taken photos of other firms' IC design and then developed their own IC products. An illustration is the Hanxin chip scandal in 2006, in which Jin Chen, the dean of Shanghai Jiao Tong University's School of Micro-Electronics, was removed from his position after his "Hanxin chips I-IV" were found to be counterfeits. However, legal and foreign pressures aside, with the growing complexity of technology, it is becoming less feasible to operate in this manner. As Weiping Liu, director and general manager of CEC Huada Electronic Design Company, indicates: "IP protection is essential to our business; more protection will help us develop our technology."[7]

An important finding from our interviews is a genuine, homegrown impetus for strong protection of intellectual property rights (IPRs). It is not derived from the lobbying of MNCs in the country, although that is going on, but rather the domestic industry, led by the systems companies, and the associated government entities who realize that stronger IPRs are necessary to upgrade skills.

Government Efforts

The government has been encouraging domestic semiconductor firms to develop their own IP as well as promoting IP protection. As Fei Teng of the ITEI vividly describes, "Chinese firms are still at a negative position in terms of IP protection. Although the Chinese IC industry is developing very fast, it is far from strong. In fact, most Chinese firms are still in the stage of imitation only. That's why we have this national campaign of education on IPR. The purpose of the campaign is to stimulate firms to innovate themselves, not just imitate. Chinese firms are mostly at the level of exchange of goods, instead of exchange of IP. They have no idea how to sell ideas and protect their IP. The whole country needs to brainstorm to understand IPR."[8] As a result, we see an IP social movement in China today, albeit at an early stage.

At the central government level, the Intellectual Property Affairs Center (IPAC), affiliated with the Ministry of Science and Technology (MOST), has played several roles. On the theoretical side, it conducts research on national patent strategy; on the practical side, it oversees MOST projects, including patent application and licensing, and helps judicial departments determine proper technology protection. An illustration of the function of the IPAC is its cooperation with Synopsys, mostly relating to software licensing. Synopsys is collaborating with the IPAC with the idea that Chinese firms who use its software will likely need further licenses for software upgrading. "It took a long time to finalize the deal," says Lincun Yang, director of the IPAC. "The contract is very carefully articulated. The licensed software can only be used by twenty firms in each of the seven specified software development bases supported by MOST.[9] The firm which will use the licensed Synopsys software cannot exceed an employment size of fifty people. Furthermore, these firms can only use the software for two years."[10]

Regional and city governments are also very active in this IP movement. For example, the municipal government of Shanghai has initiated the "Shanghai Intellectual Property Action Outline," in which it encourages domestic firms to invest in their own research and development (R&D) and to respect other companies' IP. An important initiative is government help in building a public R&D platform to lower R&D costs for small firms. The Shanghai Integrated Circuits R&D Center is such a public technology platform provider. With a budget of about $10 million per year, it has contracted with some big electronic design automation (EDA) tool companies to get software licenses for local firms to share. It also provides a multi-project wafer (MPW) to help lower small firms' R&D costs. In addition, the center assists companies in applying for IC layout IP protection at a low cost. Since China entered the World Trade Organization (WTO), there already have been five hundred applications for layout IP protection.

There is greater recognition of the need to follow international rules to improve technological skills. The No. 18 Policy, which offered special tax benefits to domestic IC firms only, is an example. After protests by the U.S. government, the Chinese government agreed to end the program. Local firms are forced to compete with global players under the same rules.

However, government efforts are not enough to make a strong IC industry. As Yang puts it, "The ironic thing about [the] Chinese IC industry is that large-scale IC manufacturing firms, where the government has put a lot of efforts (money, equipments, talents), failed, while design houses, where no effort was ever made by the government, have flourished."[11] The fact that design houses, which usually require effective IP protection to make profits, flourish under China's weak IPR protection regime is a paradox that needs to be investigated further.

Is Intellectual Property a Big Concern in the Chinese Semiconductor Industry?

While IP protection in China is weak, there are surprisingly few IP disputes in the semiconductor industry. Fei Teng, of the ITEI, states that "in 2004, there were ten big cases on IP disputes in total, two or three of which were involved with trademarks in the semiconductor industry. In China, [the] public security bureau deals with disputes on trademarks and copyrights, while patents disputes are solved by formal lawsuits."[12] According to Dr. Jing Lin, of the Shanghai Municipal Informatization Commission, "so far, no big IP disputes have been observed among domestic IC companies here in Shanghai. The reason may be that every firm is busy with its own development and sometimes cross-licensing is occurring among small firms."[13]

However, there has been one very well-publicized IP infringement lawsuit. This was between two big foundries—the SMIC and the TSMC. Having legally obtained advanced technology from Siemens and Infineon, the SMIC then lured more than one hundred employees away from the TSMC in 2002. These workers brought advanced 0.18-micron process technology to SMIC, doing so in record time. In 2004 the SMIC became the number-four foundry in the world in revenue.[14] Viewing the growing threat from the SMIC, the TSMC sued the SMIC in December 2003 in the United States for infringement of five U.S. patents and misappropriation of technical and operational trade secrets relating to methods for conducting semiconductor fabrication operations and manufacturing ICs. Earlier, the TSMC had sued the SMIC in the Hsinchu District Court in 2002 for improper hiring practices and trade secret misappropriation. In early 2005, the lawsuit was settled at the price of about $175 million to be paid by the SMIC. Outside observers believed this to be a good deal for the SMIC.[15] Calvin Chin, a manager from the SMIC, put it this way:

> The only winners in the lawsuit were the lawyers. While [the] TSMC may have filed suit to damage our IPO, once that process went through, that was gone as a motivation to continue the dispute. It became a distraction for both companies, and it was better for us to simply cross-license all of [the] TSMC's patents. Since they've been around for nearly twenty years, their patent portfolio is substantially bigger than ours. And the dispute didn't help us entice customers to either of our companies. It doesn't help the foundry industry when two of the top three companies are in a major dispute. I think we are satisfied with the terms of the settlement: $175 million over six years.[16]

This may be an exceptional case. In general, local Chinese firms are vulnerable in lawsuits. "When companies face a lawsuit, we have to pay [a] very high legal fee—as much as, [or] more than, several millions USD. This is very difficult for growing companies,"[17] observed Hongbing Peng, director of IC Product Contracts at the Ministry of Information Industry (MII).

As a result, many firms have learned to tolerate infringement. "Employees could leave and take designs with them to compete with us," says Weiping Liu, of Huada Electronic Design. "Two groups of employees have done this. It is hard to prove infringement, though, so we have not stopped them. For example, if they copy 70 percent of our design and develop 30 percent on their own, that may be enough to keep us from stopping them."[18] Lun Zhao, of the DMT, reports, "We did have an experience with Gemplus where they infringed one of our patents. They came to us and offered to exchange other patents with us in exchange for a license to that patent. But we didn't want any of these other patents. When we analyzed the situation, we found that their sales in China were small, but they had large sales of the infringing product in Europe. Our patent was in China, though, not in Europe. So ultimately we decided not to fight."[19]

Indeed, most patents are for defensive purposes. "DMT does own more than one hundred patents, but most are just for protection, not for making money from them."[20] This is consistent with earlier studies of IPRs in semiconductor firms in the United States. Such firms patent actively to reduce the risk of being held up by external patent owners and to negotiate access to technologies on more favorable terms. This is done even though firm managers don't regard patents as an effective mechanism for appropriating returns on R&D investments.[21]

A semiconductor manufacturing firm lacking a strong patent portfolio with which to negotiate licensing or cross-licensing agreements could face a severe erosion of profits at a time when the costs and risks associated with infringement have increased. One industry executive estimates that "a new manufacturer would need to spend $100 to $200 million to license what are now considered basic manufacturing principles but which do not transfer any currently useful technologies."[22]

In contrast to fabrication, on the design side, IP protection is believed to be essential. Currently, there are an estimated six hundred design houses in China. A severe market shakeout is expected. Weiping Liu says that only about one hundred of the over six hundred design houses will survive.[23] Similarly, John Deng, chairman of Vimicro, says that many existing design houses are not in a healthy condition. He explains: "In China, it is not hard to start up a fabless due to the chaotic nature of the market and information technology (IT) mania. However, a lot of these firms actually don't know anything about business models. They don't have earlier (overseas) experiences. There is only entrepreneurial spirit, no entrepreneurial success."[24] More extremely, Wayne Dai, of VeriSilicon, says, "I can see only about twenty qualified design firms in China. Half of the over six hundred design houses in China are not real; they are not doing design, but distributions."[25]

Despite the small size of the IC design industry (about RMB 10 billion[26] in 2006, or $1.46 billion), it is growing fast. A few returnee-run design houses such as Vimicro and Spreadtrum have globally comparable competence. The growth of IC firms has raised some interesting issues. "IP protection is essential to our business, and more protection will help us develop our technology," says Weiping Liu, "but IP infringement is not a big concern for us."[27] This seemingly paradoxical quote means that although the IPR protection regime in China is weak, there are not many IP infringements in this industry. Or, in other words, even if there are many infringements, it is not a big concern for innovators.

How Do Integrated Circuit Firms Manage Their Intellectual Property in China?

Effective protection of IPRs is essential for this industry. However, currently there are infringements in China that evidently do not hurt IP holders' businesses much. We encountered this puzzle when we conducted interviews in Shanghai and Beijing. So why do IC firms not care much about infringements?

Basically, as semiconductor designs require more complex technology, imitation becomes very difficult. Reverse engineering is less feasible.[28] And imitations are usually much inferior to the originals. For example, some firms reverse engineer Vimicro's designs in China and sell their products at half of Vimicro's price. However, they are very inferior, according to John Deng, chairman of Vimicro.[29] Although departing employees can take away certain technology, it is hard for them to develop similar businesses. One reason is that individual employees in a design firm usually master a certain module within the technology. "DMT is a team, not a group of individuals with their own patents," says Zhao Lun. "While individual pieces can be taken, it is hard to copy our environment."[30]

Even successful infringers face difficulty in production because they need to sell their final designs to foundries or system companies. According to Zhao Lun, "If someone were to copy our products, they would find it difficult to sell them. They would also find it difficult to sell their other products."[31] A tentative explanation is that there is a social protection mechanism in Chinese society. For instance, customers would be reluctant to buy products from "betrayers" and would try not to make their principal suppliers lose face, especially when there is a relationship between suppliers and buyers. (However, further evidence is still necessary on this argument.) As John Deng, of Vimicro, says, "It really takes a lot of resources and capabilities to be a successful firm in China—and everywhere in the world."[32]

It is well known that formal IP protection is not essential to capturing returns to innovation in many industries.[33] Patents are effective in only a few, such as chemical industries, while other mechanisms—such as lead time, learning-curve advantage, secrecy, and sales and service efforts—are better means of appropriation in some other industries. Our interviews were consistent with this argument.

According to Deng, "A good way for Vimicro to deal with [the] IP problem is to move fast and make imitators difficult to follow."[34] One manager from a start-up equipment company held that "IP won't be a big concern for us since the technology for semiconductor equipments is very complicated. It is hard to imitate. For reverse engineering, it can only maintain its usefulness for ... about six to twelve months. As long as the firm can keep introducing new products, IP infringement is not a problem."[35] Formal litigation is not a good way to protect IP for either domestic firms or MNCs in China due to its high cost.

Some companies, especially those competing globally, do make great efforts to protect IP. "We actually have to overcompensate on IP protection to compensate for the reputation China has for weak IP protection," says Calvin Chin, of the SMIC. "We have very strong policies and procedures. We have no floppy disk drives or removable hard disk drives in our computers to download files. Everything we do in IT is server-based. As a public company, we realize how damaging it would be for us to fail to protect a customer's IP." In the SMIC, every exiting employee goes through an exit interview process, where they sign IP-protection agreements. They also have production, data and network security, confidentiality and nondisclosure agreement (NDA) contracts, and continuous employee training. "I would argue that no court in the world has the capability to adjudicate many of our technologies," adds Chin. "We specify governance by U.S. courts or sometimes HK [Hong Kong] courts in the event of a dispute. Our contract language is not dissimilar to contracts I worked on in my former jobs back in the U.S."[36]

There is another way that foreign companies find it relatively cost-efficient to protect IP. Since it is expensive and almost impossible to identify all infringements in China, tracking big licensees' revenue growth can help IP holders avoid further loss. For instance, "Synopsys can monitor the revenues of the top twenty design houses to get their money," says Lun Zhao, of the DMT. "If their revenue is growing, they can assume that their receipts from those houses should also be growing. They can come down to us and ask for more royalties. We send about $700,000 in royalties to them per year."[37]

As mentioned earlier, actively developing one's own IP is a way to avoid or deal with IP disputes. A young star design firm, Spreadtrum, in Shanghai, is an example of innovation. "Although Spreadtrum licenses some IP from ARM in the UK, 90 percent of IP used is owned by itself," says Ping Wu, president and CEO of Spreadtrum. "Furthermore, the IP licensed from ARM mainly are general IP modules. We develop the core IP ourselves."[38] Its mobile-phone chip designs are leading-edge. Basically, Spreadtrum does everything from scratch. Since the product is systematically complex, it is hard for any individual leaving the firm to imitate Spreadtrum's product. However, it is not easy to keep innovating. Moreover, Spreadtrum has to know how to scale itself and to use resources more efficiently to maintain a high growth rate in this highly competitive field.

Law firms in China play various roles in IP protection. Their work can be categorized as (1) IP-formation issues, (2) IP-infringement alerts, and (3) IP-dispute resolution. On IP formation, they help firms with patent and copyright application. Domestic Chinese firms still need to be educated about this, because earlier "soft" knowledge was not considered valuable and was not well protected. And small firms may not have enough money for patent application (for example, it costs $20,000 to $30,000 to apply for patents in the United States) and look to the government for help. As for IP-infringement precautions, law firms help semiconductor firms prepare confidentiality agreements to be signed by employees upon hiring and noncompete agreements dealing with employees leaving companies.

If IP infringement has occurred, a law firm can help with legal remedies. Sometimes "many administrative bureaus within the government need to be involved, and the law firm can help with coordination," according to Hongxu Qin and Harry Su, of Coudert Brothers LLP.[39]

Intellectual Property Rights for Multinational Corporations, Compared with Chinese Firms

Table 20.1 summarizes IPR-protection issues in the semiconductor industry, with a preliminary comparison between domestic Chinese firms and MNCs such as Intel.

Other than a formal IP mechanism (patent, copyright, and so on), MNCs may use a hierarchical modular R&D structure. According to Xiaohong Quan, an MNC's internal R&D activity has its own value chain, and different actors can specialize in different activities along the chain. Furthermore, such specialization is not only across firm boundaries; it can be done within an MNC across different levels and in different locations. For example, one MNC studied by Quan carefully separated its systems design work from specific component technologies in the overall design. The systems design work was kept in the home country, while elements of the implementation were conducted in its China laboratory.[40]

Table 20.1 How Do Integrated Circuit Firms Manage Their Intellectual Property in China? Domestic Chinese Firms versus MNCs

	Domestic Chinese firms	MNCs in China
Knowledge of IP protection	Weak	Strong
Owned IPs to be protected	Small amount	Large amount
Potential threats for IP	• Employees leaving • Customers revealing information • Competitors reverse engineering • Lack in IP training, IP education	• Employees leaving • Customers revealing information • Competitors reverse engineering
Measures of IP protection	• Education, training • Employee agreements, contracts • Lead time and learning-curve advantage • Modular structure • Patents, copyrights • Social protection	• Education, training • Employee agreements, contracts • Lead time and learning-curve advantage • Modular structure • Patents, copyrights
Purpose of IP patenting	Mostly defensive	Both aggressive and defensive

One firm, a joint venture between a Japanese MNC and a local China partner, separated its semiconductor fabrication process into three distinct processes. The joint venture took care not to rotate people across the three processes, so that any departing employee would know, at most, a single process.[41]

Technological value is not the only factor determining commercial value. Perception of market need is another important variable to be combined with technological value to determine whether an R&D task is core or peripheral, and consequently, whether it is better placed in developed countries with strong IPR-regimes or under the weak IPR regimes typically found in developing countries.[42]

A New Business Model for Intellectual Property—Shanghai Silicon Intellectual Property Exchange[43]

Shanghai Silicon Intellectual Property Exchange (SSIPEX) was started in 2003 and is funded by both the Shanghai municipal government and the MII. It is one of three platforms[44] sponsored by the Shanghai government to promote regional industry by helping companies to legally access Western IP design tools and libraries. SSIPEX acts as a bridge between design houses and foundries by providing access to virtual modules of functionality for companies. The company deals with two kinds of IP: one kind is used by design institutes and houses to create chip processes, and the other kind is used by foundries for layout, process, and functions. To date, SSIPEX has collected the second-largest database of IP in the world, comprised of more than three thousand items.[45] (The IP database of Design and Reuse, also known as D&R, in France, is number one in the world, with more than four thousand items.) SSIPEX has also created a database of patents, compiled from Chinese, U.S., European, and Japanese sources.

SSIPEX assists Chinese companies in a number of ways. First, it provides them with access to a data set of 4 million patents. Second, it helps companies learn about potential infringement in advance of their design work, instead of after the fact. This also indicates that China (or at least the Shanghai government) is serious about respecting IP. SSIPEX is also a good avenue to identify patents that may be available for purchase. In fact, its director, Jason Zhu, checks with six patent attorneys every day to ask about possible opportunities for acquiring patents and to learn about transactions and trends in the market, any new infringement actions that have been filed, any new settlements, and so on.

SSIPEX has over 113 companies paying for its IP services. It also has thirty customers for its patent database. SSIPEX works with IP owners in three ways. One way is that IP owners give SSIPEX specifications, and the information is then posted for its customers. A second way is that IP owners provide more detailed technical specifications, which some companies are careful to guard. SSIPEX will sometimes act as a nonexclusive broker for these owners at their request. A third way is for SSIPEX to integrate pieces of IP from various owners and then codevelop a piece of IP. There may be the potential for IP infringement here, but according to Zhu, "We are a reputable company and we are also backed by the government, so this makes us more powerful if something does go wrong. In China, we also don't want to lose face, so that is another reason not to permit illegal copying. Some of our customers, like Hua

367

Hong, have adopted technology that counts the number of chips produced. If an IP licensing agreement entitles them to produce 500,000 chips, they will be able to count them and then stop."

However, the effectiveness of this new model is not yet proven. Although SSIPEX is funded both by member subscriptions to its IP repository and by transaction fees on IP licenses, to date, the organization has not reached the break even point. Until the licensing component becomes more significant, the organization will struggle to sustain itself. The government-ownership nature of SSIPEX is inevitably also a concern. Time is still needed to determine the extent to which a public platform like this can help with the growth of the semiconductor industry.

Notes

[1] Fei Teng, International Technology and Economy Institute (ITEI), Development Research Center of the State Council, interview, May 13, 2005; and Lun Zhao, President, Datang Microelectronics Technology (DMT), interview, May 13, 2005. "The Chinese semiconductor industry is just developing," said Lun. "We must do everything to achieve competitive power. On the other hand, the competition on the technology side is not so advanced. Ninety nanometers (nm) and 65 nm are far off for us. We are using more economical and more convenient technology. Point 18 micron is best now, and 0.15 is also used a little. But 0.15 is not so convenient, not a mainstream process. [The] 0.13 price is too high to use now."

[2] Lincun Yang, director, Intellectual Property Affairs Center (IPAC), Ministry of Science and Technology (MOST), interview, May 13, 2005.

[3] Interview, May 16, 2005.

[4] Interview, May 13, 2005.

[5] Hongxu Qin and Harry Su, Coudert Brothers LLP, interview, May 19, 2005.

[6] Shichang Zou, Shanghai IC Industry Association, interview, May 16, 2005.

[7] Interview, May 14, 2005.

[8] Interview, May 13, 2005.

[9] Since 1995 the Chinese government has established 29 national software industry development bases, in cities such as Beijing, Shanghai, Guangzhou, and Nanjing.

[10] Interview, May 13, 2005.

[11] Interview, May 13, 2005.

[12] Interview, May 13, 2005.

[13] Jing Lin, deputy division chief, Information Industry Administration Division, Shanghai Municipal Informatization Commission, interview, May 16, 2005.

[14] Data from IC insights, <http://www.icinsights.com>.

[15] Weiping Liu, director and general manager, CEC Huada Electronic Design Company, interview, May 14, 2005.

[16] Calvin E. Chin, manager, Strategic Ventures, Shanghai Manufacturing International Corp. (SMIC), interview, May 17, 2005.
[17] Hongbing Peng, director of IC Product Contracts, Ministry of Information Industry (MII), interview, May 18, 2005.
[18] Liu Weiping, Huada Electronic Design, May 14, 2005.
[19] Interview, May 13, 2005.
[20] Lun Zhao, interview at the DMT, May 13, 2005.
[21] Bronwyn Hall and Rosemarie Ham Ziedonis, "The Patent Paradox Revisited: An Empirical Study of Patenting in the U.S. Semiconductor Industry, 1979–1995," *RAND Journal of Economics* 32, no. 1 (spring 2001): 101–28.
[22] As quoted in Weston Headley, "Rapporteur's Report: The Stanford Workshop on Intellectual Property and Industry Competitive Standards," Stanford Law and Technology Policy Center, Stanford University Law School (April 17–18, 1998).
[23] Weiping Liu, Huada Electronic Design.
[24] John Deng, chairman, Vimicro, interview, May 14, 2005.
[25] Wayne Dai, chairman and CEO, VeriSilicon, interview, May 21, 2005.
[26] 1 USD = 8.08 RMB (CNY) on December 2, 2005.
[27] Liu Weiping, Huada Electronic Design.
[28] Shichang Zou, Shanghai IC Industry Association, interview, May 16, 2005.
[29] Interview, May 14, 2005.
[30] Interview, May 13, 2005.
[31] Interview, May 13, 2005.
[32] Interview, May 14, 2005.
[33] Richard C. Levin, Alvin K. Klevorick, Richard R. Nelson, Sidney G. Winter, "Appropriating the Returns from Industrial Research and Development," *Brookings Papers on Economic Activity* 3 (1987): 783–820; and Xiaohong Quan, "Multinational R&D Labs in China: Local and Global Innovation" (PhD dissertation, University of California, Berkeley, 2005).
[34] Interview, May 14, 2005.
[35] Shangzhong Luo, deputy director, MOST, Soft Technology Association, interview, September 11, 2005.
[36] Interview, May 17, 2005.
[37] Zhao Lun, DMT, interview, May 13, 2005.
[38] Ping Wu, president and CEO, Spreadtrum, interview, May 19, 2005.
[39] Interview, May 19, 2005.
[40] Quan, "Multinational R&D Labs in China."
[41] Calvin E. Chin, SMIC, interview, May 17, 2005.
[42] Quan, "Multinational R&D Labs in China."
[43] The information in this section comes mainly from interviews with Jason Zhu, director and vice president of SSIPEX, and Tracey Lewis, marketing department deputy manager, March 17, 2005.
[44] The other two platforms are the Shanghai IC R&D Center (ICRD) and the Shanghai National IC Design Industry Center (ICC). The ICRD focuses

on providing manufacturing-process platforms, and the ICC focuses on design services platforms, such as electronic design automation (EDA) tools.

[45] Jason Zhu, Shanghai Silicon Intellectual Property Exchange Center, interview, May 17, 2005.

CHINA'S NEW MEDIA: SHAPING NEW INDUSTRIES WITH NEW POLICY REGIMES

F. Ted Tschang and Seng-Su Tsang

In the industrialization process, the traditional path has been for economies to move from agriculture to manufacturing to services. The so-called East Asian "tiger" economies such as Taiwan and Singapore have fit the pattern of this progression, with export-oriented manufacturing first, followed more recently by services. With the rise of modern-technology-oriented services such as software services, the path has diverged for countries such as India, which have more or less skipped the manufacturing exports stage to directly become providers of outsourced software and other software-related services such as research and development (R&D) services, business-process outsourcing, and so on. These cases suggest that countries with relatively weaker manufacturing competencies can still move ahead to service-related forms of work. The Indian software case is well documented, but far less research has been done on other countries.

This chapter looks at a set of new media sectors and the policies that support their growth. While there has been much activity and government interest in these sectors in the advanced countries, the developing countries have also started to recognize the potential of creative sectors even during their emergent stages of industrial development. These sectors might be another instance of the "industrial leapfrogging" model. While we are ultimately interested in the Chinese situation, this chapter will also look at other Asian experiences to examine what competitive or cooperative situations exist and to see what China might learn from those experiences.[1]

The new media sectors are a subset of "creative industries" that relate to content and technology—namely, video games (including games played on dedicated gaming consoles or personal computers), animation, and mobile-device (including cell-phone) content. Creative industries include sectors as diverse as design, architecture, filmmaking, and video games.[2] The definition of a creative sector revolves around the product coming from the efforts of individuals, but in many products, such as filmmaking and video games, increasing complexity has been accompanied by increasing team sizes.

Basic Characteristics of the New Media Sectors

The three new media (or digital-content) sectors that we focus on are animation (including both "hand-drawn" two-dimensional animation and digital or three-dimensional animation), video games, and mobile-device content, which includes games as well as other Web-based content.

There are some key differences, but also similarities, in how each new media sector is organized compared to previous high-tech sectors. First, there is the cultural nature of the content. While some media are exported by means of contracts (for example, animation in India or the Philippines), or by modifying "local" content for a global or regional audience (such as Japanese or U.S. animation), local digital-content providers can also strictly dedicate themselves to a domestic market and its highly localized preferences (such as Chinese animation). China's digital-content providers are very different from its manufacturing firms and are similar to its software firms in the software inputs, especially in games. In addition, digital content is largely about culturally influenced content.

Second, the structure of the industry involves a set of intermediaries, including channel providers (organizations that provide access to TV transmission), distributors, and publishers (which can both fund and distribute products) to promote consumer adoption. Production and distribution may become separated in mature markets for content such as animation and TV programs. Governments might regulate content and grant access rights to the transmission medium. For mobile content, telecommunications carriers have the ability to control access and, therefore, content. For games, there is traditionally less control over access, but game publishers control the creation of ideas through their financing of development and their control over distribution channels.

The third aspect relates to the technical and economic characteristics of the products. While all three sectors involve content, in games the content is interactively used; that is, it is accessed in a "nonsequential" manner. Firms in all the sectors are concerned with intellectual property rights (IPRs) relating to the content, because there is no intellectual property (IP) protection for a given *type* of game. The model for revenue generation is different, however, from sector to sector. It appears that while games rely on consumer purchases as their main source of revenue, the animation sector depends on TV advertising revenue and has a far greater ability to tap into associated streams of merchandise. For the mobile-device sector, operators not only benefit from content transmission but also split the content subscription fee with content providers.

The fourth aspect relates to the technological basis of these industries. While technology is embedded almost by definition in games, the proportion of content that consists of art and animation has been increasing in recent years. In animation, while content has been the primary component, the sector has seen some wrenching changes in employment and contracts due to technology, first in the conversion of hand-drawn to computer-aided animation (particularly in what is known as "in-betweener" work, which involves drawing the cells in between the reference cells), and in a further shift, at least in the U.S. animated feature-films industry, to three-dimensional animation, which is software- and computing-intensive. Many animation companies, particularly outsourcing service providers, now operate with smaller permanent workforces and rely on a wider pool of temporary employees.

New Media in Asia

A discussion of the new media industries of the countries surrounding China may also shed light on China's situation, in part because many of these countries' trends are interlinked.

Korea

In 2004 the market value of games was estimated at about $3.8 billion (of which about 43 percent comprised revenues from Internet cafés), whereas the market value of animation was $221 million.[3] There were upwards of 119 companies developing online games and 9 companies developing console games. The most rapidly growing sectors were online and mobile games. Korean game companies that were surveyed consider online, mobile, and video games to be viable and promising platforms in the future and role-playing games to be the most promising genre. It appears that Korean animation and game companies, like Japanese companies, have successfully fused some aspects of their own heritage, as well as the general Eastern cultural heritage, with aspects of fantasy and science fiction, into new "cultural worlds" and visual styles quite different from those in Western and other markets. Given the significantly advanced state of development of Korean online games, they are the most influential in China so far.

Taiwan

In Taiwan, the PC games market was $280 million in 2004, with online games dominating at $226 million and estimated to reach $331 million in 2006.[4] There were five publicly listed firms and a larger number of smaller firms supplying games. The major weakness for game firms in Taiwan is that government support is not as aggressive or as organized as in Korea and Singapore. Two possible strengths of the industry are the skill of programmers and the industry's familiarity with the Chinese culture. Like Korea, Taiwan seeks to introduce some of its games to China.

Singapore

The Singapore government has targeted digital media (including games and animation) for government support. It announced that S$1 billion (approximately US$600 million) of support per year will be divided between the digital media and two other sectors. While the government has already set up training centers in the polytechnics and more than two hundred graduates a year come out with game- and animation-specific skills, higher education is not yet very involved. Even before this initiative, smaller pools of funding were used to develop the local industry, with about forty games and animation companies being formed, most of which are smaller start-ups. In addition, several multinational corporations, such as Lucasfilm's animation studio and KOEI Co., Ltd., have been attracted to Singapore. Two of Singapore's possible weaknesses are that it has no strong indigenous culture and it has not undergone a long period of nation-building, so it may be difficult to form IP that is rich or unique.

The Philippines

The Philippines has a long historical record of animation, although most of it has involved outsourcing to Western countries. While the country had significant artistic skills and a comparative advantage in its appreciation of Western (particularly American) culture, language, and preferences, the rise in artists' wages coupled with technology-induced productivity changes, plus a global crash in animation production in the early 2000s, hurt the industry badly, and some studios were severely cut back or bankrupted. The industry has seen a rebirth in the last few years, with several firms in operation, but firms are much leaner and have smaller capacities than before, in part because of technological change and the adoption of an organizational model involving larger temporary staffs.[5]

New Media in China

In contrast to almost all the Asian countries, China's large internal market has driven its industries in separate, domestically focused ways (although in animation, there is a strong capacity for outsourcing). Table 21.1 shows the size of each sector.

Table 21.1 Main Sectors and the Size of Revenues (in RMB)

	Animation*	Online games	Mobile gaming
2002		1 billion	
2003		1.9 billion	320 million
2004	30 billion	3.6 billion (including mobile) (24 million users)	804 million
2005	70 billion	5.8 billion**	1.44 billion** (15.2 million users)

Source: Analysys International, various reports.
Note: * Production value
** Estimate

There are as many as 5,473 animation companies in China by one estimate, of which 15 are designated as "national" (sufficiently large and strategically important in nature).[6] According to one popular observation, in animation, Jiangsu and Zhejiang provinces (including Hangzhou, the capital of Zhejiang) are becoming strong regions. Shanghai has the strongest marketing and circulation capability. In addition, strong companies or historical centers exist in places such as Shanghai, Beijing, and Changsha. Some cities and studios benefited from being designated as national "animation bases" by the government.

For online games, it appears that Beijing and Shanghai are key centers, possibly because of Beijing's science and technology (S&T) resources and Shanghai's reputation as a cultural and financial hub. Of the major operators of online games, Kingsoft and Sohu are in Beijing; Shanda, SINA, and The9 are in Shanghai; NetEase is in Guangzhou; and Tencent is in Shenzhen. There were reportedly 30 online-game-development studios in Shanghai alone, and 73 nationally, already in existence as of January 2005, although this number fluctuates constantly.[7]

In the cell-phone-game sector, many service providers as well as content providers appear to be clustered in Beijing, with most providers located between the outer third and fourth "ring roads" of eastern Beijing.

The National Character of Each Sector and the Competencies of Firms

We now turn to an examination of three new media sectors of increasing interest in China: mobile-device content (including cell-phone content, and with a focus on games), online games, and animation.

Mobile-device Games

For mobile-device games, the telecommunications infrastructure and equipment is critical to the ability of consumers to take part in mobile entertainment and other services. With the diffusion of more advanced mobile phones increasing the demand for more advanced content, the number of Chinese mobile-gaming subscribers reached 15.2 million users in 2005 and was expected to reach 19.5 million in 2006. China's market for mobile gaming is expected to have high profitability but with high competition. Because less than one-third of China's 390 million mobile-phone subscribers are mobile-gaming subscribers with handsets that can support data services, the real boon for content providers is likely to come from third-generation (3G) networks and phones. It is expected that with 3G deployment, the business and revenue models will be similar to the "service model" used in the Internet-gaming industry. Users will also have options similar to those of computer users, in that they will be able to select and install games of their choice.[8]

According to one analysis, the market value of cell-phone games grew 44 percent from 2004 to 2005—when it topped RMB 1.44 billion (approximately

$185 million according to mid-2000 exchange rates)—making it one of the fastest-growing markets worldwide.[9] The government may be less likely to open the domestic market to foreign content providers, and the monopoly afforded by the regulated nature of the airwaves can help ensure this. In 2006 only two telecommunications licenses were allocated to operators (China Mobile and China Telecom), and about three hundred licenses to service providers. In the near future, another two licenses may be issued to operators, but the structure may not change immediately because of the government's extensive regulation. As in other countries, the operators have the power to define the nature of content that is offered on their network—for example, regulating on the basis of the moral nature of the content.[10] However, this ability to regulate may change as Web portals start to offer content that is downloadable to mobile phones.

Currently, the service providers have the upper hand over content providers, with many traditional as well as newly established content providers rushing into the mobile market. According to an interviewee from an application developer, content providers and application developers concentrate on relatively narrow niches, such as push mail or MP3 music applications. In this large and intensely competitive market, just about any idea—sound or not—will have been considered by some start-up. In addition, the success of companies such as Baidu and Shanda on NASDAQ appears to have encouraged both companies and financiers, and companies appear to be generously financed by venture capitalists.

One potential issue is that the high number of programmers being trained for the broad information technology (IT) areas means that the technology barriers to entry are not high. As a result, licenses are essential to restricting access. This particular situation may not change in the coming decade due to the government's conservative attitudes toward media.

According to our interviewees, two big challenges that the industry faces are those of how to identify customers' needs and how to cultivate hits. On the demand side, although China is a huge market, the demand for content services varies substantially across provinces. Currently, there is a huge shortage of marketing and other talent to help identify these differences and their implications for services and content differentiation. Talent capable of producing more advanced kinds of content is also facing shortages. Currently, colleges that offer media programs do not have the faculties to cultivate these kinds of talents. On-the-job training is currently one of the ways to alleviate the shortage of skilled workers, and Taiwan and Hong Kong also provide skilled workers. Nevertheless, the undersupply of media talents may remain for years to come.

Online Games

Online games are a significant part of Chinese gaming culture, and earned RMB3.37 billion in the fourth quarter of 2007 (or nearly $0.5 billion at mid-2000 exchange rates).[11] These games offer more assured revenue than traditional games, since piracy rates are much lower, given the provider's control of the servers, game content, and gameplay. By contrast, conventional PC games suffer piracy rates as high as 95 percent, making them very difficult propositions for firms.[12] Even though the market is growing more crowded and competitive, and more products are appearing that resemble one another, the market continues its upward trajectory, with 60 percent growth to $1.66 billion in 2007.[13]

Until a few years ago, however, as much as 80 percent of online games in China originated in other countries, predominantly Korea.[14] Selected U.S. games came into China much later—such as the massively multiplayer online (MMO) game *World of Warcraft*, which garnered 44 percent of its worldwide audience in China one month after it launched there.[15]

There are as many as one hundred online game operators in China. Online game operators tend to range widely in their primary line of business, including software, games, and other Internet services. Some of the biggest are Shanda and The9 (both primarily online game operators), Kingsoft (a diversified software company), NetEase and SINA (both Web portals), Sohu (a Web portal including a search engine), and Optisp (a telecommunications company that became the third-largest online game operator and that has joint ventures with U.S. publisher Electronic Arts). Of China's 2007 market share, 20 percent was held by Shanda, 14 percent by NetEase, and 12 percent by The9.[16]

A key generator of online game revenue has been the Internet café, with more than 50 percent of online play occurring in the 350,000 cafés in China in 2005. While this is similar to what is happening in Korea, the high penetration of broadband in Korean homes has led to many Internet cafés closing, and China is starting to see a similar trend of home play displacing public play.[17] Increasing connectivity tends to drive the growth in online games, and in 2005, 40 million of the 107 million Internet users played them.[18]

Broadband penetration will become critical to MMO games, since players value rapid response times (in other words, reductions in the delays between their inputs to the game and the display of the eventual outcomes). Broadband connectivity in the home is expected to provide a similar stimulus to MMO games as advanced handsets and increased connectivity have done with cell-phone games.

Whereas in the United States the provider of online games is usually a game publisher or the studio itself, China's industry has involved providers such as Shanda, which may license other developers' products in addition to developing their own. Shanda Networking is one of China's success stories. Shanda was started in 1999 by Chen Tianqiao, who at one point was the third-richest man in China. It started in the online business by offering a popular Korean game, *The Legend of Mir II*, after its online animation business failed due to piracy. Shanda now has 2,500 employees and is listed on the NASDAQ in the United States. In recent years the company has come under pressure, as the number of players of online games—and accordingly, the revenues for online games—has dropped. Shanda's troubles began with poor investment strategies in nongame sectors, and were accompanied by the end of the product cycle for some of the company's earlier product successes. For instance, concurrent online users on *The Legend of Mir II* dropped from 763,000 to 630,000 at the end of 2005.[19] The challenges to the providers of online games include the need to find viable alternative business models, as the number of games goes up and as imitators start to compete on price. Shanda's recent troubles prompted it to offer two online games on a free-to-play basis (but to charge for premium "content")—a step that most companies have already copied—and to seek to remake itself again as a diverse media group.[20]

Another issue is the high standard and competition set by Korean products. The Korea Game Development and Promotion Institute (KGDI) reported that, as of the end of 2004, 51.4 percent of the online games in the Chinese market (or 154 products) were made in Korea.[21] According to a study by the Taiwanese firm FIND (Focus on Internet News and Data), the China market represented 30 percent of Korea's overseas gaming revenue in the mid-2000s. For example, Actoz Soft entered the China market in 2001 with the role-playing game *1000 Years*, on Asiagame.com. This attracted 1 million players in the first five months and generated $1 million in licensing income. In 2003, Actoz's total revenue was $46.4 million, with 71 percent of this in China.

Despite the early strength of Korean products, their market share has been dropping in recent years, plunging to around 10 percent according to one recent account.[22] In part, this is because the Korean dominance in the Chinese market made the Chinese government nervous and led to its decision to impose restrictions on foreign game firms in 2004. As a result, Korean game firms now form joint ventures; for example, Actoz joined Sea Rainbow Holding Corporation to form Oriental Interactive Inc., NHN joined Sea Rainbow Holding Corporation to form Ourgame Assets, and Webzen joined The9.com to form 9Webzen. Nevertheless, recent successes by the domestic industry suggest that Chinese gaming has gained considerable market share at the expense of Korean and other imports. Shanda has had some success with its domestic MMO game *The World of Legend*, which attracted 300,000 users and became the fourth-most-popular online game in 2003. Similarly, Kingsoft offered *JXOnline*, another game modeled on "Chinese martial arts and modern love," which cost $1.8 million to develop.

Other challenges to the industry include the critical shortage of developers. The government recently estimated that there were only about five thousand professional game developers in China but that there was a need for twenty thousand.[23]

Animation

Foreign cartoons in the past were so easily and cheaply available in China that the government became concerned that domestic content might have trouble getting a foothold in the domestic market. About 90 percent of the animation market has been occupied by producers from Japan, the United States, and South Korea.[24] Additional foreign cartoon producers are attempting to enter the market. International Television for Asia (ITA) signed an agreement with the Beijing China Cartoon Media Group Co., Ltd. (BCCM), to deliver cartoons, games, screen savers, and other content to China Mobile in 2005.[25]

Conditions for this industry's sustainable development appear to be in place. In 2006 China nearly doubled its animation output to 81,000 minutes.[26] It appears that strong Chinese animation producers do exist, although some originated in Taiwan and Hong Kong. It also appears that a combination of policy measures and the need for at least some content to appeal to domestic culture will protect the domestic producers.

The decoupling of the production and distribution networks is producing healthy growth among content providers. With the advent of local satellite TV stations and digital mobile access in the sophisticated urban areas, a more diverse set of demands may emerge for animation programs. However, reflecting global trends, consolidation in the industry may occur as several large TV stations monopolize the countrywide channels. This in turn may lead to a reduction of diversity in animated and other TV fare as the major markets become more homogeneous.[27]

Zhejiang province (especially the capital, Hangzhou) and Jiangsu province have, among others, become strong bases for animation. Internationally competitive companies include Hengdian Group and Shanda.[28] Another example is the Institute of Digital Media Technology, in Shenzhen; established by a Hong Kong group, the company made the first three-dimensional high-definition animated film in China.[29] One of the domestic companies with the strongest IP is the Hunan Sunchime Cartoon Group. The group's *Blue Cat* serial is shown on 1,020 television stations in mainland China, Hong Kong, and Taiwan; has spawned 6,600 products to date; and has helped the group to create a retail network of two thousand specialty stores. Sunchime's copyright revenue and related business income alone amounted to RMB 150 million and RMB 280 million, respectively, in 2002. In 2005 the company's revenue was RMB 1 billion or approximately $150 million.[30]

As with game developers, there is also a severe shortage of animators. While in 2004 the apparent demand for animation professionals was nearing 250,000, there were only about 10,000 college-trained professionals in China.[31] Estimates vary widely about the numbers of animation professionals. One report noted that at the end of 2005, there were 64,000 students majoring in animation who had graduated from universities, and a further 466,000 studying in colleges. The differences may be ascribable to the various definitions of animators and artists in use.[32]

Policies

All three of these sectors have recently garnered government attention and support. Current government policies fall into several categories: investments, such as development funds and infrastructure; protectionist measures, such as bans on foreign content and reserving space in the various channels for domestic content; training; and exposure, such as through exhibitions and fairs. These are represented by the Ministry of Culture's recently articulated goals for promoting the nation's animation, comic book, and gaming industries. Those goals are as follows:[33]

- Adjustment to the market environment to aid the growth of the animation and gaming industries through legal and political measures
- An increase in the share of Chinese original animations in the domestic market and acquisition of a share of the international market
- Promotion of anime and games suited for minors
- Incubation of the industry, and construction of business parks for animation and gaming companies
- Stricter import criteria, introduction of superior international games, and prevention of the import of goods that are unfit for the country
- Introduction of animation and gaming exhibits and fairs

Since investment and protectionist measures are the main purview of the national government and appear to be more widespread, they will be more closely examined here. Both kinds of measures are similar to those in other industries.

Investment and Policies

While for almost all games, the critical point is the financial viability of the company, policy has been used to influence this in marginal ways. The General Administration of Press and Publication has the primary responsibility for regulating the online games industry, including policies to help domestic producers, such as a recent plan to invest RMB 1.8 billion to develop one hundred online games and their associated IP, based on Chinese history and heroes.[34] While this is as risky as it is breathtaking (since it will be extremely hard to "pick winners"), we have not heard anything more of this in our recent interviews with several online game companies. There are, however, city-level grants, such as the Beijing Science and Technology Committee grants, that are in place.

Perhaps the bigger issue in China is that the structure of the industry is not yet mature. The challenge of making games lies in their "hits" nature (that is, the nature by which the few successes tend to garner most of the market's attention, and therefore revenue), and the interactive nature of games makes it hard to determine which basic concept will ultimately succeed. Even then the concept must be properly implemented.[35] Only proven team experience may make it a little easier to judge outcomes.

Regulations: Channel and Access Conditions

The government appears to be willing to set aside broadcast channels for the domestic industry, particularly in the case of animation. The government has set aside 60,000 minutes of programming time for animation on two thousand TV stations, but Chinese companies could only fill one-third of this time, and the rest was filled by foreign imports.[36] The State Administration of Radio, Film and Television (SARFT) has designated nine "national animation bases" and has granted three TV animation channels. Changsha has two of these bases, producing about 15,000 minutes of animated film each year.[37]

Bans and Provisions for Domestic Cultural Content: Diversity Enhancing or Not?

The state uses bans extensively, officially to protect the young from any bad influences, but this also has the effect of curtailing the indigenous development of cultural content. Bans have been imposed on certain animation-related activities as well as on video games, and make it difficult for foreign producers to dominate the market.[38] For instance, bans on foreign cartoons have been proposed. The SARFT has banned animated features or series that included live actors, arguing that these hybrid shows would "jeopardize the broadcasting order of homemade animation and mislead their development"—apparently because such programs presumably masquerade as animation while actually using existing IP or movie stars to gain audience interest.[39] In online games as well, while thirty or forty were once imported in 2004, only ten per year are now allowed.[40]

While these measures presumably free up market space for domestic cartoons to enter, the social consequences of the bans are longer-standing and harder to evaluate. One effect is that they deny the consumer exposure to a wider range of influences. In the case of games, while this might be deemed "good" in that bans might mitigate the influences from global entertainment, the risk is that if bans are too sweeping, they might deny the players exposure to a variety of influences and therefore to potentially new sources of innovation. A common observation in video-game development is that players tend to mimic what they have played as they become developers themselves. If innovative games and other media are banned, China risks raising a generation of players that is not exposed to new gameplay styles. Recent Chinese policy appears somewhat more considerate in this matter. The recent consideration by the SARFT to ban foreign cartoons suggests protectionism, but, at the same time, the Ministry of Culture's policy to introduce "superior international games" appears to be an effort to boost the cultural exposure of the populace.[41]

Local Policies

Local policies may be more effective than national policies, because many effects in the new media industries—for example, clustering and labor pooling—are local. On the production side, it appears that with strong firms, China's market can have a national character, as seen in the U.S. video-game industry, where firms sell across state borders. This is quite different from the situation of Chinese software firms, which sometimes have had trouble cross-selling to other provinces trying to protect their own firms.

Shanghai is helping its local online-gaming companies to build up their brands and offers the same preferential tax and land-leasing treatments that it offers its software companies.[42] Another example is Hangzhou, which has hosted China's animation festival, and has put in place fifteen new measures to help develop itself as an animation center.[43]

Human Resources

With a chronic shortage of resources across the board, government action in all the sectors is similar. With the fourfold shortage in game developers expected in the next five years, the central government has announced plans for gaming programs in ten universities and plans also to open a professional game development college to train senior talent.[44] Animation and media programs are receiving a similar boost.

Discussion

Core competencies in the new media lie, in part, in technology and skilled human resources, but the importance of creativity and culturally unique content suggests a need to approach human resource formation (and hence, policy toward this) differently; historical, multidisciplinary, and diverse experiences are especially important. Another core competency in the new media that may be different from traditional sectors is the need to develop and control IP, and to ensure that new content models and business models match up to this.

The Chinese market has historically seen intense, imitative, cost competition, making it hard for firms to differentiate themselves on IP and other output characteristics. Foreign content producers, particularly in online games, have occupied strong niches or held predominant market shares. Given the intense domestic and foreign competition in online games, it may be difficult for all but the largest Chinese firms to compete now. This is similar to the software industry's experience, where domestic firms were unable to match foreign multinationals at the high end of the market and were competing with one another in a self-destructive way on cost at the low end.

Policies are being reshaped to recognize that these new service industries are cultural ones. The government appears to be getting serious about IP protection for domestic content. Furthermore, government policies toward the new media sectors, as in software, are geared toward protectionism. This may work for new media better than in the technology industries, because local content is one possible product differentiator that consumers may prefer, so there is a need to allow domestic efforts to flourish. In contrast, in technological industries such as electronics, the state-of-the-art may be easily duplicable and therefore globally diffusible once it is reached—making process improvement the general differentiator after products mature. In addition, with so many consumers, the market is quite competitive, at least on the demand side. Another advantage is that local firms can produce domestic content in combination with business and application models derived from other countries. An example is koook.com, which generates its own music from sources in the Greater China region. Similarly, some companies have developed gaming and television programming capabilities that allow interactive games to be played over the television. Thus, unlike traditional manufacturing, where product choice outcomes in the market are due to cost competitiveness and the ability to improve or add features, in content industries it is all about emotive content, recognizable IP, and the business model. At the same time, the globalization of certain cultures clearly poses a challenge to local content producers, as shown by the U.S. MMO game *World of Warcraft* and Korean MMOs, which have exploited audience tastes for "foreign" products.

The size of the domestic market could also provide the government with a bargaining position to gain access to technology or to bargain for product exports. However, it appears that many domestic content providers are not yet eyeing foreign markets; they are too busy fighting for domestic market share. To the extent that they do, the Taiwanese, Korean, and Southeast Asian markets appear to be the major overseas markets targeted.

There is an issue of policy integration, with both national and state governments offering packages to firms that develop mobile games. In the case of mobile content, for instance, China's development of its own 3G standard (STD-CDMA) suggests that imported software will be disadvantaged. Some firms that we interviewed saw opportunities in this situation. They can duplicate what was done in earlier standards with the help of the new barriers and can now theoretically compete with even traditional heavyweights such as Nokia. The barriers buy time for the start-ups. As a result, the huge mobile market of China is unlikely to benefit foreign firms unless the latter work with local firms with good connections.

Most of the new start-ups in China are led by experienced engineers. Although for both games and animation it is possible for individual developers and artists to gain skills equivalent to their foreign counterparts, at the level of the organization (for example, in terms of implementation), local organizations may find themselves behind.

While online games initially made a lot of money, price pressure is making it hard to sustain this level of profit. However, since process improvement can do little to improve product success (unlike in manufacturing), innovation may be the only way out. Such a situation reflects another finding in games, which suggests the importance of creative ideas for new gameplay and technology. Since games are based on interactivity, and since they have a strong technology basis in software, developments that create new forms of interactivity or gameplay are likely to surface out of *creative* technological achievements. One possible determinant of success is that countries with deep traditions in creative work, or in the technologies that underpin and help bring such creativity to life, will foster new kinds of games and gameplay. To achieve this, the Chinese government could look at creativity and content-enhancing policies (for example, Singapore's various competitions for content).

Conclusion

Like other technological sectors, the new media sectors are rapidly growing in China and are focused mainly on the domestic market. While there is a strong challenge from foreign firms, especially in online games, the government is using a variety of policy instruments similar to those it has applied in other industries to protect and gestate domestic firms. Unlike in other industries, where protectionism may foster only uncompetitive firms, with the new media, the uniqueness of the cultural content and the direct sales to users may make these policies worthwhile.

Notes

[1] This chapter relies on secondary and primary sources of data. Our primary sources include a set of interviews with about five mobile-phone operators, including content providers, and a PC-game-producing company. These interviews were conducted between January 2006 and May 16, 2006.

[2] United Kingdom, Department for Culture, Media and Sport (DCMS), *Creative Industries Mapping Document 1998*, <http://www.culture.gov.uk/reference_library/publications/4740.aspx).

[3] Korea Game Industry Agency (KOGIA), *2004 White Book: The Rise of Korean Games*, 2004, <http://www.kogia.or.kr/english/information/01white_2004.jsp>.

[4] Taiwan, The Digital Content Industry Promotion and Development Institute (DCIPDI) "2005 Digital Content White Paper," 2005, <http://www.digitalcontent.org.tw/index.php#Scene_1>.

[5] F. T. Tschang, "The Philippines IT-Enabled Services Industries," Report to the World Bank/Foreign Investment Advisory Service and Board of Investments, 2005, <http://siteresources.worldbank.org/INTPHILIPPINES/Resources/Tschang-word.pdf>.

[6] "China Doubles Animation Output," Animation World Network, <http://news.awn.com/index.php?ltype=cat&category1=Business&newsitem_no=18797>.

[7] "China's Online Gaming Sector Makes Progress," *People's Daily*, January 21, 2005, <http://english.people.com.cn/200501/21/eng20050121_171418.html>.

[8] Analysys International, "China Mobile Gaming Market Development 2005," June 20, 2005, <http://www.marketresearch.com/map/prod/1191203.html>.

[9] Analysys International, "Analysys International Says China's Mobile Gaming Industry Will Enter High Profit Stage Soon," press release, January 23, 2006, <http://english.analysys.com.cn/3class/detail.php?id=152&name=report&daohang=产业分析&title=Analysys%20International%20Says%20China's%20Mobile%20Gaming%20Industry%20Will%20Enter%20High%20Pro20Profit%20Stage%20Soon>.

[10] T. H. Nguyen, E. Liew, S. Pittet and T. J. Hart, "Industry Must Overcome Hurdles for Mobile Gaming to Succeed" (Gartner Research Report, September 12, 2005), 9.

[11] Analysys International, "Analysys International Says China Online Gaming Market Reached CNY 3.367 bln in Q4 2007," February 20, 2008, <http://english.analysys.com.cn/3class/detail.php?id=351&name=report&daohang=%B2%FA%D2%B5%B7%D6%CE%F6&title=Analysys%20Internatio nal%20Says%20China%20Online%20Gaming%20Market%20Reached%20 CNY%203.367%20bln%20in%20Q4%202007>.

[12] J. Gaudiosi, "China Caught in Web of Games," *Hollywood Reporter*, April 29, 2005 <http://www.hollywoodreporter.com/hr/search/article_display.jsp?vnu_content_id=1000874842>.

[13] Pearl Research, "Games Market in China Grew 60% to $1.66 billion in 2007, Expected to Exceed $3.4 billion in 2010," press release, March 19, 2008, <http://www.pearlresearch.com/products/Cn08.html>.

[14] CBS News, "Web Games All the Rage in China," December 8, 2003, <http://www.cbsnews.com/stories/2003/12/08/tech/main587365.shtml>.

[15] "Chinese Government to Invest $1.8 billion in Video Games," *The Inquirer*, August 1, 2005, <http://www.theinquirer.net/?article=25045>.

[16] Analysys International, "Analysys International Says China Online Gaming Market Reached CNY 3.367 bln in Q4 2007."

[17] Gaudiosi, "China Caught in Web of Games."

[18] CBS News, "Web Games All the Rage in China."

[19] P. Loughrey, "Shanda Adopts Free-play MMORPG Model," *Games Industry*, December 2, 2005, <http://gamesindustry.biz/news.php?aid=13374>.

[20] "Games Operator Opts for Variety," *People's Daily*, March 1, 2006, <http://english.people.com.cn/200603/01/eng20060301_246925.html>.

[21] Korea Game Industry Agency (KOGIA), *2004 White Book: The Rise of Korean Games*, 12.

[22] "Korean Game Firms Failing in China," *Chosun Ilbo*, November 20, 2007, <http://english.chosun.com/w21data/html/news/200711/200711200016.html>.

²³ "China's Online Gaming Sector Makes Progress," *People's Daily,* January 21, 2005, <http://english.people.com.cn/200501/21/eng20050121_171418.html>.

²⁴ J. Landreth, "China Bans TV Toons That Include Live Actors," *Reuters,* February 23, 2006.

²⁵ "International Cartoon Producers Vie for Chinese Cell-Phone Animation Market," *People's Daily,* November 8, 2005, <http://news.awn.com/index.php?ltype=cat&category1=Internet+and+Interactive&newsitem_no=15340>.

²⁶ Animation World Network, "China Doubles Animation Output," January 4, 2007, <http://news.awn.com/index.php?ltype=cat&category1=Business&newsitem_no=18797>.

²⁷ Analysys International, "China TV Industry 2005" (Report, October 2005), <http://www.researchandmarkets.com/reports/317630/china_tv_industry_2005> 3.

²⁸ Analysys International, "China TV Industry 2005," 3.

²⁹ "Platform for Animation Industry to Be Built in South China City," *People's Daily,* January 20, 2006, <http://english.people.com.cn/200601/20/eng20060120_237015.html>.

³⁰ Q. Tian, "Innovation Key to Success of Cartoon Industry,"*China Daily,* November 15, 2006, <http://www.chinadaily.com.cn/bizchina/2006-11/15/content_733487.htm>.

³¹ "China Opens Toon Industry to Private Investors," Animation World Network, August 20, 2004, <http://news.awn.com/index.php?ltype=date&newsitem_no=11747>.

³² "China Doubles Animation Output," Animation World Network, January 4, 2007, <http://news.awn.com/index.php?ltype=cat&category1=Business&newsitem_no=18797>.

³³ "China to Expand Its Anime Market," Anime News Network, February 11, 2005, <http://www.animenewsnetwork.com/article.php?id=6152>.

³⁴ Olivia Chung, "China Online Games Sector Battles Foreign Domination," *The Standard,* February 18, 2005, <http://www.thestandard.com.hk/archive_news_detail.asp?pp_cat=&art_id=3371&sid=&con_type=1&archive_d_str=20050218>.

³⁵ F. T. Tschang, "Videogames as Interactive Experiential Products and Their Manner of Development," *International Journal of Innovation Management* 9, no. 1 (2005), 103–31.

³⁶ "China Opens Toon Industry to Private Investors," Animation World Network, August 20, 2004.

³⁷ "China, ROK Join Hands in Animation Industry," *People's Daily,* November 7, 2005, <http://english.people.com.cn/200511/07/eng20051107_219480.html>.

³⁸ China Bans 50 Electronic Games, *China Daily,* January 27, 2005, <http://www.chinadaily.com.cn/english/doc/2005-01/27/content_412706.htm>.

³⁹ J. Landreth, "China Bans TV Toons that Include Live Actors."

[40] Olivia Chung, "China Scores in Online Games Battle," *China Business*, May 19, 2007, <http://www.atimes.com/atimes/china_business/ie19cb02.html>.

[41] "Foreign Cartoons May Be Banned from Prime Time," *China View*, June 9, 2005, <http://news.xinhuanet.com/english/2005-06/09/content_3062253.htm>.

[42] "China's Online Gaming Sector Makes Progress," *People's Daily*, January 21, 2005 <http://english.people.com.cn/200501/21/eng20050121_171418.html>.

[43] "New Measures Help Hangzhou to Become Animation Design Capital," *People's Daily*, September 12, 2005, <http://english.people.com.cn/200509/12/eng20050912_208086.html>.

[44] "China's Online Gaming Sector Makes Progress," *Peoples Daily*, January 21, 2005, <http://english.people.com.cn/200501/21/eng20050121_171418.html>.

ABOUT THE CONTRIBUTORS

Erik Baark is professor and acting dean at the School of Humanities and Social Science, Hong Kong University of Science and Technology. He received a PhD in information and computer science from the University of Lund, Sweden, and was awarded a D.Phil. in history from the Faculty of Humanities, University of Copenhagen, Denmark. His primary research interests relate to innovation systems and policies in China and other East Asian countries. He has published numerous books, including *Lightning Wires: Telegraphs and China's Technological Modernization 1860–1890* (1997). His articles have appeared in leading international area studies journals, such as *China Quarterly*, and in innovation research journals, such as *Research Policy* and the *International Journal of Technology Management*. Baark has also worked as a consultant for international agencies, such as the United Nations Development Programme and the World Bank.

Cong Cao is a researcher with the Levin Graduate Institute of International Relations and Commerce at the State University of New York, and the University of Oregon. He is interested in the social studies of science and technology, with a focus on China. He is the author of *China's Scientific Elite* (2004), a study of Chinese scientists who hold elite membership in the Chinese Academy of Sciences, and *China's Emerging Technological Edge: Assessing the Role of High-End Talent* (with Denis Fred Simon, forthcoming).

Yuan-chieh Chang is an associate professor at the Institute of Technology Management, National Tsing Hua University, Taiwan. He received a B.A. in electrical engineering from the National Changhua University of Education, Taiwan; an MBA in the management of technology from the Asian Institute of Technology, Bangkok, Thailand; and a PhD in science and technology policy from the University of Manchester, England. Currently, Chang teaches MBA and EMBA courses in the management of technology, technological forecasting and evaluation, science and technology policy, and innovation management. He has served as a research advisor for various leading research institutes in Taiwan. Chang's research interests include geographical systems of innovation, university-industry links, academic entrepreneurship, and the globalization of R&D. He has published journal papers in *R&D Management*, *Technology in Society*, *Research Technology Management*, *Technological Forecasting and Social Change*, and the *International Journal of Technology Management*.

Shin-horng Chen is director of the International Division at the Chung-Hua Institution for Economic Research in Taipei, Taiwan. From 2003 he has served as an editorial member of the *International Journal of Technology and Globalization*, published at Harvard University. Chen has taught at the National

Taiwan University and National Central University, Taiwan, and his professional experience includes research assistantships at the Center for Urban and Regional Development Studies at Newcastle University (UK) and the Industrial Economics Research Center at the Industrial Technology Research Institute (Taiwan). He received B.A. and M.A. degrees in economics from National Taiwan University and earned his PhD from the ICT Programme at the University of Newcastle-upon-Tyne (UK).

Chen's recent publications include "Raising R&D Intensity: A Strategy for Taiwan" (with Meng-chun Liu, Jia-Zhen Chen, and Chun-Hung Kuo) in *Asia Pacific Tech Monitor* (2005); "Taiwanese IT Firms' Offshore R&D in China and the Connection with the Global Innovation Network," in *Research Policy* (2004); and "International R&D Deployment and Locational Advantage: A Case Study of Taiwan" (with Meng-chun Liu) in *International Trade in East Asia*, edited by Takatoshi Ito and Andrew K. Rose (2005).

Bangwen Cheng is a professor at the School of Management, Huazhong University of Science and Technology, China. A council member of the China Society, he is also an expert on science and technology indicators with special government clearance. He holds a master's degree in engineering from the Huazhong University of Science and Technology. His research focuses on science and technology statistics and indicators for China's Ministry of Science and Technology.

Henry Chesbrough is executive director of the Center for Open Innovation at the Haas School of Business at the University of California, Berkeley. Previously, he was an assistant professor of business administration and the class of 1961 fellow at the Harvard Business School. He holds a PhD in business administration from the University of California, Berkeley, an MBA from Stanford University, and a B.A. from Yale University, summa cum laude.

Chesbrough's research focuses on managing technology and innovation. His book, *Open Innovation* (2003) articulates a new paradigm for organizing and managing R&D. In this new approach, companies must access both external and internal technologies, and take them to market through internal and external paths. This book was named a "Best Business Book" by *Strategy & Business* magazine and the best book on innovation by National Public Radio's *All Things Considered*. His most recent book, *Open Business Models* (2006), extends his analysis of innovation to business models, intellectual property management, and markets for innovation. This title was named one of the ten best books on innovation in 2006 by *BusinessWeek*, and is being translated into six languages.

Chesbrough's work has been published in *Harvard Business Review, California Management Review, Sloan Management Review, Research Policy, Industrial and Corporate Change, Research-Technology Management, Business History Review*, and the *Journal of Evolutionary Economics*. He contributes a

About the Contributors

monthly column on innovation to BusinessWeek.com and serves on the editorial board of *Research Policy* and the *California Management Review*.

Yih-Luan Chyi received a B.A. in economics from National Taiwan University and a PhD in economics from Northwestern University. In 1998 she became a professor in the department of economics at National Tsing Hua University. Funded by the National Science Council in Taiwan, she visited the London School of Economics and Political Science from 1998 to 1999.

Since 2002 Chyi has served as an associate editor for *Taiwan Economic Forecast and Policy* and *Taipei Economic Inquiry*. She has published articles in the *Review of International Economics*, the *Review of Development Economics*, *Applied Economics*, *Applied Financial Economics*, *Economic Modeling*, *Asian Academic Journal*, *Academia Economic Papers*, and the *Taiwan Economic Review*.

Her current work focuses on indicators of high-tech industrial clusters, knowledge spillovers in clusters and regions, the contribution of venture capital to the industry performance, and knowledge-based service industries. In 2006 she was elected the first director of the Economic Research Center on Globalization, College of Technology Management, National Tsing Hua University.

Paul Duo Deng is a doctoral candidate of economics at Brandeis University. His main research interests are economic growth, macroeconomics, and the evolution of institutions, including the economic transition of China. His recent research finds that foreign entry induces a "creative destruction" effect on Chinese domestic firms' productivity and innovation behavior. He is currently a lecturer in the economics of development at Brandeis International Business School. He holds a master's degree in economics from Baylor University and bachelor's degree from the Zhengzhou Institute of Aeronautics in China.

Dieter Ernst is senior fellow at the East-West Center in Honolulu. He was senior adviser to the Organisation for Economic Co-operation and Development (OECD); research director at the Berkeley Roundtable, University of California, Berkeley; and professor of international business at the Copenhagen Business School.

Ernst has cochaired an advisory committee of the U.S. Social Science Research Council to develop a new program on innovation, business institutions, and governance in Asia. He has also served as scientific advisor to several institutions, among them the OECD, the World Bank, the United Nations Conference on Trade and Development, and the United Nations Industrial Development Organization.

Ernst's current research focuses on offshore outsourcing through global production and innovation networks, global markets for knowledge workers, and implications for industrial and technology policies. He has published numerous books, as well as articles in leading journals on the internationalization of innovation and the determinants of its diffusion within and across countries. His recent books include *Innovation Offshoring: Asia's Emerging Role in Global*

Innovation (2006), *International Production Networks in Asia: Rivalry or Riches?* (2000), and *Technological Capabilities and Export Success: Lessons from East Asia* (1998).

Douglas B. Fuller is a lecturer in the Department of Management of King's College, University of London. He holds a PhD in political science from the Massachusetts Institute of Technology and has been researching technology and economic policymaking in East Asia for over a decade. He has spent approximately six years living in Greater China and has also conducted field research in Korea, Japan, and Malaysia. His current research interests include the economic prospects for developing countries under globalization and the comparative political economy of East Asia. He has contributed to two recent books, *Global Taiwan* (2005) and *How We Compete* (2005), and has published a number of articles in academic journals, including *Industry and Innovation*, *Asian Survey*, and the *Journal of Contemporary China*. He has served as a research fellow at the Stanford Program on Regions of Innovation and Entrepreneurship, which is part of Stanford University's Walter H. Shorenstein Asia-Pacific Research Center, and MIT's Industrial Performance Center.

Fuller is currently working on a manuscript about how China's successful technological development fits neither the prescriptions for development offered by revisionist political economists nor those of the mainstream economists adhering to the Washington Consensus.

Marguerite Gong Hancock is the associate director of the Stanford Program on Regions of Innovation and Entrepreneurship (SPRIE). For SPRIE, she manages project research, conferences and seminars, publications, and oversees the project's affiliated academic and government research partners in six countries in Asia. She leads SPRIE's China team, guiding research on high-tech regions, case studies of information technology companies, and new work on the globalization of research and development and high-tech leadership in Greater China.

She is coeditor, with Henry S. Rowen and William F. Miller, of *Making IT: Asia's Rise in High Tech* (2006) and *The Silicon Valley Edge* (2000). She continues to be an active member of Stanford's Entrepreneurship Task Force and a speaker to university and business leaders, including presentations for executive education and conferences in Silicon Valley and Asia.

A specialist on government-business relations in the development of information technology, Hancock has worked as director of Network Research for the Stanford Computer Industry Project at the Graduate School of Business; as a research associate at the East Asia Business Program of the University of Michigan; and as a company consultant in Boston and Tokyo. She holds a B.A. in humanities and East Asian studies from Brigham Young and an M.A. from Harvard in East Asian studies. While pursuing a PhD at the Fletcher School of Law and Diplomacy, she focused on computer industry development in China.

About the Contributors

Stefan Hennemann is a lecturer in economic geography at Justus Liebig University, Giessen, Germany. He received his PhD from the University of Hanover, Germany, in 2005 for research on technological change and endogenous innovation capacity in China. In 2006 he was postdoctoral research fellow and university planning consultant at the Higher Education Information System, Hanover. After focusing on the relevance of the educational sector for innovation systems, Hennemann is now working on developing methods and instruments for regional policy intelligence systems. His general research interests focus on applied economic geography in general, and the modeling of regional development capabilities in China and Germany in particular.

Albert Guangzhou Hu is an associate professor of economics at the National University of Singapore. He received his B.A. from Nankai University in 1991 and his PhD from Brandeis University in 1999. His research interests include the economics of technological change, international economics, and the economy of China. He has written and coauthored papers published in a number of journals, including *China Economic Review*, the *Journal of Comparative Economics*, *International Journal of Industrial Organization*, *Research Policy*, and *Review of Economics and Statistics*. He has also served as a short-term consultant for the Asian Development Bank and the World Bank.

Chih-Young Hung is an associate professor at the Institute of Technology Management, National Chiao Tung University, Taiwan, where he has taught since 1991. He earned his PhD from Texas Tech University in the United States. He received his bachelor's degree in 1979 from the Department of Electrical and Control Engineering at the National Chiao Tung University, Taiwan, with a focus on Taiwan's high-technology industries. He has published a series of papers in studying industry evolution from the finance perspective, and also several professional books on finance-related subjects. His current research interests include financial strategies, business valuation, value-based management, technology valuation, and venture capital. He has also served two terms as chairman of the Chinese Society of Business Valuation, which promotes academic studies and professional practices of business valuation in Greater China.

Gary Jefferson writes about institutions, technology, economic growth, and China's economic transformation. At Brandeis University, Jefferson teaches courses on the economics of innovation and institutions and China. He also teaches a course on the political economy of China at Tufts University's Fletcher School of Law and Diplomacy.

Jefferson's publications include *Enterprise Reform in China: Ownership, Transition, and Performance* (2000); "What is Driving China's Decline in Energy Intensity?" in *Resource and Energy Economics* (2004); "An Investigation of Firm-Level R&D Capabilities in East Asia," in *Innovation and Production Networking in East Asia*, ed. Shahid Yusuf (2004); "R&D and Technology Transfer: Firm-Level

Evidence from Chinese Industry," in *Review of Economics and Statistics* (2005); and "Privatization and Restructuring in China: Evidence from Shareholding Ownership" in the *Journal of Comparative Economics* (2006).

Jefferson's research is supported by the Department of Energy and the National Science Foundation. A graduate of Dartmouth College (A.B.) and Yale University (PhD), Jefferson has lived and taught at the Chinese University of Hong Kong and at Wuhan University in China and frequently travels to China.

Yee-Man Lai is a second-year M.A. candidate and Chinese overseas student in Taiwan. After completing high school in Hong Kong, she earned her B.A. in economics from National Tsing Hua University in Taiwan. Lai received an overseas student exchange scholarship that enabled her to study as a short-term exchange student in China. She conducted five weeks of research under Huang Zhi-Xiao at Fudan University in Shanghai, looking at the relationship between journalism and economics.

Lai is currently writing her master's thesis, "The Effectiveness of Knowledge Spillovers on Firm's Performance: An Empirical Study from Taiwan's Hsinchu Science-based Industrial Park in Taiwan."

Loet Leydesdorff holds an M.S. in biochemistry, an M.A. in philosophy, and a PhD in sociology. He currently researches science and technology dynamics at the Amsterdam School of Communications Research at the University of Amsterdam. He has published extensively on systems theory, social network analysis, scientometrics, and the sociology of innovation

In 2006 he published *The Knowledge-Based Economy: Modeled, Measured, Simulated*. Previous monographs include *A Sociological Theory of Communication: The Self-Organization of the Knowledge-Based Society* (2001), and *The Challenge of Scientometrics: The Development, Measurement, and Self-organization of Scientific Communications* (1995). With Henry Etzkowitz, Leydesdorff initiated a series of workshops and conferences about the triple helix of university-industry-government relations. He received the Derek de Solla Price Award for Scientometrics and Infometrics in 2003, and held the City of Lausanne Honor Chair at the School of Economics, Université de Lausanne, in 2005.

During the period 2007–2010, Leydesdorff is serving as a visiting professor at the Institute of Scientific and Technical Information of China (ISTIC), and as an honorary fellow of the Science and Technology Policy Unit (SPRU) of the University of Sussex. He has also been an honorary fellow of the Virtual Knowledge Studio of the Netherlands Academy of Arts and Sciences since 2006.

Zheng Liang now serves as research fellow and assistant director of the China Institute for Science and Technology Policy at Tsinghua University (CISTP). The CISTP was established jointly by the Ministry of Science and Technology of China and Tsinghua University in order to study science and technology policy and strategy. Liang also serves as assistant professor of the School of Public

Policy and Management, Tsinghua University. Previously, he served as associate professor of the Business School at Nankai University.

Liang received his PhD in economics from Nankai University and subsequently published his doctoral research on the economics of science, technology, and innovation in 2004. Liang's current research focuses on a comparison of science and technology policies in different countries, and the globalization of R&D. He has published nearly thirty papers in academic journals and coauthored five books, including the controversial *The Criticism of Top 10 Economists in China*, which has been distributed in Japan, Australia, and elsewhere. Recently Liang has worked on several important strategic projects, notably on China's Medium- and Long-tern Science and Technology Development Plan.

Ingo Liefner is professor of economic geography at Justus Liebig University, Giessen, Germany. He graduated in 1996 from the University of Hanover, Germany, where he wrote a thesis on the regional economic impact of innovations in gene technology. In 2000 he received his PhD from the University of Hanover. His research focuses on theoretical and applied economic geography as well as development economics, and his research interests include higher education systems and the regional economic effects of technical progress in developing countries. He has published books and articles on topics ranging from university funding and research performance, university-industry linkages, and foreign direct investment in China, to knowledge transfer within multinational companies, and the innovation systems of developing countries.

Bao-shuh Paul Lin is the vice president and general director of the Information and Communications Research Laboratories (ICL), at Taiwan's Industrial Technology Research Institute (ITRI). He also directs the Committee of Communication Industry Development in Taiwan's Ministry of Economic Affairs. Under his leadership, the ICL pioneered significant technical innovations and earned medals from the government in 1996, 2003, 2004, 2007, and 2008.

After earning a PhD from the University of Illinois at Chicago, Lin worked for several high-tech firms, including AT&T's Bell Labs and Boeing, before joining ITRI in 1991. He was later promoted to deputy general director of ITRI. From 1998, Lin served as senior vice president of Philips Research and president of Philips Research East Asia, based in Shanghai, where he established the company's research and development capacity and position.

A fellow of the Institute of Electrical and Electronics Engineers, and an influential participant in the industrial technical policy arena, Lin has been actively involved in launching the National Technology Development Programs in Taiwan. He has won numerous awards and honors during his long career, and has published more than 120 technical papers and reports in fields related to information and communications technology.

Tzung-Pao Lin is a consultant to the Information and Communications Research Laboratories (ICL) at Taiwan's Industrial Technology Research Institute (ITRI). He joined ITRI in 1992 and held deputy director, director, and deputy general director positions within ICL/ITRI until 2006. In these capacities, he led various government-sponsored communication technology development and applications projects, during which time he also owned 10 patents and published more than 60 papers.

Prior to joining ICL/ITRI, Lin worked at AT&T Bell Laboratories, where his activities included exploring state-of-the-art new services equipment and formulating overall new residence and small business services and applications. At AT&T, he led a team to design the software architecture and interprocesses of the Automated System for Performance Evaluation of Networks (ASPEN).

In 1991 Lin returned to Taiwan as chief engineer for Tailyn Communication Co., where he eventually became vice president of the Customer Premises Equipment (CPE) Business Unit, responsible for the R&D, marketing, and sales of the CPE unit.

Lin received his B.S. degree in electrical engineering from National Taiwan University in 1977, and his M.S. and PhD degrees in electrical engineering from the University of Washington in 1982 and 1985, respectively.

Meng-chun Liu has been a research fellow and deputy director of the Mainland China Division at Chung-Hua Institution for Economic Research, Taipei, since 2002. He received his PhD in economics from Monash University, Australia. He also teaches at Yun-Zu University, focusing on foreign direct investment and regional market analysis.

Based on a previous, coauthored paper titled "International R&D Deployment and Locational Advantage: A Case Study of Taiwan" (2005), Liu is currently researching a book titled *Regional Innovation System in China and R&D Investments by FDI Firms*, which empirically considers Taiwanese firms' influences on R&D networking and investment in China.

Liu's recent coauthored publications include (with Gee San) "Social Learning and Digital Divides: A Case Study of Internet Technology Diffusion" (*Kyklos*, 2006), and (with Monchi Lio) "ICT and Agricultural Productivity: Evidence from Cross-country Data" (*Agricultural Economics*, 2006). He has also published in *World Development*, *Economic Record*, and *Pacific Economic Review*, among other economics and policy journals.

William F. Miller has spent about half of his professional life in business and about half in academia. He was the last faculty member recruited to Stanford University by the legendary Frederick Terman, who was then vice president and provost of Stanford. Miller himself later became vice president and provost of Stanford. He conducted research and directed many graduate students in computer science.

About the Contributors

In 1968 Miller also played a role in the founding of the first Mayfield Fund (venture capital) as a special limited partner and advisor to the general partners. As president and CEO of SRI International (1979–1990), Miller opened SRI to the Pacific region, established the spin-out and commercialization program at SRI, and founded the David Sarnoff Research Center (now the Sarnoff Corporation) as a for-profit subsidiary of SRI. He also became chairman and CEO of SRI.

In 1982 Miller was appointed to the National Science Board. He has served on the boards of directors of several major companies, such as Signetics, Firemans Fund Insurance, First Interstate Bank (and later) Wells Fargo Bank, Pacific Gas and Electric Company, Varian Associates, WhoWhere? Inc. (chairman), and Borland Software Corp. (chairman).

He cofounded SmartValley, Inc. and aided the formation of CommerceNet, where he also serves on the board of directors. Miller was a founding director of the Center for Excellence in Non-profits. He currently serves as chairman of the Board of Sentius Corporation and is a founder and chairman of Nanostellar, Inc. Miller is a life member of the National Academy of Engineering, a fellow of the American Academy of Arts and Science, a fellow of the American Association for the Advancement of Sciences, a life fellow of the Institute of Electric and Electronics Engineers, and a member of the Silicon Valley Engineering Hall of Fame.

Claudia Müller has worked as an advisor to the German executive director at the World Bank since 2006. Her responsibilities include reviewing World Bank documents on economic policy across a broad range of issues, including on sub-Saharan Africa, gender, agriculture, and health, and preparing discussions of these issues for shareholders. She received a diploma (2001) and a PhD (2006) in economics from the University of Cologne. In her PhD, Müller investigated whether and how the reverse brain drain to China contributes to China's innovative capacity. Her research was based on approximately 100 face-to-face interviews with Chinese entrepreneurs who returned to China from overseas in order to start up high-tech companies.

From 2001 to 2006, she worked as a researcher at the University of Cologne, as a lecturer on China's economic development at the Cologne Business School, as a consultant to a regional German government on the effective promotion of high-tech start-ups by universities (jointly with Professor Rolf Sternberg), and as a consultant for the German Development Institute in Bonn.

Haoyuan Qin is affiliated with the College of Management at Huazhong University of Science and Technology. He is a coauthor of both *China Science and Technology Indicators 2004* (2006) and *China Science and Technology Indicators 2006* (2008).

Xiaohong Quan is an assistant professor of entrepreneurship in the College of Business at San Jose State University. She holds a PhD from the University of California, Berkeley. She was a fellow at the Stanford Program on Regions of Innovation and Entrepreneurship (SPRIE) at Stanford University from 2005 to 2006. Quan has published in both English and Chinese academic journals and books. Her research interests include technology and innovation management, entrepreneurship, international business, and regional economic development. Her current research involves topics related to multinational corporations' research and development labs in China, the Chinese software industry, and social networks and entrepreneurship in Silicon Valley. Quan earned bachelor's and master's degrees in economics, both from Peking University, China.

Henry S. Rowen is a senior fellow at the Hoover Institution, a professor of Public Policy and Management emeritus at the Graduate School of Business, and a senior fellow emeritus of the Walter H. Shorenstein Asia-Pacific Research Center, all at Stanford University. Rowen is an expert on international security, economic development, and high-tech industries in the United States and Asia. His current research focuses on the rise of Asia in high technologies.

From 2004 to 2005, Rowen served on the Presidential Commission on the Intelligence of the United States Regarding Weapons of Mass Destruction. From 2001 to 2004, he served on the Secretary of Defense Policy Advisory Board. Rowen has served as Assistant Secretary of Defense for International Security Affairs in the U.S. Department of Defense (1989–1991); chairman of the National Intelligence Council (1981–1983); president of the RAND Corporation (1967–1972); and assistant director, U.S. Bureau of the Budget (1965–1966).

Rowen most recently coedited *Making IT: The Rise of Asia in High Tech* (2006). He has also coedited *The Silicon Valley Edge: A Habitat for Innovation and Entrepreneurship* (2000); *Behind East Asian Growth: The Political and Social Foundations of Prosperity* (1998); and *Defense Conversion, Economic Reform, and the Outlook for the Russian and Ukrainian Economies* (1994), coedited with Hoover fellow Charles Wolf and Jeanne Zlotnick. Among his articles are "Kim Jong-Il Must Go" (*Policy Review*, 2003); "The Short March: China's Road to Democracy" (*National Interest*, 1996); and "Inchon in the Desert: My Rejected Plan" (*National Interest*, 1995).

Rowen earned a bachelor's degree in industrial management from the Massachusetts Institute of Technology in 1949 and a master's in economics from Oxford University in 1955.

Jihong Wu Sanderson is a leading expert in the areas of globalization, strategy, and innovation in China. She is the founding executive director at the Center for Research on Chinese and American Strategic Cooperation at the University of California, Berkeley, which seeks to enhance cooperation between China and the United States through research, exchange, and educational programs.

About the Contributors

Sanderson is a faculty member at the Haas School of Business at the University of California, Berkeley, where she teaches the graduate course "MOT, Doing Business in China" and "IT in China" at the School of Information. Sanderson also teaches "The Strategy of China's Globalization" at executive leadership programs in both the United States and China. Her book, *Next Step: The Options for Chinese Firms Going Global*, was published in Chinese in 2006.

Sanderson has extensive executive experience in an investment bank, at an electronics manufacturer, and within a conglomerate in China. In the United States, she works as a strategic consultant, helping firms initiate and optimize their businesses at the global level. She serves as board advisor for many companies and organizations, including the U.S. Women's Chamber of Commerce and the Silicon Valley China Wireless Technology Association.

Adam Segal is the Maurice R. Greenberg Senior Fellow in China Studies at the Council on Foreign Relations. An expert on Chinese domestic politics, technology development, foreign policy, and security issues, he has a PhD and B.A. in government from Cornell University, and an MALD in international relations from the Fletcher School of Law and Diplomacy, Tufts University. Previously, he was an arms control analyst at the Union of Concerned Scientists and a visiting scholar at the Center for International Studies, Massachusetts Institute of Technology. He has taught at Vassar College and Columbia University and been a visiting scholar at the Shanghai Academy of Social Sciences and Tsinghua University in Beijing. Segal has written a book—*Digital Dragon: High-technology Enterprises in China* (2003)—as well as several articles on Chinese technology policy. His work has recently appeared in the *International Herald Tribune*, the *Financial Times*, *Foreign Affairs*, the *Los Angeles Times*, the *Asian Wall Street Journal*, and *Washington Quarterly*.

Kristy H. C. Sha joined the Industrial Technology Research Institute (ITRI) in 1990. She is currently special assistant to the general director of the Information and Communications Research Laboratories (ICL), one of ITRI's core labs. Sha graduated from National Taiwan University and received her PhD from the Institute of Management of Technology, National Chiao Tung University; her thesis was titled "Three Cases of ITRI's Innovations in Value-Based Management." Her research interests are innovation management and industrial development strategy. She has published nearly thirty papers, both in referred journals and international conferences. She has also coauthored a chapter of the book *The Information Society in the Asia Pacific Region: Diffusion, Access and Socio-economic Impact* (2004).

Bing Shi is head of the Evaluation Division at the Bureau of Planning and Strategy, Chinese Academy of Sciences (CAS), and deputy director and associate professor at the Evaluation Research Center, CAS. She received her bachelor's degree in law from Renmin University of China, and her master's degree from the University of Science and Technology, China. From 2003 to 2004, she was a visiting scholar at the University of Oregon.

Chintay Shih is currently special advisor and former president of Taiwan's Industrial Technology Research Institute. He recently joined National Tsing-Hua University as dean of the College of Technology Management. Previously, he was a distinguished visiting scholar at the Walter H. Shorenstein Asia-Pacific Research Center, Stanford University. Shih serves as science and technology advisor of Taiwan's Executive Yuan, chairman of the Asia Pacific Intellectual Property Association, managing director of the Taiwan Electrical and Electronics Manufacturer's Association, chairman of the Chinese Business Incubation Association, and chairman of the Asia-Pacific Industrial Analyst Association. Shih is also a member the Economic Advisory Committee of the President's Office, and of the boards of directors for the Taiwan Semiconductor Manufacturing Company and Vanguard Semiconductor. Honored as a fellow of the Institute of Electrical and Electronics Engineers in 1992, Shih also received the Engineering Medal of the Chinese Institute of Engineering in 1995 and the First Medal of the Ministry of Economic Affairs in 2003. He served as president of the Chinese Institute of Engineers between 1998 and 2000, and chairman of the Taiwan Semiconductor Industry Association from 1996 to 2000.

Shih holds a B.S. from National Taiwan University, an M.S. from Stanford University, and a PhD from Princeton University.

Denis Fred Simon is professor of international affairs at Pennsylvania State University and director of the Program in U.S.-China Technology, Economic and Business Relations. Previously, he served as founding provost and vice president for academic affairs of the Levin Graduate Institute of International Relations and Commerce, State University of New York (2004–2008). Prior to joining the Levin Institute, he served as dean of the Lally School of Management and Technology at Rensselaer Polytechnic Institute in Troy, New York. He also has served as professor of international business strategy and technology management at the Fletcher School of Law and Diplomacy, Tufts University (1987–1995), and as the Ford International Professor of Management and Technology at the Sloan School of Management at the Massachusetts Institute of Technology (1983–1987).

Before joining Rensselaer, Simon was president of Monitor Group (China), where he helped to drive overall business development and provided high-level management support and intellectual leadership for Monitor's strategy engagements in China. Prior to joining Monitor, he was managing director of the Business Strategy and Architecture Innovation Center for Scient Corporation, Singapore; general manager and associate partner at Andersen Consulting China, where he directed the China Strategy Group; and president of China Consulting Associates (Boston). Simon also has served as a private consultant to numerous Fortune Global 500 firms on their business entry strategy and operations in China.

ABOUT THE CONTRIBUTORS

Simon has written and lectured widely on innovation, high-technology development, foreign investment, and corporate strategy. His most recent publications include (with Cong Cao) *China's Emerging Technological Edge: Assessing the Role of High-End Talent* (2008); *Techno-Security in an Age of Globalization* (1997); *Corporate Strategies towards the Pacific Rim* (1996); and *The Emerging Technological Trajectory of the Pacific Rim* (1995). He is currently writing a book about the development of the computer industry in China.

Simon received his M.A. in Asian studies in 1975 and PhD in political science in 1980, both from the University of California, Berkeley. He received his B.A. in Asian studies from the State University of New York in 1974.

Rolf Sternberg is an economic geographer who received his PhD (1987) and habilitation (1994) from the University of Hanover, Germany. Between 1995 and 1996, he worked as a professor for regional geography at the Department of Geography, Technical University of Munich, Germany. Between 1996 and 2005, he was a full professor (chair) of Economic and Social Geography at the University of Cologne in the Faculty of Economics and Social Science. In 2005 he moved to the University of Hanover to become head of the Institute of Economic and Cultural Gography.

Sternberg's main research interests are in the field of spatial consequences of policy activities (for example, technology policy instruments, such as innovation centers and science parks); the impact of networks among innovative actors (firms, research institutions, politicians) on regional development; new firm formation processes and effects; and regional-sectoral clusters and their economic impacts. In addition to five books, he has published about 160 articles in many notable journals, including *Research Policy, Regional Studies, Economic Geography, European Planning Studies*, and *Small Business Economics*. Since 1998 he has led the German team of the Global Entrepreneurship Monitor (GEM) and been a member of the GEM Research Committee.

Jian Su is an associate professor in the School of Economics at Peking University. He received his PhD degree in economics from Brandeis University. His current teaching and research fields are macroeconomics and the Chinese economy. His recent publications include "Privatization and Restructuring in China: Evidence from Shareholding Ownership," *Journal of Comparative Economics* (2006); "The Sources and Sustainability of China's Economic Growth," Brookings Papers on Economic Activity (2006); and a series of papers in Chinese.

Richard P. Suttmeier is professor of political science, emeritus, at the University of Oregon. He received his bachelor's degree from Dartmouth College and his PhD from Indiana University. He has written widely on Chinese science and technology affairs and is currently working on projects dealing with the role of science and technology in China-U.S. relations and with China's strategy for the development of technical standards.

Seng-Su Tsang is assistant professor of strategy, policy, and technology in the Department of Business Administration at National Taiwan University of Science and Technology. He received his B.S. in chemical engineering from Tsinghua University, Taiwan; his MBA from Sun Yat-sen University, Taiwan; and his PhD in public policy and management from Carnegie Mellon University. His research interests include innovation measurement in services, technology licensing and business franchising, and government policy for clusters activation. His current applied research emphasizes innovations for business services, a four-year project sponsored by the Ministry of Economic Affairs, Taiwan. Tsang is engaged with a research team to develop business models for digital content provision in mobile areas. He has been a consultant on market analysis for CPC Corporation, Taiwan, since 2003.

F. Ted Tschang is an assistant professor of economics and technology on the faculty of the Lee Kong Chian School of Business, Singapore Management University. He has also been a visiting scholar at the Asian Development Bank Institute in Tokyo, a research associate at the United Nations University/Institute of Advanced Studies in Tokyo, and a research fellow at the Harvard Kennedy School of Government. He received his PhD in public policy analysis and management from Carnegie Mellon University and also holds graduate degrees in economics and electrical engineering. His research is in the areas of technology and innovation management, industrial development, and technology policy. Together with collaborators, he has researched the software industry in China and India, and the research and development and innovation system in Singapore. Tschang is currently working on a major study of creativity, product development, and industry formation in the U.S. computer games industry and the IT-enabled services industry in the Philippines.

Kung Wang is currently director and professor at the Graduate Institution of Industrial Economics at National Central University (NCU), Taiwan. He was general director of Industrial Economics and Knowledge Center of the Industrial Technology Research Institute (ITRI) from 2000 to 2004; he also established ITRI's market analysis and consulting services division. Wang has held numerous Taiwanese governmental positions, including advisor for national science, director-general of the Science-based Industrial Park Administration, and Commissioner of Fair Trade Commission–The Executive Yuan. In academia, he has served as the director and chairman of the Graduate Institution of Industrial Economics and the Department of Business Administration at NCU. He has also served on boards of directors for state-owned enterprises and venture capital funds, and as an advisor to the Taiwanese government and industrial and commercial associations. He holds B.S. and M.S. degrees from National Taiwan University and a PhD from the Massachusetts Institute of Technology.

About the Contributors

Yi-Ling Wei is a senior industrial and science and technology researcher at the Industrial Economics and Knowledge Center (IEK) in the Industrial Technology Research Institute (ITRI), Taiwan. Since 1996 she has performed market and industrial research in the automotive, machinery, and semiconductor equipment industries, and has disseminated those results in publications and seminars. Currently, Wei's major study, supported by the Ministry of Economy Affairs, Taiwan, focuses on innovation and entrepreneurship in high-tech industry clusters and public research and devlopment policy evaluation. She received a B.A. in mechanical engineering from Tamkang University (1992) and an MBA from Providence University (1994).

Poh Kam Wong currently holds joint professor appointments at the National University of Singapore (NUS) Business School and Lee Kuan Yew School of Public Policy. Concurrently, he serves as director of the NUS's Entrepreneurship Centre, where he spearheads the university's entrepreneurship education program and various university technology commercialization and spin-off support programs, including incubation, seed funding, and mentoring.

Wong has published extensively on innovation policy and technology entrepreneurship in journals such as *Organization Science*, the *Journal of Business Venturing*, and *Research Policy and Entrepreneurship Theory & Practice*. He has consulted widely for many international organizations, government agencies, and high-tech firms in Asia. Aactive angel investor, Wong chairs the Business Angel Network Southeast Asia and is a partner in an angel fund (BAF Spectrum) coinvested in by the Singapore government.

Wong has two B.S. degrees, an M.S. and a PhD, all from the Massachusetts Institute of Technology.

Zhaohui Xuan is a researcher in the Institute of Science and Technology Statistics and Analysis at the Chinese Academy of Science and Technology for Development. He is responsible for developing statistics related to regional and enterprises' science and technology activities. He is a coauthor of both *China Science and Technology Indicators 2004* (2006) and *China Science and Technology Indicators 2006* (2008). He holds a master's degree in regional economics and a bachelor's degree in urban and regional planning from Peking University.

Lan Xue is professor and executive associate dean of the School of Public Policy and Management, executive vice president of the Development Research Academy for the Twenty-first Century, and director of the China Institute for Science and Technology Policy, all at Tsinghua University. He is also an adjunct professor at Carnegie Mellon University and a fellow of the IC2 Institute at the University of Texas, Austin. Xue's teaching and research interests include public policy analysis and management, science and technology policy, higher education policy, and crisis management. He has published widely in these areas and serves on the editorial or advisory boards of several international academic journals.

Xue has led many research projects, which have been funded, variously, by the U.S. National Science Foundation, the U.S. Department of Commerce, the Australian Government, the Ford Foundation, and the National Natural Science Foundation of China. He has consulted for the World Bank, the U.S. Department of Commerce, the International Development Research Center (Canada), and SRI International. In the 2003–04 Strategic Research for the China's National Medium- and Long-term Science and Technology Development Plan, Xue served as deputy chair of the National Innovation System Group. He is on the advisory boards of many government agencies in China and, in April 2003, was invited to give a lecture to the top Chinese leadership on issues of science and technology policy.

Trained as an engineer, Xue has two master's degrees, in technological systems management and public administration, both from the State University of New York, Stony Brook. At George Washington University (GWU), he taught engineering administration and international affairs, and also served as a faculty associate at GWU's Center for International Science and Technology Policy. In 2001 he received a National Distinguished Young Scientist Award in China. Xue holds a PhD in engineering and public policy from Carnegie Mellon University.

Ping Zhou is a senior researcher at the Institute of Scientific and Technical Information of China (ISTIC), and an associate researcher for scientometrics at the Steunpunt O&O Indicatoren (Policy Research Center for R&D Indicators) of the Katholieke Universiteit Leuven (Catholic University of Leuven). She has a B.S. in chemistry and an M.A. in the study of historical documents. She is currently working on a PhD at the Amsterdam School of Communications Research, University of Amsterdam, under the supervision of Professors Loet Leydesdorff and Wolfgang Glänzel. She has coauthored a number of publications with these scholars.

Recent and Forthcoming Publications of the Walter H. Shorenstein Asia-Pacific Research Center

Books
(distributed by the Brookings Institution Press)

Donald Macintyre, Daniel C. Sneider, and Gi-Wook Shin, eds. *First Drafts of Korea: The U.S. Media and Perceptions of the Last Cold War Frontier.* Stanford, CA: Walter H. Shorenstein Asia-Pacific Research Center, forthcoming 2009.

Steven R. Reed, Kenneth Mori McElwain, and Kay Shimizu, eds. *Political Change in Japan: Electoral Behavior, Party Realignment, and the Koizumi Reforms.* Stanford, CA: Walter H. Shorenstein Asia-Pacific Research Center, forthcoming 2009.

Donald K. Emmerson, ed. *Hard Choices: Security, Democracy, and Regionalism in Southeast Asia.* Stanford, CA: Walter H. Shorenstein Asia-Pacific Research Center, 2008.

Gi-Wook Shin and Daniel C. Sneider, eds. *Cross Currents: Regionalism and Nationalism in Northeast Asia.* Stanford, CA: Walter H. Shorenstein Asia-Pacific Research Center, 2007.

Stella R. Quah, ed. *Crisis Preparedness: Asia and the Global Governance of Epidemics.* Stanford, CA: Walter H. Shorenstein Asia-Pacific Research Center, 2007.

Philip W. Yun and Gi-Wook Shin, eds. *North Korea: 2005 and Beyond.* Stanford, CA: Walter H. Shorenstein Asia-Pacific Research Center, 2006.

Jongryn Mo and Daniel I. Okimoto, eds. *From Crisis to Opportunity: Financial Globalization and East Asian Capitalism.* Stanford, CA: Walter H. Shorenstein Asia-Pacific Research Center, 2006.

Michael H. Armacost and Daniel I. Okimoto, eds. *The Future of America's Alliances in Northeast Asia.* Stanford, CA: Walter H. Shorenstein Asia-Pacific Research Center, 2004.

Henry S. Rowen and Sangmok Suh, eds. *To the Brink of Peace: New Challenges in Inter-Korean Economic Cooperation and Integration*. Stanford, CA: Walter H. Shorenstein Asia-Pacific Research Center, 2001.

Studies of the Walter H. Shorenstein Asia-Pacific Research Center
(published with Stanford University Press)

Jean Oi and Nara Dillon, eds. *At the Crossroads of Empires: Middlemen, Social Networks, and State-building in Republican Shanghai*. Stanford, CA: Stanford University Press, 2007.

Henry S. Rowen, Marguerite Gong Hancock, and William F. Miller, eds. *Making IT: The Rise of Asia in High Tech*. Stanford, CA: Stanford University Press, 2006.

Gi-Wook Shin. *Ethnic Nationalism in Korea: Genealogy, Politics, and Legacy*. Stanford, CA:Stanford University Press, 2006.

Andrew Walder, Joseph Esherick, and Paul Pickowicz, eds. *The Chinese Cultural Revolution as History*. Stanford, CA: Stanford University Press, 2006.

Rafiq Dossani and Henry S. Rowen, eds. *Prospects for Peace in South Asia*. Stanford, CA: Stanford University Press, 2005.